The Space Shuttle Program

Technologies and Accomplishments

Davide Sivolella

The Space Shuttle Program

Technologies and Accomplishments

 Springer

Published in association with
Praxis Publishing
Chichester, UK

Davide Sivolella
Aerospace Engineer
Hemel Hempstead
United Kingdom

SPRINGER PRAXIS BOOKS IN SPACE EXPLORATION

Springer Praxis Books
ISBN 978-3-319-54944-6   ISBN 978-3-319-54946-0   (eBook)
DOI 10.1007/978-3-319-54946-0

Library of Congress Control Number: 2017942940

© Springer International Publishing AG 2017
This work is subject to copyright. All rights are reserved by the Publisher, whether the whole or part of the material is concerned, specifically the rights of translation, reprinting, reuse of illustrations, recitation, broadcasting, reproduction on microfilms or in any other physical way, and transmission or information storage and retrieval, electronic adaptation, computer software, or by similar or dissimilar methodology now known or hereafter developed.
The use of general descriptive names, registered names, trademarks, service marks, etc. in this publication does not imply, even in the absence of a specific statement, that such names are exempt from the relevant protective laws and regulations and therefore free for general use.
The publisher, the authors and the editors are safe to assume that the advice and information in this book are believed to be true and accurate at the date of publication. Neither the publisher nor the authors or the editors give a warranty, express or implied, with respect to the material contained herein or for any errors or omissions that may have been made. The publisher remains neutral with regard to jurisdictional claims in published maps and institutional affiliations.

Cover design: Jim Wilkie
Project Editor: David M. Harland

Printed on acid-free paper

This Springer imprint is published by Springer Nature
The registered company is Springer International Publishing AG
The registered company address is: Gewerbestrasse 11, 6330 Cham, Switzerland

Contents

Acknowledgements

I have always been an insatiable reader. In my teen years, I used to fantasize that one day I would write a book, possibly on the subject of space exploration. That dream became a reality when in the summer of 2013 Springer-Praxis published *To Orbit and Back Again: How the Space Shuttle Flew in Space*. Little did I know that a few years later, I would be able to repeat the endeavor. Shortly after having delivered that manuscript, Clive Horwood of Praxis suggested I take up this new project. For this reason, I will always be indebted to him for his confidence and trust. And I can never praise enough space historian David M. Harland, who played an instrumental role in making this book readable and in improving my writing skills. I am also immensely grateful to Maury Solomon and her staff at Springer in New York for assisting me in transforming the original idea into a finished book.

Writing a book such as this takes a lot of time, particularly if you have a full-time day job and family. No one knows this better than my wife Monica. Her continuous support and patience, in particular in the final months of writing when I really had to sacrifice a lot of family time, were of paramount value for the success of this project. This book would not have become a reality if it were not for my parents, Pasquale and Maria, who, since I was a young boy, encouraged me to pursue my passion for space exploration. Several friends, close and far, gave me their most heartfelt moral support. I will name Vincenzo Gallo, Brigida Marica Corrado and Giuseppe Pelosi, although there were many more whose kind words of encouragement, approval, and interest I shall always remember.

Unless otherwise stated, the illustrations are taken from www.nasa.gov website or associated websites such as www.nasaimages.com, www.spaceflight.nasa.gov and www.ntrs.nasa.gov. As such, they are freely available under NASA terms. For all pictures from other sources, written permission has been obtained for their use in this book.

1

A Remarkable Flying Machine

THE SPACEPLANE

Even before the last Apollo crew left the Moon on December 14, 1972, the American aerospace industry had already begun to assign their design offices to the next space program. Although it would travel to low Earth orbit only, and therefore might look less exciting than reaching Earth's natural satellite, the new spaceship promised to be something radically different from its predecessors. In fact, it would be the chance to make a long-standing dream come true. The idea had been conceived several decades earlier, even before the Space Age began.

By the end of the 18th century, those few sci-fi novels or stories that attempted to discuss space flight relied on rather impractical means of transportation. For instance, in Edgar Allan Poe's short story *The Unparalleled Adventure of One Hans Pfaall*, the main character engages in a voyage into space using a revolutionary balloon that enables him to reach the Moon in ninety days. In Jules Verne's *From the Earth to the Moon*, three men travel to the Moon on a bullet-like capsule that is shot into space using an enormous cannon. It was only when people such as the Russian Konstantin Tsiolkovsky, the American Robert H. Goddard, and the German Werner Von Braun became involved that real physics was brought into play to determine a plausible and efficient propulsion system for space travel. Within the first two decades of the 19th century these men were able to demonstrate, either using theoretical studies or field experimentation (or both) that in terms of Isaac Newton's third law of motion rockets are the most viable means of reaching space and moving within its emptiness. At the same time, the burgeoning aircraft industry had created the firm impression that any flying machine worth its name really ought to have wings. As rocketry was seen as a specialization of aeronautical engineering, and because most of the engineers with an interest in space travel had an aeronautical background, it became natural to marry rockets and wings. Thus was born the concept of a spaceplane.

In the aftermath of World War II, the notion of winged rockets flying into space was reinforced by both sci-fi literature and the cinema industry. The winged rockets of the 1950 George Pal movie *Destination Moon* or the Walt Disney animations *Man in Space*

© Springer International Publishing AG 2017
D. Sivolella, *The Space Shuttle Program*, Springer Praxis Books,
DOI 10.1007/978-3-319-54946-0_1

and *Man on the Moon* broadcast in 1955 are clear examples. The concept became so strongly embedded into the aerospace community that even the brightest minds whole-heartedly embraced it and deliberately neglected its most obvious pitfall: wings are not required for space travel, as there is no air for an aerodynamic surface to react against in order to produce lift. Likewise, wings are not needed for ascent as the sooner a rocket can clear the atmosphere the better it is able to save on propellant and thus increase the useful payload. Wings become active only during atmospheric re-entry, to steer the craft towards the chosen landing site and perhaps also to reduce stress loading. Considering that a typical atmospheric re-entry lasts 30 to 45 minutes, it is readily apparent that for 99% of the mission wings are simply dead weight. That mass would be better devoted to improving the haulage of the vehicle. However, the ability of a spaceplane to land on a runway rather than splash down in the ocean was very attractive.

It was therefore natural to look at a spaceplane-based configuration when the first plans were being drawn up to send a man into space. The first was the Boeing X-20 Dyna-Soar. Funded by the US Air Force and initiated on October 24, 1957, the plan called for a small single-man spaceplane that could perform military tasks including reconnaissance, dropping nuclear bombs from space, and disabling enemy satellites. For almost a decade starting in 1959, NASA operated the North American Aviation rocket-powered aircraft X-15 to test hypersonic flight and to "push the envelope" toward the fringes of space. One proposed evolution of such a marvel of aerospace engineering would be an aircraft capable of orbital flight.

But neither the X-20 nor the orbital X-15 left the drawing board. Quite apart from the issue of funding, the determining factor in the abandonment of these projects was time. By the close of the 1950s, the Cold War was swiftly taking a new turn with the Soviet Union successfully delivering Sputnik into orbit on October 4, 1957. The next step would be to send a man into space and use that capability as a strategic military asset. Spaceplanes were elegant, but they presented engineering challenges that could be resolved only by costly and time consuming testing and analysis. The alternative was to put an astronaut into a conical capsule that could be recovered from the ocean after splashing down at the end of a ballistic atmospheric re-entry. Apart from being simpler, this would require significantly less research and development. Both sides of the Iron Curtain therefore set out to beat the other to being the first to send a man into space.

The single orbit of the Earth by Yuri Gagarin on April 12, 1961, won the race for the Soviet Union. By the time John Glenn matched this feat for the USA with a three orbit mission on February 20, 1962, the Soviets had surged far ahead with Gherman Titov spending an entire day in space.

All manned space programs of the 1960s, whether American or Soviet, employed capsules of ever greater size, complexity, and capability. For much of this time, the two sides were competing in a race to be first to land a man on the Moon. This effort was won by the Apollo program in July 1969. With this program drawing to a close, NASA set itself new challenges for the coming decade.

Maxime A. Faget, the chief mastermind behind the Mercury capsule, says that at that point, "A space station seemed like an obvious thing to do, to have a laboratory in space where people could work. We could understand microgravity effects both on people and on physical processes a lot better with a laboratory than we could in these tiny capsules. So there was a significant effort at the beginning, and in the middle of the Apollo program, and increasing as we went along, towards building a rather large laboratory for flight…As we got deeper and deeper into the space station program, it became obvious that we would need a better vehicle…Gemini was too small, Apollo was too expensive and almost looked like overkill. The Saturn launch vehicles were quite expensive."

As aircraft are inherently built for a long service life, which requires reusability, it was clear that NASA would be attracted to a spaceplane.

SPACE SHUTTLE 101

On January 31, 1969, NASA hired McDonnell Douglas, North American Rockwell, Lockheed, and General Dynamics to make Phase-A studies of what was then called the Integrated Launch and Re-entry Vehicle (ILRV). This would later be renamed the national Space Transportation System and commonly referred to simply as the Space Shuttle. In April, NASA set up the Space Shuttle Task Group to define the layout and requirements of this new vehicle. In October 1969, less than three months after the triumph of Apollo 11, Dr. George E. Muller, Associate Administrator for Manned Space Flight at NASA HQ, delivered a lengthy paper to an IATA[1] conference in Amsterdam, Holland, on manned space flight in the post-Apollo era. After praising the Apollo program for its achievements, he reminded his audience of the profound impact of space technology on daily life, such as telecommunications and weather forecasting. It being an aviation conference, he recalled the numerous spinoffs that space technology had already injected into the design of modern aircraft, including materials for airframes and engines and avionics. "Yet we are only on the threshold of space exploration. We have yet to reap the benefits of the capability to operate in this new territory that mankind has acquired. Those benefit will be great, beyond our ability to imagine." He cited as examples the exploration of the solar system, monitoring the Earth's resources and environment, weather control, development of new innovative materials unattainable on Earth owing to gravity, and so on. The key to such development would be the construction of a space station that would be manned from the start by at least twelve astronauts. It would be modular in design to allow expansion to address new developments and research requirements. But to do such things, it would be necessary to make access to space as cheap as possible and to make space operations as flexible as those of an airline. The primary goal of the Space Shuttle, for which Phase-A was already underway, was to make access to space routine.

Muller emphasized the key features of the Space Shuttle as reusability and low-cost operations. The total system would consist of a booster and an orbiter vehicle. Both would be propelled by liquid hydrogen and liquid oxygen, which are the most powerful combination of propellants for chemical rockets. Regarding the simplicity of ground operations, he

---

[1] The International Air Transport Association (IATA) is a trade association of most of the world's airlines to support aviation with global standards for airline safety, security, efficiency, and sustainability.

went so far as to forecast that the Shuttle would be fueled by trucks, akin to how an aircraft is fueled on the apron. The rocket engines would be so reliable that only a couple of hours of servicing would be necessary between one flight and the next, and the time between overhauls would be of the same order of magnitude as for a conventional jet engine. Advances in electronics would allow low-cost and miniaturized control boxes and computers to manage the systems and every aspect of flight. One obvious consequence would be the simplification of the flight instrumentation available to the pilots, so much so that it would resemble the flight deck of a modern airliner.[2] Sophisticated computer technology and software would create the conditions for ground operations similar to those at an airport. As every onboard function would be controlled and checked by the computers, all that would be left to do in terms of ground operations would be to board the passengers and cargo.

The orbiter vehicle would be capable of all-weather re-entry and landing, again drawing the similarity to an airliner. The thermal protection system would be made reusable by employing heat-resistant materials which would form both the primary load-bearing structure of the airframe and the heat shield. The vehicle would also have a propulsion system for powered landings to enhance safety and to allow self-ferry in the case of landing away from the launch base. Landing visibility would be at least as good as that provided by a high-performance jet or the planned SST. The landing characteristics and handling would be similar to those of an airliner.

Finally, Muller foresaw that the Shuttle would be so successful and the benefits so great that other countries would seek to buy outright or rent them for their own space programs. The result would be a global space transportation system, possibly managed at multi-company and multi-national level. In short, his paper, like many others that were written in the following decade, envisaged the Space Shuttle as a high-performance aircraft.

In the meantime, Boeing and Lockheed joined forces after realizing the project was so complex that it would not be possible for a single firm to do the whole job. Trans World Airlines (TWA) joined the team. This was a strategic and wise move, because if the Space Shuttle was to operate like a commercial aircraft then only an airline could provide the necessary first-hand knowledge of aircraft maintenance, logistics, and the selection and training of crews. The airline would also prove to be valuable in defining launch, recovery, and logistics operations on a scheduled basis. In the division of labor, Boeing would lead the design of the booster and the main engines and Lockheed would work on the orbiter vehicle and avionics.

Similarly, McDonnell Douglas partnered with Martin Marietta, Thompson Ramo Wooldridge (TRW), and Pan American World Airways. They too investigated a two-stage, fully recoverable and reusable configuration involving a booster and an orbiter. In this case Martin Marietta would be responsible for the booster, McDonnell would design the orbiter, and TRW would work on the avionics.

Finally, a third alliance was made by Grumman, General Electric (responsible for electronics integration and the thermal protection system) and Northrop (the orbiter, due to the

---

[2] In all previous capsules, the astronauts were surrounded by control panels full of switches and dials.

company's experience in testing "lifting body" research aircraft such as the M2-F2 and HL-10), with the participation of Easter Airliners.

Following submission of the Phase-A proposals on November 1, 1969, the space agency announced on May 12, 1970, that McDonnell Douglas and North American Rockwell had been selected to carry forward their studies into Phase-B. However, the very next month NASA funded all the losing companies to extend Phase-A in order to review additional configurations. It took several years and design reviews, mostly fueled by a series of budget reductions for research and development, to define what became the Space Shuttle comprising an orbiter, an external tank, and a pair of solid rocket boosters. On July 27, 1972, North American Rockwell was awarded a multi-billion dollar contract to design and develop the orbiter vehicle. Development of the solid rocket boosters was assigned to Morton Thiokol. By now, every aspect of the Space Transportation System was radically different from the concept that Mueller had presented to IATA just a few years earlier.

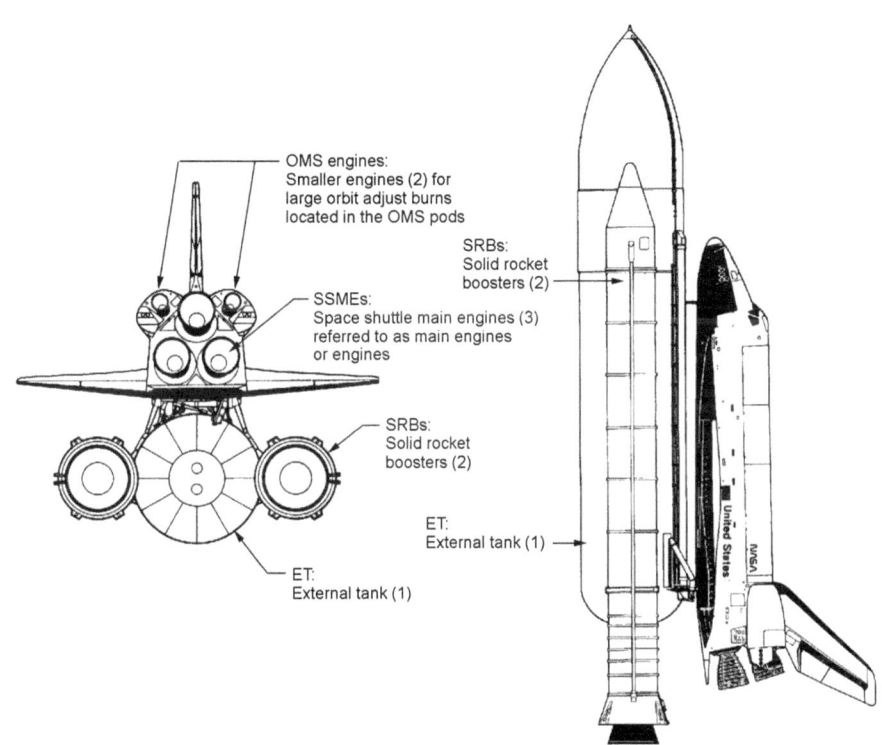

The Space Transportation System, otherwise known as the Space Shuttle.

SOLID ROCKET BOOSTERS

The first element of Mueller's vision to fall afoul of the budget cuts was the reusable manned booster vehicle. In May 1971 the Office of Management and Budget, which draws up the President's budget proposal to Congress, said that NASA would receive enough money to continue development of the orbiter vehicle but not the booster in the form that it envisaged. So the idea of an entirely reusable two-stage system had to be discarded. Everyone was sent back to the drawing board, this time with the task of designing an expendable booster.

It was felt that an expendable booster could be developed relatively rapidly and cheaply in order to get the program running. When further funding became available later on, this could be replaced by an entirely reusable one. Boeing proposed that the orbiter vehicle be installed on top of a modified version of the first stage (S-IC) of the Saturn V launcher. Such a solution was of keen interest for the company, which was eager to keep its S-IC production facility open as the Apollo program drew to a close. Likewise, Martin Marietta proposed using a larger version of its Titan III, designated the Titan III-L, to boost the Shuttle. It would combine a new liquid-fueled core with a diameter of 16 feet and up to six strap-on solid-fuel rockets, each 10 feet in diameter. A third proposal was a joint venture between Thiokol, Aerojet, United Technologies, and the Lockheed Propulsion Company, all of which made solid rocket motors. They suggested a mixed cluster of their various rockets. Finally, NASA's Marshall Space Flight Center proposed a pressure-fed expendable booster whose great attraction was the elimination of the complex turbomachinery which would normally pressurize the propellant on its way to the engines, thereby making a saving on the developmental costs. The downside was that the structure of the propellant tanks would require to be much thicker to withstand the internal pressures. On the whole, the increase in weight and resultant reduction in performance of the booster was deemed a reasonable way to minimize its development costs and thereby allocate more funding to cutting-edge technologies such as the reusable main engines of the orbiter vehicle. This approach made even more sense if the expendable booster was only an interim solution. At the same time, the designers started to consider externally mounting the propellant tanks of the orbiter vehicle.

When it was realized that with such a configuration the "staging velocity" could be cut to 5,000 (perhaps only 4,000) feet per second, NASA did an about-face and reasserted its wish for a fully reusable booster. In fact, for that staging velocity, the booster would be far smaller than was initially envisaged, and so would not require sophisticated thermal protection. With renewed hopes, Marshall kept proposing its pressure-fed booster, arguing it could be made reusable even without wings. Its thick walls would create a structure that was robust enough to survive a parachute-assisted splashdown in the ocean and the ensuing perils afloat. Furthermore, because the thick walls would serve as a heat sink during re-entry, this would eliminate the additional weight of a thermal protection system. Owing to its inelegance in comparison to a winged booster, this concept was unofficially referred to as the "Big Dumb Booster."

Boeing also stuck to its initial proposal of a converted S-IC stage, this time fitted with wings, a tail, a nose housing a flight deck, and ten jet engines to enable it to fly back to the launch site. NASA was very interested in this idea because it represented the agency's aspiration for a fully recoverable fly-back booster. But there were some practical issues to be addressed, such as the booster's thermal protection system and how the pilot would escape in the event of an abort.

A fresh idea was a Thrust Assisted Orbiter Shuttle (TOAS) in which the external tank carried sufficient propellant to enable the main engines of the orbiter vehicle to operate all the way from the ground into orbit, not just after staging from the booster. In this design the orbiter vehicle clearly could not ride on top of the booster. In fact, this "parallel staged" configuration required a booster to be mounted on each side of the big external tank. After providing additional thrust at lift-off, the boosters would be discarded at the proper staging velocity. Although the performances for both solid and liquid propellant boosters were somewhat similar, a solid booster was judged to be at least $1 billion cheaper to develop than its liquid counterpart, and hence it was selected as the winner.[3]

The Solid Rocket Boosters (SRB) were the largest solid rockets yet built, the first to be specifically designed for recovery, refurbishment, and reuse, and the only ones rated for use with a human crew.[4] From a performance point of view, the two SRBs provided most of the thrust required to lift a Shuttle off the ground and ascend to an altitude of about 150,000 feet or 24 nautical miles. Furthermore, the entire weight of the fully loaded external tank and the orbiter vehicle and its payload was transmitted through the SRBs to the mobile launcher platform. Each booster produced a sea level thrust of about 3,300,000 pounds, and the pair provided at least 70% of the overall thrust required for lift-off.

Each SRB was 149.16 feet long and 12.17 feet in diameter, with an approximate weight of 1,300,000 pounds at launch; 1,100,000 pounds of it was the propellant and the rest was the structure. Due to its sheer size, the booster could not be built as one monolithic piece. Instead, the casing was divided into four major subassemblies that were called the forward casting segment, the center casting segment (of which there were two in each booster), and the aft casting segment. Each of these was made of a number of smaller segments, for a total of eleven. In particular, the forward segment had a forward dome segment and two cylindrical segments. The two center castings segments each consisted of a pair of cylindrical segments. The aft casting segment consisted of an attachment segment, two stiffener segments, and the aft segment. The four major subassemblies together formed the part of the booster that was responsible for generating thrust and hence constituted a solid rocket motor.[5]

---

[3] A solid rocket is simpler than a liquid-fueled one because it does not require tankage, a propellant feeding system, thermal insulation to maintain the propellants at cryogenic temperatures, engine turbomachinery, etc.

[4] The US Air Force had intended to launch a Gemini spacecraft as part of the Manned Orbital Laboratory (MOL) using a Titan-IIIM rocket which had twin solid boosters which were smaller than those developed for the Space Shuttle, but this project was canceled in 1969 before any missions were flown.

[5] For this reason, the four major subassemblies were also called solid rocket motor segments.

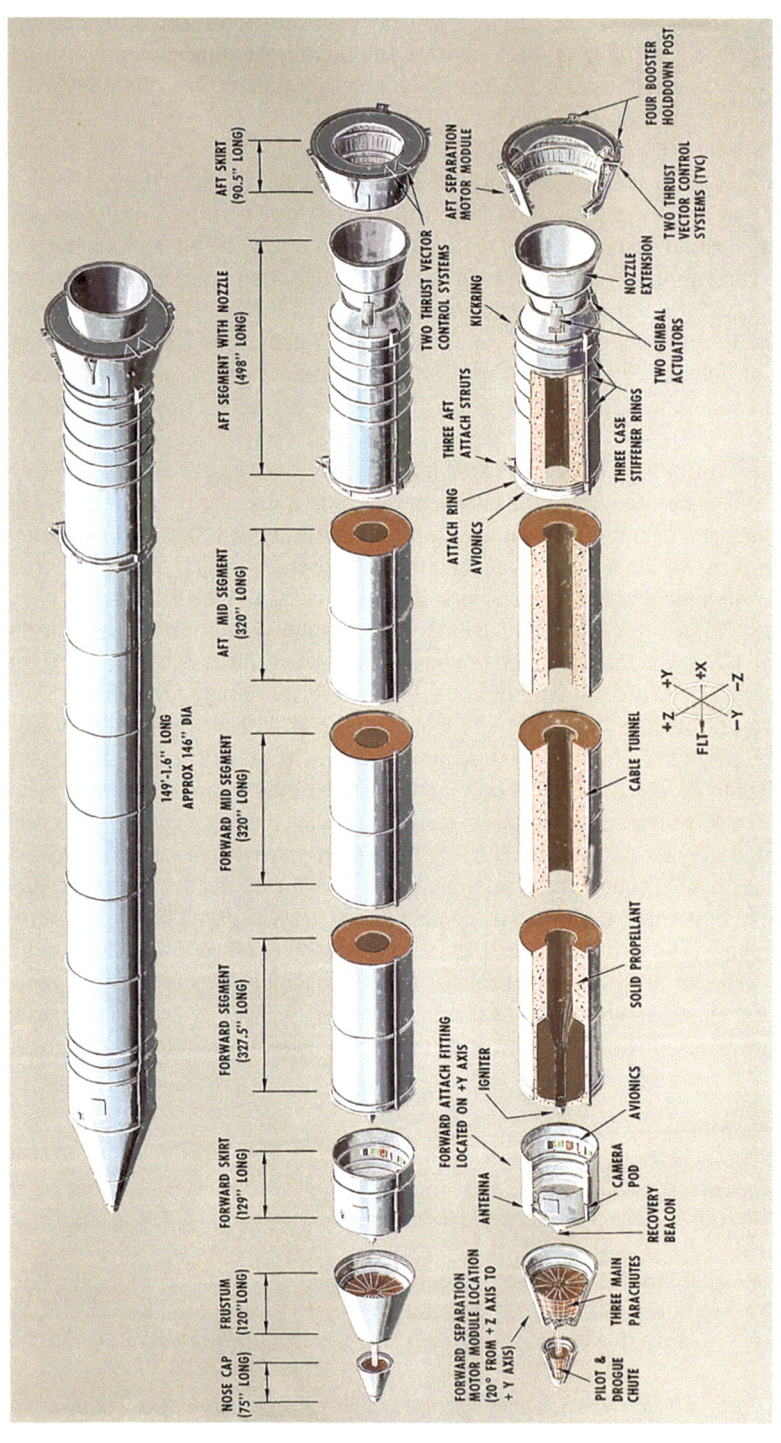

A technical view of the Solid Rocket Booster.

To Max Faget such a configuration was a mistake, "The mistake we made on the solid rockets, it was a major mistake, was that we decided that we'd build the solid rockets in Utah, and that limited the size of the solid rockets. It also meant that they needed to be segmented. But the size limitation was probably more serious than the segmenting. The diameter of the solid rockets had to be less than twelve and a half feet [to ride on] the existing rail transportation, which had to go through a number of tunnels and so forth. We just simply couldn't make it any bigger around than that. What was also a factor in the design of solid rockets was that you can't make them so long. The length-to-diameter ratio is a limiting design factor, and consequently we limited the total amount of impulse, thrust times time. The total amount of propellant in the solid rockets was limited by that, which meant that the solid rockets would only be able to help during the first couple of minutes, as opposed to being attached longer. We staged those solid rockets at only a little over 4,000 feet a second, which really meant that the Shuttle and its tank had a big job to do, and it really limited the performance and inhibited any growth in the payload capability, which the program has had to suffer with all this time."

The solid rocket motor was combined with other elements to create a complete solid rocket booster. At the very top of the booster, a nose cone assembly housed the recovery system. This assembly was split into a nose cap section and a frustum, with the former housing both the pilot and the drogue parachutes while the latter held the three large main chutes, flotation devices, and handling hardware for water recovery. Attached to the external surface of the frustum were four solid motors to separate the booster from the external tank at the appropriate point during the ascent. A forward skirt was placed below the nose cone assembly and on top of the forward solid rocket motor segment to serve as the structure which would bear the parachute loads during deployment, descent, and towing, and to mount the internal elements of the electrical and instrumentation subsystem, the rate gyro assembly, and the range safety panels. On the aft casting segment at the bottom of the solid rocket motor there was a conical aluminum and steel structure named the aft skirt to provide aerodynamic and thermal protection and mountings for the thrust vectoring subsystem, other components of the electronic subsystem, and four separation motors. This skirt also had four attachment points to the mobile launch platform to support the Shuttle in all conditions prior to booster ignition.

However, the simplicity of the design was stained by a reduced safety factor for the crew. In fact, unlike a liquid rocket engine, a solid rocket can only be shut off at propellant depletion. In the case of malfunction, all you can do is to press the button for self-destruction. As former Space Shuttle Program Manager Arnold D. Aldrich said of the SRBs, "There is not a way out. There's no way to get off. They have to burn for 2 minutes. Then at 2 minutes they separate after they stop thrusting, and they fall away. But if you try to separate them, there was no way to shut them down, and if you try to separate them while they're burning, the momentum of the thrust holds them in their places. They're connected. So even if you blow the bolts to separate them, they won't separate because the thrust keeps them where they are. I was quite alarmed that one failure could cause a catastrophe. There would be no way to recover from a failure [of an SRB] during the first 2 minutes of flight."

While the Space Shuttle community ended up accepting the situation and trusting the basic reliability of solid rockets, Aldrich and his team spent some time contriving ways in which "you could maybe shut a solid rocket booster down, and you can blow something off the top, and it will dissipate the thrust. But no system like that had been developed that was operational, that you could use. The analysis of it was that it would take some period of time after you blow it for the thrust to phase down, and it would cause an [unacceptable thrust] imbalance. In the end, it was thought that that was probably not something that was really reliable to implement, that would be any safer than just trying to do the rockets right and keep them like they were."

EXTERNAL TANK

The design of the orbiter vehicle underwent a drastic mutation, too. As it was meant to be a spaceplane, for the initial configuration no one raised an eyebrow when it was required to have internal tanks for the liquid hydrogen and oxygen which would feed its engines. Once again, visions of it as just a high-performance aircraft took the lead over engineering judgment.

These propellants were chosen due to their high efficiency in a rocket engine, but hydrogen has a drawback. As the most basic course in chemistry will readily show, hydrogen is the lightest element in the universe and therefore has a very low density; around one-fourteenth that of water. Although it would constitute only about one-seventh of the propellant load in terms of weight, the hydrogen would occupy nearly three-fourths of the volume. Being low in density and hence light in weight, to carry the hydrogen within the orbiter vehicle would have required very large tanks, which would have two major consequences. Firstly, despite being made of aluminum, these tanks would add a great deal of weight at the expense of payload capacity and they would also increase the complexity of the thermal protection system required for re-entry. Secondly, even supposing it would have been possible to fly such a monster, the development costs were well beyond the funding that even the most enthusiastic members of Congress were willing to allocate.

This led to consideration of partially reusable configurations in which the orbiter vehicle would carry its liquid hydrogen externally in an expendable aluminum shell. This would reduce the structure of the orbiter vehicle and greatly reduce the surface area that would require thermal protection. Also, new studies showed that with this configuration it would be possible to reduce the staging velocity at which the booster was jettisoned. That would ease the development of both vehicles and increase the payload capacity.

The next logical step was to stretch the external tank to carry the liquid oxygen as well. This offered a two-fold benefit. Firstly, it would further reduce the size of the orbiter vehicle to only that necessary to perform the mission. Secondly, with all the propellant carried externally, the spacecraft could achieve a standard design that was independent of the tank size. The tank could be enlarged to further reduce the staging velocity. In turn, this would reduce the size of the booster, thereby cutting the cost of the program. This reasoning led directly to the familiar configuration of an orbiter vehicle attached to the enormous expendable external tank, alongside which were the two reusable solid boosters.

Having a length of 153.8 feet, a diameter of 27.6 feet, and a capacity of 1,385,000 pounds of liquid oxygen and 231,000 pounds of liquid hydrogen, the External Tank (ET) was the largest and heaviest (when loaded) component of the Shuttle stack. As the central, integrating structural element, it was subjected to static loads from the attached orbiter vehicle when on the pad and to thrust loads from the SRBs and the orbiter vehicle's main engines in flight. The structural configuration chosen for this colossal piece of hardware was simple: two tanks, one for each of the two propellants for the main engines, joined by a strong cylindrical structure known as the intertank. This eliminated the conceptual and operational complexity of a common bulkhead between the two propellant tanks. Given the higher density of liquid oxygen, its tank was the smallest and it was located on top for vehicle controllability. The two tanks were an aluminum monocoque consisting of a fusion-welded assembly of preformed chem-milled gores, panels, machined fittings, and ring chords. Internally, anti-slosh and anti-vortex provisions in the bottom part minimized liquid residuals and damped out destabilizing oscillations of the fluid that otherwise could have induced instability in the delivery of fluid to the feed lines to the engines.

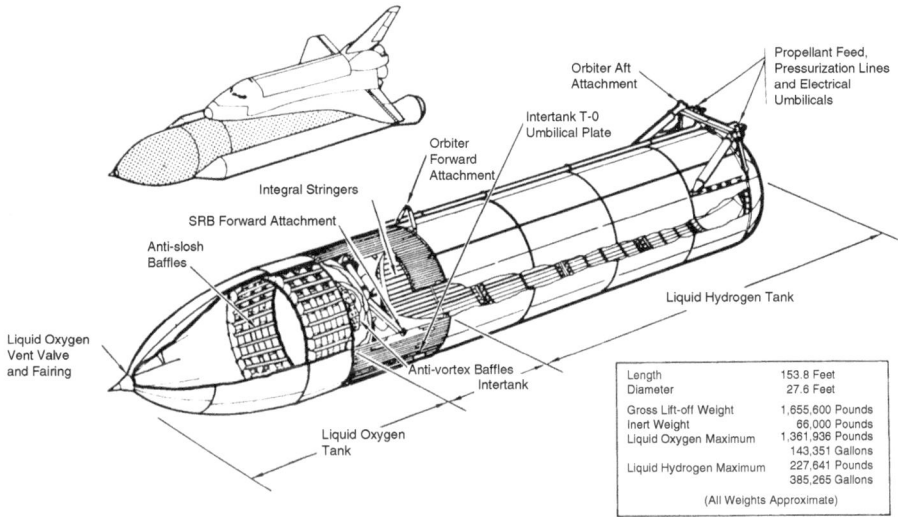

The major components of the External Tank.

ORBITER VEHICLE

The primary requirement of the Space Shuttle was reusability. This translated into each vehicle having an operational life of at least 10 years, during which it was to fly one hundred missions.

The engineers called upon to design the Shuttle were the same as had made Apollo feasible. They addressed the problem head-on by deciding which materials to use for the structure of the spaceplane. Reusability implied the Orbiter Vehicle (OV), usually referred to simply as the Orbiter, had to be able to survive the fiery environment of atmospheric re-entry repeatedly without impairing the structure. In effect, this required designing a thermal protection system capable of protecting the structure from melting during re-entry.

The ablative shield that was installed on the base of the Apollo capsule, although very simple in concept and very efficient in performance, was immediately ruled out because it would have been too difficult to create an ablative shield for the enormous aerodynamic surface of the Orbiter which could be refurbished during a turnaround between missions of just two weeks. And apart from the issue of reusability, it would have been difficult to install an ablator on such a large surface and the weight penalty would have been enormous.

Attention turned to a so-called "hot structure" where the primary components are made of a material that is capable of withstanding high temperatures without losing its load-carrying capabilities; one which can not only withstand mechanical loads but also be integrated into the thermal protection system. However, even such a structure would not be able to be directly exposed to the heat of re-entry. It would still require a thermal protection system in the form of shingle-like plates installed in the style of the tiles on the roof of a building. The materials chosen for these shingles could not have any heat-sinking capability, the absorbed heat had to be re-radiated to create an equilibrium state between the heat that was being absorbed and that which was being lost by radiation. Of course, heat would still be transferred to the primary structure, but only very slowly due to the low heat-sink characteristics of the material and to an insulator installed beneath the shingles to slow the inward flow of heat. By the time heat began to penetrate the primary structure, the vehicle would have passed through the period of peak heating and the structure would have remained within the limits of its materials. Other design requirements were that the shingles be able to expand and contract to accommodate the changes in temperature without working loose, and that they withstand the aerodynamic loads and prevent flutter. Unlike an ablative shield, the shingles would not be consumed during re-entry. Their reusability would further enhance the operating efficiency of the vehicle. And maintainability was guaranteed because a damaged shingle could easily be replaced during the turnaround between missions.

The highly successful X-15 and SR-71 "Blackbird" were both designed to have a "hot structure," and they proved this concept in a real high-temperature environment. In particular, the structure of the SR-71 was fabricated from titanium, which had the incredible capability of retaining its mechanical properties at 340°C. It was, therefore, the material of choice for building the Orbiter's hot structure. Since the upper surface of the wings and fuselage would be subjected to much lower temperatures during re-entry than the belly, no additional thermal protection would be needed in these areas. The belly would have to withstand 1,400°C, therefore the titanium structure was to be covered by shingles that promised not only to be robust but also to save a good deal of weight in the thermal protection system because this would be needed only on the underside of the Orbiter.

Although the hot structure concept was eagerly adopted by the companies which were competing for the contract to develop the Orbiter, it soon became apparent that there was a major flaw. The materials from which the shingles were to be fabricated, molybdenum and columbium, are very susceptible to oxidation at high temperatures, and this would severely impair their mechanical properties. Anti-oxidation coatings were available, but even the tiniest scratch on a single shingle surface could lead to a catastrophic failure during re-entry.

Another issue was that at that time the aerospace industry had limited experience in the use of titanium; essentially just the SR-71, and it would not have been easy to gain access to this highly classified military project in order to apply the techniques to a civilian project. And, of course, given the limited budget, cost was an important argument against using an exotic metal in the structure of the Orbiter.

Since the early 1930s, aircraft have been constructed using aluminum alloys that provide lightweight but strong structures in an inexpensive way. With the prospect of creating a hot structure for the Orbiter diminishing, attention turned to whether the vehicle could be made of aluminum, with which the aerospace industry had a wealth of experience. An affirmative answer came on the day that a new thermal protection system made its appearance, a Reusable Surface Insulation (RSI) in the form of tiles. As would a shingle, a tile would re-radiate the incoming heat to create an equilibrium between incoming and outgoing heat fluxes with such efficiency that the underlying primary structure would be subjected only to a small thermal load. In fact, it would be possible to build the entire airframe using aluminum, a metal that has excellent mechanical properties but a very low resistance to heat. Like a shingle, a tile would not be consumed during re-entry, making it reusable. And it could be replaced when damaged. Further, since the tile material was already oxidized there was no need to apply a protective coating. In addition, the RSI material promised to be lightweight.

NASA therefore decided to construct the Orbiter structure from aluminum and to protect it with tiles during re-entry.

The Orbiter structure used a reinforced shell known as a semi-monocoque. This kind of structure was recognized very early on by aircraft designers as the best for a flying machine. It consists of an external skin (the shell) reinforced with longitudinal elements called *stringers* linked by transversal elements called *frames* (for fuselage sections) or *ribs* (for wings). If one considers how easy it is to pierce an egg using a sharp object, it is easy to understand why such structures are configured this manner. A pure shell structure (such as an egg) is excellent at withstanding a load distributed over an area (such as the hydrostatic pressure inside a vessel) but is unable to resist compression and shear due to concentrated loads. An aerospace structure is subjected to a variety of loads applied to specific points, such as the attachment points for the landing gear to the fuselage or the attachment structure of a spacecraft atop a rocket. Stringers, ribs, and frames enable the structure to accept and distribute all manner of loads, making it strong, stiff, and lightweight.

Obviously, with an Orbiter developed as a reusable spaceship capable of landing on a runway, its configuration would resemble an aircraft with a fuselage, wings, and empennage. The fuselage itself comprised the forward fuselage, mid fuselage, and aft fuselage.

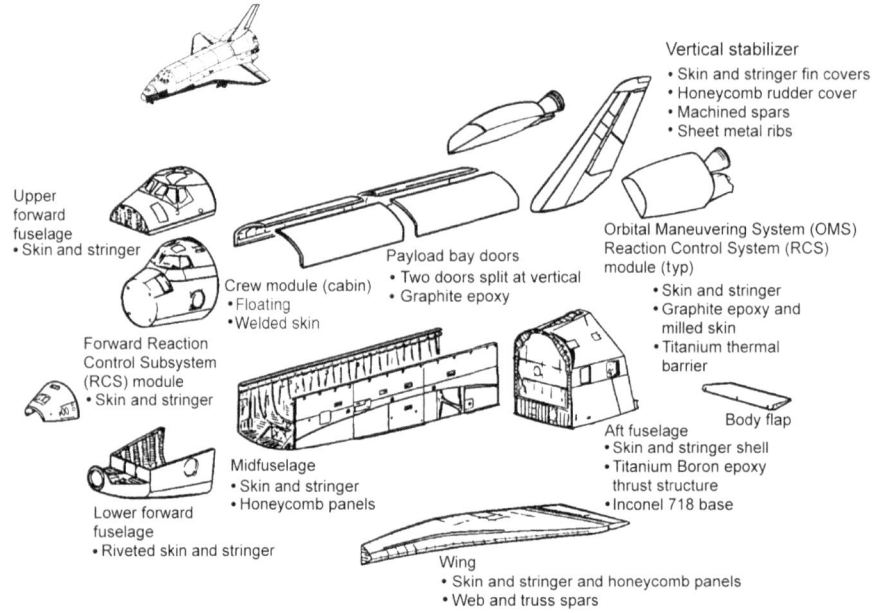

Vertical stabilizer
• Skin and stringer fin covers
• Honeycomb rudder cover
• Machined spars
• Sheet metal ribs

Upper
forward
fuselage
• Skin and stringer

Orbital Maneuvering System (OMS)
Reaction Control System (RCS)
module (typ)
• Skin and stringer
• Graphite epoxy and
  milled skin
• Titanium thermal
  barrier

Crew module (cabin)
• Floating
• Welded skin

Payload bay doors
• Two doors split at vertical
• Graphite epoxy

Forward Reaction
Control Subsystem
(RCS) module
• Skin and stringer

Body flap

Aft fuselage
• Skin and stringer shell
• Titanium Boron epoxy
  thrust structure
• Inconel 718 base

Midfuselage
• Skin and stringer
• Honeycomb panels

Lower forward
fuselage
• Riveted skin and stringer

Wing
• Skin and stringer and honeycomb panels
• Web and truss spars

The structural elements of the Space Shuttle Orbiter.

Orbiter Forward Fuselage

The forward fuselage design was dictated not only by the requirement to incorporate the crew compartment module into the structure, but also to improve the hypersonic pitch trim and directional stability and to reduce re-entry heating on the body sides. It also had to react to both primary body-bending loads[6] and nose landing gear loads. From a construction point of view, the fuselage was built in upper and lower sections to accommodate the crew module in the manner of a sandwich. This configuration was selected on the basis of crew safety. In fact, had the crew module been built as a conventional aircraft fuselage, any damage to the forward fuselage such as a crack or a hole in the skin that depressurized the module would seriously jeopardize the safety of the astronauts. The crew module was to be a simple pressure vessel that sat within the fuselage and was attached to it at certain hard points. This was an elegant solution that allowed the fuselage to bend and warp without transferring any loads to the crew module, which had only to withstand its pressurization.

In terms of layout, the crew module was arranged as a flight deck, a middeck, and a lower equipment bay.

The flight deck was the top level of the crew module. Positioned up front and side by side were the workstations for the mission commander and the pilot. These had controls

---

[6]A bending load is a tendency to change the radius of a curvature of the body.

and displays for maintaining autonomous control of the vehicle throughout all mission phases. There were seats for two mission specialists in the aft portion of the flight deck, which also had several stations for executing attitude or translational maneuvers during rendezvous, station-keeping, and docking operations, for opening and closing the doors of the payload bay, for deploying and retrieving payloads, and for closed-circuit television operations.

An aperture in the left side of the flight deck floor led to the middeck, where there were three avionics equipment bays; two on the forward side and one on the aft part of that deck. Depending upon the mission requirements, bunk sleep stations, exercise equipment, and a galley could be installed on the middeck. There were also a number of modular lockers for storing the crew's personal gear, personal hygiene equipment, experiments, and apparatus required for the mission. During launch and re-entry, four crew members would be seated on the flight deck and the others on the middeck.[7]

The middeck floor gave access to the lower equipment bay, which housed major components of the waste management system and the environmental control and life and support systems, such as pumps, fans, lithium hydroxide canisters, absorbers, heat exchangers, and miscellaneous ducting.

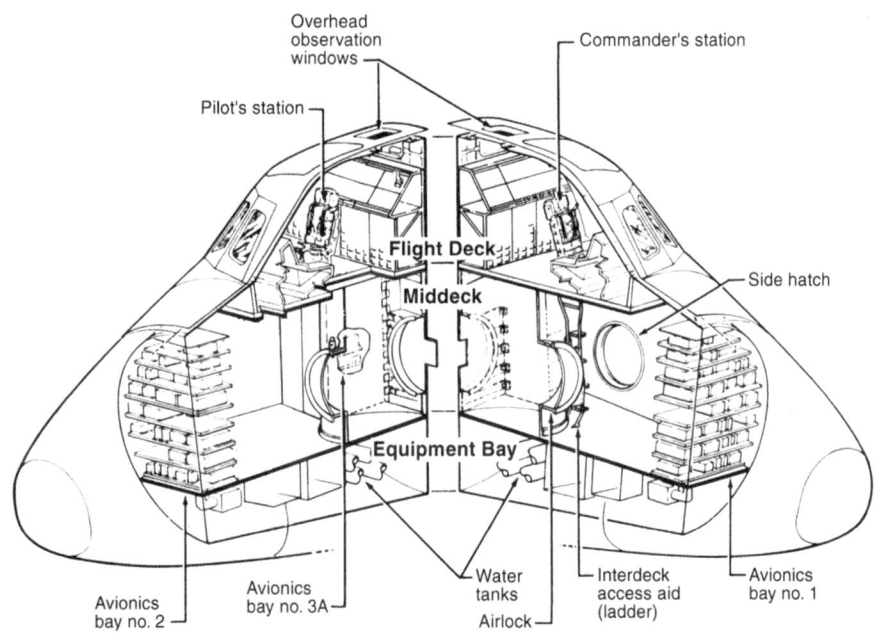

The crew compartment.

---

[7] The largest crew of a single Shuttle mission had eight astronauts.

Orbiter Mid Fuselage

The mid fuselage was the largest structural section of the Orbiter, having a length of 60 feet, a width of 17 feet, and a height of 13 feet. Built in conventional aluminum, it was open at both ends and interfaced with the forward fuselage, aft fuselage, and the wings. It supported the payload bay doors, door hinges, and various other parts of the vehicle.

This section was divided into thirteen bays, each made of vertical side machined elements and horizontal elements made of boron/aluminum tubes which had bonded titanium end fittings. Its external contour was defined by numerical-control machined skin panels strengthened by longitudinal and vertical stiffeners. Two machined sills and longerons ran its entire length to respectively accept the bending and longitudinal loads imparted by the payload and the payload bay doors. Another vital component was a lateral trunnion support structure located on the side wall forward of the wing carry-through structure, which reacted to main landing gear loads.

Two historical conditions led to such a large cargo bay. The design of the Shuttle was undertaken in an era when NASA was sending men to the Moon and there were proposals for a large space station on the drawing board. Already under development was a modification of the third stage of the mighty Saturn V rocket to accommodate facilities and experiments. Named Skylab, this "orbital workshop" was launched in 1973 and successively hosted three crews of three astronauts. However, even before the end of the Apollo program, the production line for the heavy-lift Saturn V had been shut down. As a result, NASA planners started to think of a future space station as a large structure that would be assembled on-orbit piece by piece. To achieve this, a vehicle capable of transporting the various components would be required. For this reason, NASA wanted the Shuttle to have a payload bay 17 feet in diameter. At the same time, to overcome Congressional opposition to the Shuttle development, NASA sought the support of the US Air Force, which agreed but demanded some changes to the overall configuration of the Orbiter. In particular, that being the Cold War era, the Air Force wished to place into orbit ever larger and heavier spy satellites. This would require the payload bay to be considerably enlarged. In fact, the Air Force demanded that it be at least 60 feet in length.

To give the Shuttle the incredible versatility which it demonstrated throughout the program, it was necessary primarily to design a proper way for retaining the payloads in the cargo bay. The issue was that owing to the variety of missions the Shuttle was expected to conduct, the requirements for payloads were either vague or did not exist. Nevertheless, the payload bay had to be designed.

As Thomas L. Moser, who led the structural design of the Orbiter, remembers, "We looked at multiple types of payloads, different orientations, different centers of gravity, different positions, different weights, everything else. We designed the mid fuselage to accommodate 10 million types of payloads. The Orbiter has never had a problem accommodating any payload."

To avoid having a very stiff payload bolted onto the mid fuselage structure, the attachment fittings were designed in such a way that they could slide. This meant that a payload could be designed independently of the Orbiter, and vice versa. For non-deployable payloads intended to be kept in one place during the entire mission, so-called passive fittings were utilized, over which neither Mission Control nor the crew had any control. For deployable payloads, active motor driven retention mechanisms were used. These were activated by the crew to release a payload or to secure it into the bay. Both retention mechanisms could be attached to the longerons by cross-bay bridges and to the floor of the bay by keel bridges.

With a length of 60 feet, a diameter of 20 feet, and a total area of around 1,600 square feet, the payload bay doors remain the largest fairing ever built for a launcher. They played the two-fold role of preventing contamination of the payload from the terrestrial environment and of providing a non-turbulent airflow over the fuselage to reduce drag when in the atmosphere. Structurally speaking, a payload bay door was comprised of five skin panels, each constructed as a honeycomb sandwich where the faceplates and the enclosed honeycomb were made of graphite-epoxy and Nomex respectively. The panels were connected using shear pins and stiffened by means of an internal framing of ribs, intercostals, and longerons, all of which were made from solid graphite-epoxy laminates.

The payload bay doors were to be opened shortly after the vehicle achieved orbit, and left open until several hours prior to the de-orbit maneuver. For the full duration of the orbital flight, the thermal stresses of passing into and out of the Earth's shadow would cause expansion and contraction, warping the doors in a more or less marked way. This warping of the doors would not impair the mission on-orbit, but it might prevent the doors from closing for the return to Earth. If they could not be adequately sealed, then the hot plasma of re-entry would penetrate the Orbiter, with potentially catastrophic results. To alleviate this, Moser and his team came up with the expedient of making "the payload bay doors very flexible, so that once they start to close on-orbit you can zip them closed." They were like a shirt with a zip closure. This was achieved by having three sets of latching mechanisms, one on the aft bulkhead of the forward fuselage, another on the forward bulkhead of the aft fuselage, and the others along the centerline of the doors.

Orbiter Aft Fuselage

The aft fuselage was the most densely packed part of the Orbiter, housing all of the actuators, turbopumps, and propellant lines needed to feed and gimbal the three main engines. It also had the three auxiliary power units and the associated hydraulics, the ammonia boiler and flash evaporator for active thermal protection during ascent and re-entry, and the engine and vehicle avionics. It supported and interfaced with the two aft orbital maneuvering system/reaction control system pods, the wings, the body flap, the vertical tail, the launch umbilicals, and the three main engines.

This section of the fuselage bore the immense stress of the 1.5 million pounds of thrust from the main engine cluster and transmitted it to the remainder of the fuselage without catastrophic failure. To accomplish this, the 28 truss members of the internal thrust structure of what was essentially a box 18 feet long, 22 feet wide, and 20 feet high, were made of titanium and fabricated by a process called diffusion bonding. As Moser remembers, "What we did, is we would take two pieces of titanium that were going to go together, and instead of welding them we'd put them in high temperature in a vacuum and push them together until they bonded. At the molecular level they became one and the same." The result was a single hollow, homogeneous mass that was lighter but stronger than a forged piece. For selected areas of the structure, the diffusion bonded truss elements were reinforced with boron/epoxy in order to further reduce weight and increase strength.

Orbiter Wings

At the start of the Shuttle program, it was envisaged that the whole system would be reusable and be capable of landing on a conventional runway. This required the new spaceship to be capable of generating lift when re-entering the atmosphere. This was easier said than done though, because defining the correct shape, dimension, position, and airfoil of a wing is one of the most difficult issues in designing an aircraft. And all the more so for a vehicle that was to leave Earth as a rocket, accomplish its orbital mission as a spaceship, and return to Earth as a glider.

Engineers at NASA and the companies competing for the contract to develop the Shuttle started out with the concept of a "lifting body." But this proved to be a dream and nightmare for aerodynamicists. The main advantage of such a configuration was that it blended the wings and fuselage. The advantage would be a smaller structure, compared to a conventional machine of the same performance. It also eliminated the irritating aerodynamical wing-body interfaces that give rise to considerable drag. But there were also several drawbacks. For instance, a lifting body does not generate as much lift as a winged aircraft, and the increased drag raises the landing speed, which is an undesirable handling quality. Also, the sophisticated aerodynamics of a lifting body are such that the smallest departure from the correct configuration can seriously affect the whole design. And to safely pilot a lifting body, it is necessary to develop sophisticated software to enable an onboard computer to stabilize a machine which is intrinsically unstable. Although appealing as a new and innovative design, the idea of making the Orbiter in this way was soon abandoned for a more conventional concept with a fuselage and wings. However, the prospect of achieving an easy design with a traditional layout was complicated by the need to address issues imposed by its role in the Cold War.

In the late 1960s and early 1970s, the two superpowers aimed to maximize their control over "outer space." In a James Bond-like fashion, one possible role for the Shuttle was for it to be launched and promptly rendezvous with and snatch an enemy satellite, returning home 90 minutes later after making a single orbit. Many military satellites travel in polar orbits, meaning that the plane of the orbit is perpendicular to the equator. For the Shuttle to match such orbits it would be necessary to launch from Vandenberg Air Force Base in California. This site had the additional advantage that it was not accessible to members of the public, meaning that launch operations could be conducted in secret. But launching from Vandenberg introduced a new parameter into the Shuttle program, namely the cross-range capability of the Orbiter during re-entry. The Air Force wanted it to return to the launch site at the end of a single polar orbit, but the rotation of the Earth would have carried that site 1,100 miles east of the orbital track. This required the Orbiter to be capable of gliding during all phases of re-entry in order to steer towards a landing site. Also, NASA and the Air Force were concerned that a Shuttle might have to abort its mission and return home as soon as possible. For launches from the Kennedy Space Center in Florida, heading eastward, this would not pose a problem since it would be feasible to find a good selection of emergency landing sites. But for missions in polar orbit, the only place to land, even in an emergency, was Vandenberg itself. Thus a cross-range of up to 1,100 miles was deemed essential.

This meant the wings could not be short and straight as NASA had desired for its original concept, but a delta shape capable of producing a considerable amount of lift at hypersonic speeds in order to glide far to either side of the initial track. But flying with a delta wing to achieve cross-range would expose the Orbiter to thermal heating for a longer period, and this, in turn, would require the thermal protection system to cover the entire exposed surface. And since a delta wing would have a much greater area than a small straight wing, a delta wing would further increase the weight of the thermal protection system. Nevertheless, the delta wing was chosen to satisfy the Air Force's need for cross-range because, without Air Force support, the program would have stalled in Congress.

If the shape of the wing was determined mainly by the Air Force's need for cross-range and for vehicle control throughout the entire flight regime, then the size of the wing was largely determined by landing conditions. Aircraft possessing swept wings, such as a delta wing, have to land with a high angle of attack in order to increase the lift at low speed. Since this angle is also a function of the sweep of the leading edge of the wing, the greater the required angle of attack the longer the landing gear struts must be to prevent the rear of the aircraft from scraping on the runway. The longer the struts, the heavier the main gear and the greater the volume which is required to stow it in flight. By combining all of these requirements, the optimized shape proved to be a "double delta" wing that had an 81° sweep angle for the forward section and a 45° sweep angle for the main section. It was about 60 feet in length at its intersection with the fuselage and had a maximum thickness of 5 feet. Each wing had four major subassemblies: the forward wing box, the intermediate section, the torque box, and the wing-elevon interface.

The forward wing box had a conventional configuration made of aluminum ribs, aluminum tubes, and tubular struts.[8] The upper and lower skin panels were stiffened aluminum and the leading edge spar utilized corrugated aluminum. The intermediate section consisted of aluminum multi-ribs and tubes with upper and lower skin panels comprised of aluminum honeycomb. Almost half of the volume of this section was occupied by the main gear well. All torsion and bending loads imparted on the wing were accepted and distributed to the rest of the vehicle structure using the torque box section. This consisted of four corrugated aluminum spars plus a series of tubular struts, with upper and lower skin panels of stiffened aluminum. Two pairs of elevons provided for lateral maneuverability as the Orbiter was passing through the lower and denser atmosphere during re-entry.

Orbiter Thermal Protection System

When the final decision was made to use RSI materials, Lockheed was already well ahead in studying the characteristics of this new material, as well as in experimenting with techniques for efficient fabrication. In fact, these studies started as early as 1957 and the company investigated a broad range of reusable surface insulation materials. These included all-silica systems, zirconium compounds, and alumina and aluminum silicates called mullites. Improvements in materials production resulted in a new all-silica material called

---

[8] Generally speaking, for all sections of the Orbiter's wing these tubular struts acted as ribs to maintain the airfoil of the wing, whereas in a conventional aircraft the ribs are holed webs. The tubular strut configuration was chosen to save weight.

Li-900 that had a density of just 9 pounds per cubic foot and an amorphous fiber silica possessing an astonishing purity of 99.8%. Two kinds of tiles were created from the same material for use on different parts of the Orbiter. High Temperature Reusable Surface Insulation (HSRI) was black, and Low Temperature Reusable Surface Insulation (LSRI) was white. White tiles were used for those areas of the Orbiter that were mainly exposed to the cold of space and, being thin because they did not have to cope with the intense heat of re-entry, needed a high reflectivity in order to reflect the heat of the Sun. However, the HRSI were painted with a black coating to increasing the reflection of re-entry heat. These were applied to the entire belly of the Orbiter.

   To understand this concept, it is necessary to recall that silica fiber, by itself, is white, and its low thermal emissivity makes it a poor radiator of heat. It would be necessary to make thicker and heavier tiles to accommodate extremely high surface temperatures. But applying a coating that would turn it black in order to achieve a high emissivity would allow it to radiate heat efficiently and thus remain cool. The coating was a borosilicate glass, with silicon carbide to further raise the emissivity. This was called reaction cured glass. The fact that silica and glass were both based on silicon dioxide ensured that the coefficients of thermal expansion of the coating and the substrate matched, preventing the coating from cracking under the thermal stress of re-entry. In fact, the glass coating could soften at very high temperatures and "heal" minor nicks or scratches on the surface of the tile, enabling it to survive repeated heating cycles to 2,500°F and thereby attain the reusability desired by the Shuttle designers.

   Rigid tiles were not the only form of thermal protection. When it was realized that the use of tiles which had the required minimum thickness of 0.5 inch would have overprotected surfaces that were not expected to exceed 700°F, these tiles represented unnecessary weight. For example, the upper surfaces were adequately shielded from maximum heating during re-entry by the 40° angle of attack, but they were exposed to maximum solar radiation on-orbit. Even before *Columbia*'s maiden flight, NASA engineers had started to develop a flexible insulation material. This Felt Reusable Surface Insulation (FRSI) consisted of a waterproofed Nomex fibrous pad that had a thickness in the range 0.16–0.40 inch. It was used to protect the parts of the Orbiter that faced temperatures of 350– 700°F. And when *Discovery*, *Atlantis* and *Endeavour* emerged from the Rockwell factory in Palmdale, most of the FRSI initially used for *Columbia* and *Challenger* had been superseded by another type of flexible insulation called Advanced Flexible Reusable Surface Insulation (AFRSI) which consisted of a low-density fibrous silica batt made up of highly pure silica and 99.8% amorphous silica fibers.

   In re-entry some parts of the Orbiter, most notably its nose cap and the leading edges of its wings, had to withstand temperatures as high as 2,300°F; far in excess of the thermal limit of the black HRSI tiles. These parts were protected by another type of reusable surface insulation that was a composite material in which both the matrix and the fibers were made of carbon. This Reinforced Carbon-Carbon (RCC) retained the properties of graphite in bulk by being lightweight, tolerant of high temperatures and, with a suitable coating, resistant to oxidation. Its strength and low coefficient of thermal expansion ensured excellent resistance to thermal shock, and to the stresses imparted by changes in temperature. It also had much better damage tolerance than graphite and was readily shaped. While the nose cap consisted of a single large piece, the protection for the leading edge of a wing had

22 panels that were mechanically attached by a series of 22 "T-seal" floating joints made of RCC to minimize loading on the panels due to wing deflections. This segmentation was necessary not only to accommodate the high-temperature fabrication process but also to allow for thermal expansion during re-entry and prevent large gaps from opening. In addition, the T-seals prevented the hot boundary layer gases from penetrating the leading edge cavity of the wing during re-entry, where they could cause catastrophic damage.

SPACE SHUTTLE: WHAT FOR?

In September 1969, several months after the Apollo 11 lunar landing and initiating Phase-A of the Space Shuttle development, President Richard M. Nixon established the Space Task Force to chart NASA's path for the decades to come. In particular, by the end of the century the agency was expected to have landed men on the surface of Mars. Although this mandate was not issued with the same degree of urgency that President John F. Kennedy had stirred up in 1961 when he committed NASA to land a man on the Moon by the end of the decade, it was just what the space agency had been seeking. At that time, plans were being drawn up for a large space station that would be assembled from components delivered by Saturn V class launchers. Orbital space tugs would aid in constructing the outpost, as well as nuclear-powered vehicles by which to reach the Red Planet. Tugs would also retrieve satellites from high Earth orbit and deliver them to the space station for essential maintenance, and then return them to their orbital slot. At the same time, other stations would be built to jumpstart a space-based manufacturing industry that would improve life on Earth by producing highly pure pharmaceuticals and innovative alloys impossible to create on Earth due to the disruptive influence of gravity. The key to this vision was the development of a reusable "space truck" to cheaply and routinely carry to and from orbit astronauts and miscellaneous payloads, and assist whenever appropriate in the in-space construction activities and satellite maintenance.

This imaginative vision was shattered less than six months later when, on January 13, 1970, NASA Administrator Dr. Thomas Paine announced at a press conference in Washington that, owing to funding cuts, only the reusable Space Shuttle could be afforded. There would be no space station, no return to the Moon, and no missions to Mars. For the foreseeable future, man would be confined to low Earth orbit.

The great promise of the Space Shuttle was it would reduce the cost of access to space, but as an economic analysis demonstrated, this could be achieved only if *every* payload was assigned to the Shuttle. To make the Shuttle a genuinely low cost space transportation system, the flight rate would have to be increased in such a way that there would not be room for any other expendable launch vehicle; these would have to be phased out. This condition provided the stimulus for the definition of mission requirements for which the Shuttle would be designed.

After returning from the Apollo 16 mission, Thomas K. Mattingly II was quickly involved in drawing up the requirements for typical missions that the Shuttle might execute, "We knew the Shuttle was going to last for decades, and we knew nobody was smart enough to define what the missions that would come after we started were going to evolve into. So we took great pride in trying to define the most stressful missions that we could."

For this reason, three broad mission categories were put on the table. "One was to be acting as a laboratory, and we laid out all the requirements we could think of for a laboratory – the support and what the people needed to work in it, and all that kind of stuff. Then there was another mission that was defined as deploying a payload on-orbit, and that had the manipulator arm and cradles and all of the things to do that. Then there was a polar mission. The laboratory mission was high inclination, the deployment mission was a due east azimuth launch…And each of these was sized to stress the vehicle to its maximum."

Although there never was a polar mission, 30 years and 134 flights[9] flown by five Orbiters[10] proved that indeed the Space Shuttle was a remarkable machine capable of executing a wide variety of missions. Satellites, telescopes and probes were deployed from the payload bay as the first step towards achieving their final destination. In the decades when the International Space Station existed only in artistic renderings and technical drawings, the Space Shuttle took the role of a space-based research facility. It executed complicated and risky satellite retrievals and refurbishments, chief among the latter being the five servicing missions to the Hubble Space Telescope. When the time came, the Shuttle made on-orbit assembly of a space station a reality.

It can be argued that despite the flaws in the original design imposed by an ever shrinking budget, versatility is what made the Space Shuttle such a remarkable flying machine. In the ensuing chapters we shall examine what technology was created to enable the Shuttle to exhibit this multi-talented and resourceful character while, at the same time, expanding its flight envelope and pushing its performance boundaries beyond the initial limits.

---

[9] The Space Shuttle program tallied 135 flights, but STS-51L in January 1986 did not make it to orbit because it suffered a catastrophic failure some 72 seconds after lifting off, destroying the *Challenger* and killing its seven crew members.

[10] In order of appearance they were called *Columbia*, *Challenger*, *Discovery*, *Atlantis*, and *Endeavour*.

2

Launch Platform

"A TALE OF TWO UPPER STAGES"

In the minds of its designers and advocates, the Space Shuttle was to be the first part of an infrastructure that would extend from low Earth orbit all the way to the Martian surface. It would be nothing more than a truck for special deliveries. A truck with such a large cargo bay that it would be capable of carrying into orbit the components for a modular space station in low orbit, satellites destined for geosynchronous orbits, modules for outposts on the Moon and Mars; in fact, everything else the American aerospace industry might want to place in space. As remarkable and innovative as it promised to be, one serious pitfall of this space truck was its lack of altitude. In fact, it was realized early on in its development that the operational ceiling would be about 300 nautical miles, while most of the payloads it was to haul would need to operate at higher altitudes, such as geostationary ones at 22,236 miles. Some payloads would be robotic probes to explore the uncharted territories of the solar system; destinations far and beyond the Shuttle's reach.

The solution was to create a "space tug" that would deliver these payloads to the appropriate destinations. The idea was straightforward and elegant. Once placed in space, the tug would serve a number of Shuttle missions. Upon reaching low orbit, a Shuttle would be approached by the tug. The Shuttle's payload would be offloaded to the tug, which would deliver that object to its intended operating orbit. Over time, the tug would routinely commute back and forth between low orbit and higher orbits. It would also be able to retrieve satellites and deliver them to the Shuttle, and then after the crew had repaired and refurbished them the tug would return them to their former positions. The tug would also be employed to transfer astronauts from the Shuttle to a space station and vice versa.

In the last decade of Space Shuttle operations, we grew accustomed to the vehicle docking at the International Space Station. However, because the initial concept was a delivery truck for goods and people in low Earth orbit, its creators did not include a docking capability with a space station; even if the Shuttle was the primary means of assembling a space station, it was to be the space tug that would transport everything destined for the station, including its crew.

© Springer International Publishing AG 2017
D. Sivolella, *The Space Shuttle Program*, Springer Praxis Books,
DOI 10.1007/978-3-319-54946-0_2

The space tug was to be assembled in low orbit from components carried aboard the Shuttle. Its modular design would allow the tug to be configured for each type of mission. For example, after the establishment of an outpost either orbiting around or on the surface of the Moon, or indeed both, space tugs would transport hardware and astronauts to and fro. Satellites could be repaired directly in their operating orbits, as opposed to being dragged down to the Shuttle in low orbit. In this case, the space tug would be complemented with a crew module, and a cargo compartment for the tools and materials required for the servicing activity.

Conceptual studies started as early as 1970, with the intention of having the space tug in service by the time the Space Shuttle was introduced. Unfortunately, NASA's grand plans were soon slashed. In fact, before the final Apollo lunar mission, both the Moon and Mars had been relegated to the indefinite future as likely destinations for missions involving the Shuttle. Worse, the Earth-orbiting space station dropped down the list of priorities as budget cuts were imposed by a Congress skeptical of expanded space exploration. The American space program was becoming a program limited to Earth orbit. Inevitably, doubts arose concerning the need for a space tug. NASA was already struggling to sustain the Shuttle program by insisting it would be the launcher of choice for both the nation's civilian and military space programs. The plan was to commission the Shuttle in the late 1970s. A space tug, if required at all, would not be needed until the mid-1980s, when the agency hoped it would be able to start to build a modular space station. It was considered important to keep the idea of tug alive, to be able to respond when operators expressed their wish to retrieve and refurbish their orbital assets. However, motivated by economic interests, operators did not share this vision. They felt that refurbishing a satellite would add to its overall operational costs and hold back the application of newer and better technologies with new generations of satellites.

Eventually, NASA was able to sell the Space Shuttle as the only launcher that the nation should rely on, but the agency still had to find a way to overcome the vehicle's altitude restrictions.

Mothballing the space tug, in 1974 the agency opened talks with the Department of Defense (DoD) seeking financial backing to develop an expendable Interim Upper Stage (IUS). While the IUS would be much less capable than the space tug because it would be expendable and would not be able to retrieve satellites, it would provide a viable short term solution. The Space and Missile Systems Organization (SAMSO) of the Air Force analyzed the upper stages that were already in use with other launch vehicles to determine whether they could be upgraded and made compatible with the Shuttle to function as the IUS. Safety and financial considerations soon dictated that the IUS must be a solid rocket motor.

In December 1975 the DoD opened an industrial competition for the IUS. The proposals were received in March 1976, and in August Boeing was announced as the winner with a $50 million contract to develop the stage and begin mass production as early as 1978. The company anticipated a requirement for 300 vehicles in the coming decade.

The DoD was skeptical of NASA's argument that the Shuttle should become the nation's sole means of accessing space. They wanted a redundant launcher system in case the Shuttle was unavailable for some reason when a launch was needed. So they elected to keep the Titan family of launchers in service. In order to simplify logistics and inventory, they requested that the IUS be compatible with the Titan III launcher.

When it became evident that the space station would not be funded any time soon and therefore that the space tug might not be needed for another decade, the adjective "Interim" was replaced by "Inertial," as a reference to the inertial navigation system of the new vehicle. This was only a semantic change, but by clearly signaling that the IUS was to become a long term solution it dealt a fatal blow to the space tug.[1]

The IUS was meant to deliver payloads of up to 5,000 pounds to geosynchronous orbit, but in order to be competitive with any other launcher and in particular with the increasing European competition that was already selling services using their Ariane rocket, the Space Shuttle/upper stage combination required to be economically viable for smaller payloads.

"George Low indicated some concerns when he was Deputy Administrator about whether the Shuttle could compete with the Delta launch vehicle or not," remembers Hubert P. Davis, who worked on the space tug concept. "So he said, 'Let's see what we can do with a Delta class of payloads, and how you'd propose to do that.'" Davis carried out a study which he dubbed the "Delta Killer." The result was a request by NASA for a Solid Spinning Upper Stage (SSUS)[2] capable of serving as a "perigee kick motor" to insert a payload of 2,380 pounds into geosynchronous transfer orbit. This was the same capability as the Delta rocket. NASA also requested studies for an SSUS for a payload of 4,300 pounds to match the capability in this role of the Atlas Centaur launcher.[3]

On realizing that the Space Shuttle was intended to write off all other launchers in the American stable, the Space Division of McDonnell Douglas, eager for a slice of whatever conventional launch capability was compatible with the Shuttle, proposed a design that was approved by NASA by the end of 1976. But the aerospace giant was not going to get even a dime for its efforts to develop and manufacture the SUSS; or at least not yet. Cash would flow in response to purchase orders placed by NASA and possibly other users only after the Shuttle was in regular orbital service. In exchange, NASA assured the company it would be the sole supplier of Delta class upper stages. What McDonnell Douglas designed for the Shuttle was heavily based on the Payload Assist Module (PAM), an upper stage that it was already supplying for Delta rockets. The upgraded version would come to be known as the PAM-D, where D indicated a payload capability identical to that of the Delta version. To maximize the profitability of the Shuttle, the PAM-D was made small enough for as many as four to be carried in the payload bay.

The operational advantages of deploying payloads from the Shuttle facilitated an entirely new approach to planning and executing launches. For instance, the design of payloads would benefit from the more benign launch environment afforded by the Shuttle,

---

[1] The space tug resurfaced as part of NASA's post-Shuttle planning.

[2] The term "spinning" refers to the common practice of spinning a spacecraft around its long axis during orbital maneuvering in order to increase directional stability. In fact, it is such an effective means of stabilizing the trajectory that it greatly reduces the need for the attitude control system to fire to keep the spacecraft on the right course. Thus the fuel saving is rather substantial.

[3] Strictly speaking, the IUS was to be a two-stage propulsive system. The first stage would boost out of the Shuttle's low orbit to attain a highly elliptical "transfer orbit" and then the second stage would circularize at geosynchronous altitude. In contrast, the SSUS would make only the first of these maneuvers, and the satellite would have to circularize its orbit. Hence the SSUS is also referred to as a perigee kick motor.

permitting the use of lighter and relatively more fragile structures. This was due to the 3-g maximum acceleration that the Orbiter could withstand during ascent to limit the loads placed on the vehicle and to ease the stress on astronauts and space flight participants[4] who lacked experience of the forces endured by jet pilots. Also, in the event of a launch abort the payload would not be lost, it would return with the Shuttle and, once refurbished, could be assigned to the next available flight.[5] Once on-orbit, and prior to deployment, a satellite could be checked out and if damage or a failure was found to have occurred during launch, the deployment could be canceled and the payload safely returned for repair and relaunch, thus preventing the complete loss of a valuable payload and its operational capability.

The IUS and PAM-D were carried into orbit using a reusable aluminum support structure called the Airborne Support Equipment (ASE). There were two models, one for each type of upper stage. The ASE would serve not only as a mechanical support infrastructure to accept the ascent loads exchanged between the Orbiter structure and the upper stage, it would also serve as an in-space launch pad. In both cases, the ASE would also provide services to the upper stage and its payload, such as commands, protection, safing, power distribution, communications, and so forth. Both ASE types were designed to be as self-contained and autonomous as possible, to require only the barest of attention from the crew. Similarities between the two upper stages were also shared in the way in which they would operate. As the Shuttle was being maneuvered into the desired orbit and attitude for satellite deployment, the astronauts would run a pre-deployment checklist to verify that the ascent had not harmed the upper stage and its payload, and to prepare them for the upcoming actions. Next, the ASE would be readied to bring the upper stage to the appropriate configuration. Last-minute updates of the Shuttle's orbital parameters and position would still be possible for upload into the upper stage's navigation software, to refine the precision of the upcoming journey to geosynchronous orbit. If the astronauts were satisfied and all checkouts were given the green light, the upper stage would be spring-released from the ASE to drift free of the Shuttle.

Despite having departed with a small translational velocity, the upper stage would remain in essentially the same orbit as the Shuttle. If the Shuttle were to be in close proximity at the time of engine ignition it would be blasted by the rocket's exhaust, with easy-to-imagine consequences, in particular for the delicate cabin windows. So ignition was scheduled some 45 minutes later, to allow the Shuttle time to withdraw to a safe distance. For further protection from the exhaust, the Orbiter would face its belly toward the rocket.[6]

---

[4] As it will explained in Chapter 6, these were the so-called payload specialists.

[5] During the ascent four so-called intact abort modes could be invoked based on altitude, velocity and severity of the emergency. For each abort mode the Orbiter was expected to be capable of safely landing either back at the Kennedy Space Center or at an airport in Europe or Africa. Other abort modes were possible for even more serious conditions such as loss of all three main engines. These were called contingency modes and they entailed the Shuttle trying to reach an emergency landing strip on the US East Coast or just reach a safe altitude from where the astronauts would bail out. For a contingency abort mode the chance of survival of the Orbiter and its payload were much reduced if compared to an intact abort mode.

[6] This is akin to placing your hands in front of your face to shield it from a nearby explosion.

Yet exposing thousands of delicate tiles to something that could conceivably cause real damage to the vital thermal protection system sounds like insanity! Considering that there was no capability to inspect the condition of the thermal tiles on the belly, let alone any means to repair any damage,[7] how were the crew to have known of any flaws? We shall never know because, fortunately, such a scenario never arose.

PAYLOAD ASSIST MODULE MAIDEN FLIGHT

The Space Shuttle's first satellite deployment occurred on November 11, 1982, when *Columbia* carried two commercial communication satellites. It was a pivotal moment in the Shuttle program. In fact, as the previous four test flights had proved the vehicle operational, *Columbia* was now ushering in the era of cheap and reliable commercial operations. As NASA saw it, the program was open for business, ready to receive a steady cash flow from its customers with the overall goal of pushing aside all of the competitors.

Each satellite was tucked inside the protective cocoon of the ASE that housed its PAM-D. In this case, the ASE was an open truss structure cradle made of machined aluminum frame sections with chrome plated steel longeron and keel trunnions. The cradle supported the spin table and associated drive system that formed the mounting platform for the vertical installation of the PAM-D/satellite. The nominal envelope for the installation was constrained to a cylindrical volume 7.16 feet in diameter and 8.4 feet in height on the centerline. The envelope size was chosen in order to provide commonality with the Delta launch vehicle. In this way, should the Space Shuttle not be available for some reason, the payload could be transferred to a Delta as a backup. Once confidence in the Shuttle was sufficient to retire the expendable Delta, the extra volume available within the Orbiter's payload bay could be exploited to expand the nominal envelope to 9 feet in diameter inside the cradle and 10 feet above it.

To protect the cradle and the upper stage from the fluctuating temperatures of the orbital environment, the external surface was covered with multi-layered insulation thermal blankets. Atop the cradle was a tubular frame supporting a sunshield made of Mylar insulation. The sunshield panels were fixed on the side, and stationary, but the portion that covered the top of the spacecraft formed two clamshell-like sections that were closed in order to protect against thermal stresses when the payload bay doors were opened soon after the Shuttle achieved orbit.

The PAM-D itself was rather simple, using a Star-48 solid-fueled motor made by Thiokol that had a maximum propellant load of almost 4,273 pounds. The Payload Attach Fitting (PAF) on top of the motor casing made the mechanical interface with the satellite. It had retractable fittings to deal with spacecraft-to-cradle lateral loading during ascent, and a spring-loaded system to provide the impetus required to separate from the cradle. The PAF also housed the electrical interface connectors between the solid motor and the spacecraft, the redundant safe-and-arm device for motor ignition, telemetry components, and S-band transmitters.

---

[7] In fact, both capabilities would be added many years later following the *Columbia* accident.

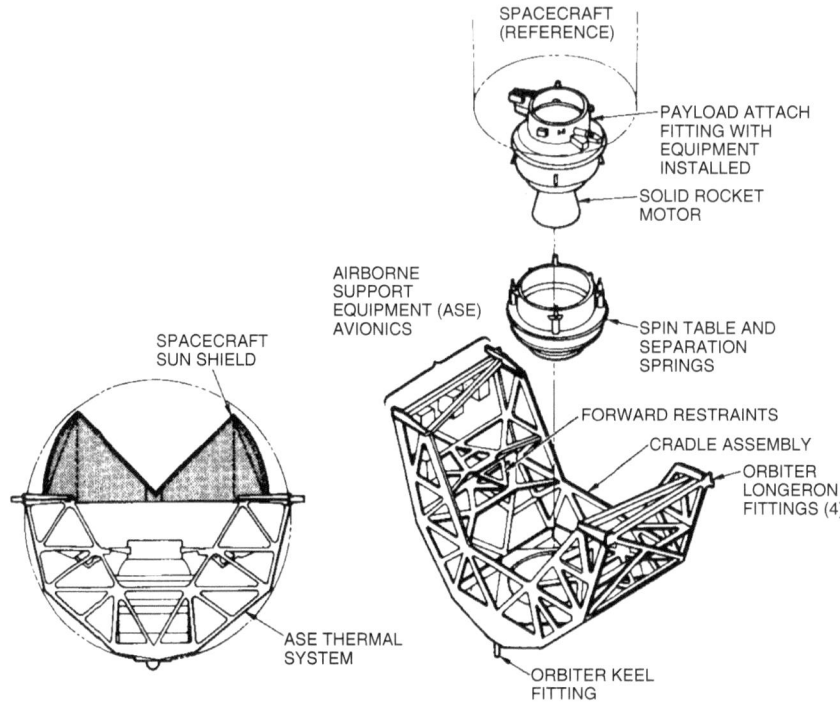

SPACECRAFT
(REFERENCE)

PAYLOAD ATTACH
FITTING WITH
EQUIPMENT
INSTALLED

SOLID ROCKET
MOTOR

AIRBORNE
SUPPORT
EQUIPMENT (ASE)
AVIONICS

SPACECRAFT
SUN SHIELD

SPIN TABLE AND
SEPARATION
SPRINGS

FORWARD RESTRAINTS

CRADLE ASSEMBLY

ORBITER
LONGERON
FITTINGS (4)

ASE THERMAL
SYSTEM

ORBITER KEEL
FITTING

Space Shuttle PAM-D system hardware.

Within 8 hours of lifting off, STS-5's crew were ready to make *Columbia* the first orbiting manned launching pad. The first to benefit from this exclusive infrastructure was SBS-3, which was a privately owned communications satellite.[8] To prepare for the deployment, the crew went through a series of checks and payload configuration procedures. Meanwhile, *Columbia* was maneuvered so that its open payload bay was facing the direction desired for firing the upper stage.

"The concept of the satellite in itself is simple. They are meant to be deployed by spinning," notes STS-5 mission specialist Joe Allen. "And the way you do it is just put the satellite on a table that will spin, like you put a record on a record player. You put it down and then you cause it to spin. In the case of the communications satellite, it is mounted on the table prior to launch, then you go to orbit…Once there you cause the table to spin and you point the Shuttle in exactly the right direction, and then at precisely the right part of the orbit you release arms that are holding the satellite to the table top. When you release the arms, springs on which the satellite sits expand to give it a very gentle push out, spinning very beautifully."

Allen's crewmate William B. Lenoir was in charge of the deployment. He coined the term Orbital Launch Director. At that early point in the Shuttle program, the crew were in

---

[8] Hughes Aircraft Company had built the spacecraft for Satellite Business System, a private communications company which was owned by subsidiaries of Aetna Life and Casualty, COMSAT General Corporation, and IBM.

communication with Mission Control for only the 15–20% of an orbit which was covered by the limited number of NASA ground tracking stations. But orbital mechanics required that the satellite be deployed when the Shuttle was crossing the equator, which would occur during one of the communication blackout zones. This meant the crew needed to decide for themselves whether to proceed with or abort the deployment. As Lenoir explains, the astronaut in charge of the satellite operations, "really was the launch director, the final say for whether you launched it out of the Shuttle or not."

With the PAM-D carrying SBS-3 drifting out the payload bay at 3 feet per second and spinning at 50 rpm, the astronauts' responsibility for this payload was complete.

The deployment of SBS-3. In the background, the sunshield doors are closed prior to deployment of the Anik C-3 satellite later in the mission.

An iconic picture of *Challenger* during STS-7, showing the typical arrangement of multiple PAM-Ds in the payload bay.

In accordance with the plan, mission commander Vance DeVoe Brand and pilot Robert F. Overmyer performed a series of burns to move *Columbia* out the way, to avoid the exhaust from the PAM-D's engine. Some 45 minutes later, the clock on the stage fired the rocket motor to boost the satellite on its way to geosynchronous orbit. As *Columbia* was facing its belly towards this activity, the astronauts were not able to witness the event with their eyes.

The next day, November 12, the show was repeated with the Canadian Anik C-3 communications satellite playing the leading role. With the approval of orbital launch director Joe Allen, the satellite was successfully deployed.

Releasing satellites in space was not new, because it was a necessary part of every mission by an expendable launch vehicle. And, after all, a key aspect of designing the PAM-D for the Shuttle was to retain compatibility with the Delta rocket as a backup option. Nevertheless, in space nothing can be left to chances, even a task as mundane as launching a satellite. As Lenoir remarks, "It was another case of never been done before, hasn't been invented." In fact, there had been concerns about the capability of the Shuttle to serve as an orbiting launch pad, particularly about whether it would be as stable as a launch pad on the ground. These concerns were answered by the arrival of the two new satellites in geosynchronous orbit several days later.

SPACE SHUTTLE INERTIAL UPPER STAGE DEBUT

The successful launches of SBS-3 and Anik C-3 were evidence that indeed the Space Shuttle/PAM-D combo was a winning team. Would the same prove true for the IUS? In the early afternoon of April 4, 1983, STS-6 set sail to answer this question. With a length of 17 feet, a diameter of 9.5 feet and a mass of 32,500 pounds, the cylindrical upper stage was an impressive sight. Furthermore, the payload, TDRS-A, was much larger than the commercial satellites deployed by the previous mission.

Physically, IUS consisted of an aft skirt, aft (first) stage, an interstage, a forward (second) stage, and an equipment section. Both stages employed solid propellant and movable nozzles for thrust vectored control. Redundant electromechanical actuators permitted up to 4° of steering on the large first stage motor and 7° on the smaller second stage motor. A hydrazine monopropellant reaction control system allowed attitude control in the coasting and thrusting phases, as well as impulses for accurate orbit injection. The majority of the avionics and control subsystems were housed in the equipment section, which also included a 9.84-foot-diameter interface mounting ring and an electrical interface connector segment to mate the spacecraft to the upper stage. Access doors allowed for replacement of components even with the spacecraft on the upper stage. A number of advanced features distinguished the IUS from other upper stages. It had the first completely redundant avionics system developed for an unmanned space vehicle, to enable it to correct in-flight features within milliseconds. It also had a redundant computer system, the second computer of which was capable of taking over the functions of the primary computer if necessary. Finally, the carbon composite nozzle throat allowed the high temperature rocket motor to operate for a long duration.

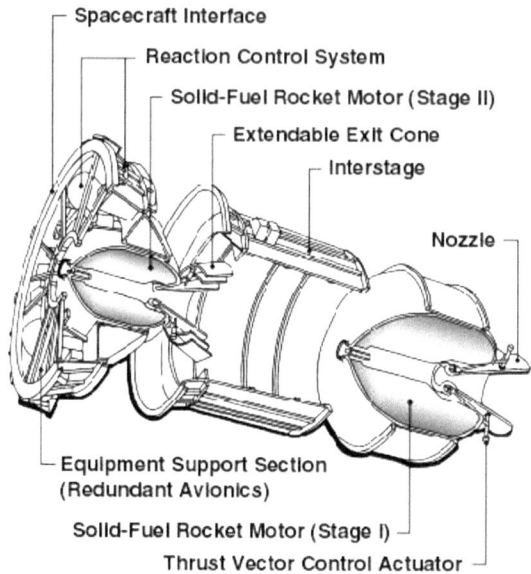

An exploded view of the Inertial Upper Stage.

As an IUS/spacecraft combination would be very heavy and would occupy almost the full length of the payload bay, the airborne support equipment was split into an aft and forward frame. The aft end of the IUS was inserted through the ring structure of the ASE aft frame, installed at the rear of the bay. Likewise, the IUS forward end was restrained by the semi-circular ASE forward frame. To reduce the ascent loads imposed on the IUS and to protect its aluminum skin-stringer construction, dampers were installed on the attachment points of the aft frame to the payload bay and on the forward frame. The ASE also offered housing for the batteries, electronics, cabling, and communication services to support the vehicle and its payload.

On attaining orbit, a sequence of actions similar to those performed on STS-5 was executed. After pre-deployment checks, the Orbiter adopted the right attitude for IUS release. Once again, the release required flipping just a few switches. However, the mechanics of deployment required a different sequence of actions since the IUS was designed for payloads much larger than those carried by the small PAM-D. In fact, STS-6 was tasked to deploy the first member of a constellation of Tracking and Data Relay Satellites (TDRS) for a brand new communications system that would replace the network of ground stations in support of the Shuttle and miscellaneous scientific missions.

An Inertial Upper Stage during ground processing. Its large size is evident from the technicians working on it.

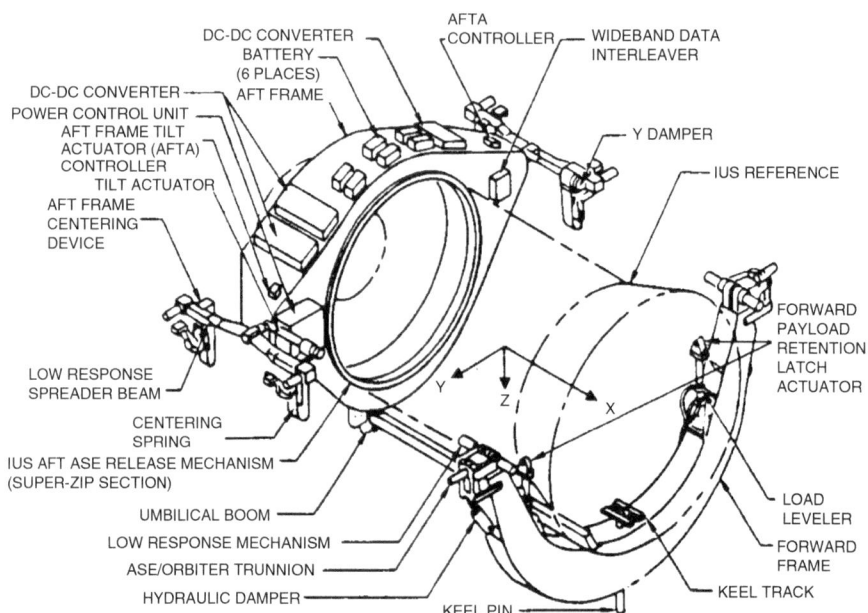

The Inertial Upper Stage Airborne Support Equipment hardware.

Almost three-quarters of the length of *Challenger*'s payload bay was occupied by the IUS/TDRS-A combination.[9] How was such a long payload to be deployed? The robotic arm could not be used, because the IUS had not been designed to be handled by it. Instead, the ASE was to perform the task. The aft frame could be tilted upward to elevate the front of its payload above the crew cabin of the Orbiter. On completion of the pre-deployment checks, the retention latches on the forward frame of the ASE were commanded to release. Then an electromechanical actuator tilted the aft frame to an angle of 29°. The TDRS-A was now just above the payload bay, and ready for the ground to verify its telemetry. Had a fault been found in the satellite that canceled its deployment, the ASE would swing down again to restore the IUS/satellite in the bay. Since the checks were satisfactory, the crew raised the aft frame to 58°, into the so-called deployment position. With separation imminent, the attitude control system of the Orbiter was inhibited to ensure stability during the process.

Just short of 10 hours after lift-off, the crew of STS-6 enjoyed the awesome sight of the massive complex placidly passing over their cabin at a rate of just 0.4 feet per second. The Orbiter then performed a maneuver to open the range to a safe distance, and some 45 minutes after its release the IUS fired its first stage.

---

[9] In fact, this would be typical for any Shuttle-based IUS deployment.

The IUS/satellite combination took full advantage of the Orbiter's payload bay. Visible in this case is the Chandra X-Ray Observatory atop the last IUS ever to be deployed by the Shuttle. The telescope was so long that there were only a few inches between it and the forward bulkhead of the payload bay.

An Upper Inertial Stage and its payload raised to the deployment position.

The Upper Inertial Stage and its payload begin their journey to geosynchronous orbit. Clearly visible in the background are the forward and aft frames of the airborne support equipment. Both are covered with gold-coated thermal insulation blankets.

With the IUS gone, the payload bay looked rather vacant. The aft frame of the ASE was returned to its normal position and locked in order to withstand the loads of re-entry.

Between 1982 and early 1986, the Shuttle routinely delivered various commercial satellites, logging a total of twenty successful deployments with PAM-D and two by IUS.[10] The *Challenger* accident on January 28, 1986, had significant repercussions on how NASA would continue to operate the Shuttle. One of them was that it would no longer be available to commercial customers since it was unreasonable to gamble the lives of astronauts to deploy a satellite that could be as effectively launched using a conventional expendable rocket. But there were exceptions such as the deployment of the Syncom IV on STS-32,

---

[10] This include the deployment of the Magnum electronic signals intelligent (ELINT) satellite by STS-51C, the first Space Shuttle mission flown for the DoD.

the delivery of five TDRS satellites, and several other satellites for the Department of Defense. The final use of an upper stage by a Shuttle was STS-93 in July 1999, when *Columbia* deployed an IUS which maneuvered the enormous Chandra X-Ray Observatory into an elliptical orbit inclined at 77° to the equator with an altitude that ranged between 6,000 miles and 86,400 miles.

DEATH STAR

Even before the IUS entered service, both the USAF and NASA knew that it would not be too long before they would need to develop a new, more powerful upper stage. Although spacecraft mass was not expected to exceed the IUS's 5,000-pound lifting capability to geosynchronous orbit until at least the mid-1980s, both institutions had projections showing that satellites were bound to grow steadily in mass by the end of the decade. For instance, the Department of Defense was already approving projects which, by 1987, would weigh up to 5,500 pounds, grow to 6,200 pounds in 1988, and reach the range 8,000–10,000 pounds by the end of the decade and start of the 1990s. Likewise, the commercial communications satellite industry, which NASA was eager to tap into as a source of revenue for the Shuttle, was projecting a 12,000-pound mass requirement by the mid-1990s.

There are several advantages in increasing the mass of a spacecraft. For instance, a larger satellite can carry a greater load of propellant with which to extend its life in terms of station-keeping and attitude control, thereby requiring fewer replacements to be launched, resulting in significant cost savings to its customers or owners. A bigger spacecraft also provides more room for payload hardware,[11] and hence increases the services or capabilities that can be offered to mission planners or customers. Though counterintuitive, it can also be cheaper. In fact, when mass is a constraint, spacecraft designers have to rely on expensive miniaturized avionics and system hardware. This obliges them to devote considerable effort to devising the optimal way to squeeze the multitude of onboard systems and payloads into the available volume. Interference between components of a system, or between different systems, is also more likely if they are in close physical proximity, and overcoming this requires expensive ground testing and delays production and launch. However, the development of very large or very heavy satellites is clearly pointless if there is no means of launching them.

This was the situation in which the American satellite industry, both civilian and military, was being forced to face in the mid-1970s. As retirement plans were being drawn for all of the existing expendable launchers in anticipation of the introduction of the Space Shuttle, satellite builders and customers became reticent to develop or require larger spacecraft, as they were aware of the lifting capability limitation of the Space Shuttle/IUS combination. Unless a more powerful upper stage that would be capable of delivering the requisite performance was approved for development, then it would make no sense to plan for spacecraft which would not be able to leave their clean rooms. The risk was that they

---

[11] While a spacecraft is considered as the payload for a launcher, the payload of a spacecraft is the equipment designed to perform the desired mission. The remaining mass consists of propellant and the hardware needed to operate the spacecraft and its payload.

might turn to the competition and require launch services onboard the fledgling European launcher Ariane, with a resulting severe loss of revenue for NASA. Besides a more powerful upper stage would also benefit the agency's planetary exploration missions such as Galileo which was to orbit Jupiter and the International Solar Polar Mission (ISPM)[12] that was to investigate our parent star.

Generally speaking, deriving the final requirements for this type of mission is an iterative process in which the launch lifting limitations are accommodated in several ways. For instance, low energy trajectories for specific launch opportunities may be assisted by planetary fly-bys, even though this will significantly increase the journey time, maintenance costs, and likelihood of a malfunction occurring before the craft reaches its destination. Another alternative consists of splitting the original probe into two or three smaller spacecraft that are launched separately as soon as an opportunity presents itself. This is also akin to reducing the mission's objectives, simplifying and making the spacecraft lighter at the expense of the scientific return and the value of the program. Often two or more alternatives are combined. As a case in point, for the Galileo mission, which comprised an orbiter for Jupiter and a probe to penentrate the atmosphere, NASA considered dispatching them by separate Shuttle/IUS launches, with each vehicle proceeding independently to its target. Along with Galileo and the ISPM, other planetary projects were in the pipeline, particularly a Saturn orbiter with probes for the atmospheres of the planet and its largest moon Titan, probes for the atmospheres of Uranus and Neptune, and a mission to rendezvous with a number of asteroids.[13]

In 1982 Congress told NASA and the USAF to collaborate in the development of a form of the Centaur upper stage, already used with the Atlas launcher, for the Space Shuttle. This was disappointing because those institutions were seeking to develop a brand new ad-hoc booster with inbuilt potential for future growth, such as to become a space tug or an orbital transfer vehicle that could be either manned or automated. However, as that project would have required funding on the order of $1,000 million compared to the estimated $250 million cost of adapting the Centaur for the Shuttle, it is easy to understand why Congress chose the latter. Adapting the Centaur had the additional advantage of bolstering the know-how of the American aerospace industry in cryogenic engine technology, enabling it to compete more effectively with the new European Ariane launcher.

The chief difference between the Centaur and the IUS was their propellants. In general terms, one way to increase a rocket's performance is to make its tanks larger and hence able to carry more propellant for longer burns. The type of engine is also a factor, as different propulsion configurations can yield significantly different thrusts. Despite the old saying that size doesn't matter, it did in the case of an upper stage for the Shuttle because the propulsive stage and its payload had to fit within the payload bay. At 60 feet, the length of the payload bay was particularly constraining. It is easy to appreciate that the longer the upper stage, the smaller the volume available for the payload it was to carry, thereby penalizing the mission objectives and capabilities. At the same time, mass was a key limiting factor on the size of an upper stage. In fact, the Orbiter and upper stage/payload combination could not exceed 65,000 pounds. It is also easy to appreciate that the heavier the

---

[12] The International Solar Polar Mission was later renamed Ulysses.

[13] At that time, no asteroid had been subjected to a close fly-by and they were entirely mysterious.

propulsive stage, the small the mass that it could carry. Both the size of the payload bay and the mass limitations of the Shuttle had the potential to seriously threaten the payload mass growth envisaged by NASA and the USAF, which was the driving reason for seeking a replacement for the IUS in the first place.

The only strategy for maintaining both size and mass within acceptable limits was to switch to an upper stage with cryogenic hydrogen and oxygen propellants, because this combination offers the greatest energy density. The Centaur upper stage used on the Atlas was that type of vehicle. Although a Shuttle Centaur would serve the needs of both NASA and USAF, it was soon realized that a one-size-fits-all approach was impractical as the civilian and military missions had radically different requirements. So there would have to be two versions of the Centaur for the Shuttle. The Air Force configuration named Centaur-G had to be capable of placing a 40-foot-long, 10,000-pound payload into geo-synchronous orbit and it would be roughly 19.5 feet long and 14.2 feet wide. It would be a shorter but fatter version of the Centaur in use with the Atlas, which was 30 feet long and 10 feet wide. The NASA variant, the Centaur-G-prime, was to be 29.1 feet long and 14.2 feet wide in order to escape from Earth and place a payload on an interplanetary trajectory. Both versions would be carried in the Shuttle's payload bay in a manner similar to the IUS, tilted to the proper angle prior to deployment, and ejected by a mechanism called the Centaur Integrated Support System.

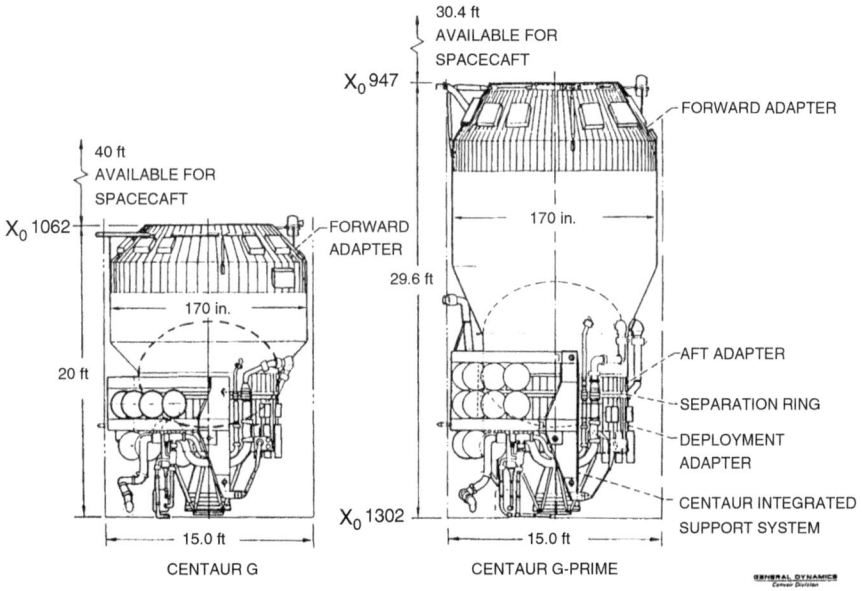

Space Shuttle Centaur configurations.

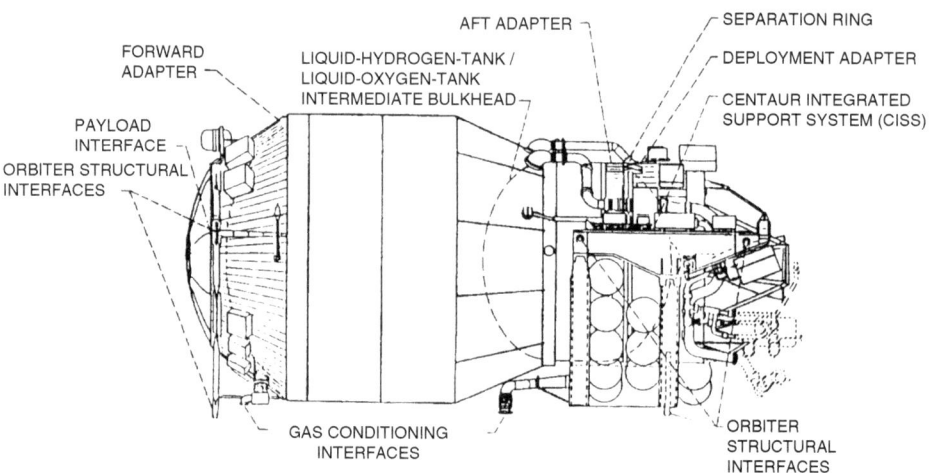

AFT ADAPTER
SEPARATION RING
FORWARD ADAPTER
DEPLOYMENT ADAPTER
LIQUID-HYDROGEN-TANK / LIQUID-OXYGEN-TANK INTERMEDIATE BULKHEAD
CENTAUR INTEGRATED SUPPORT SYSTEM (CISS)
PAYLOAD INTERFACE
ORBITER STRUCTURAL INTERFACES
GAS CONDITIONING INTERFACES
ORBITER STRUCTURAL INTERFACES

Space Shuttle Centaur-G-prime configuration and major structural interfaces.

Since NASA had already conducted feasibility studies for a Shuttle version of the Atlas Centaur, the space agency was assigned overall management responsibility for both new variants and a joint NASA/Air Force Joint Program Office would supervise their development. As General Dynamics was the designer and manufacturer of the Centaur for the Atlas, it was only reasonable to select them as the prime contractor for the new versions. Another sensible choice was to use the same dual RL-10 engine configuration as the existing variant. And the engine maker, Pratt & Whitney, would make improvements to squeeze even more thrust out of the engine nozzle. Finally, Honeywell and Teledyne were to develop the sophisticated avionics needed to safely and accurately navigate and control such a high performance vehicle.

By 1985, the new Centaur development was already struggling to negotiate what was becoming a steeply uphill path. Its budget was rapidly rising due to higher-than-anticipated costs arising from multiple working shifts and issues of quality control.[14] Other delays were due to NASA changing the redundancy requirements on complex systems which required their redesign. Integration with the Orbiter structure was also proving difficult, as the support structure within the payload bay needed to be altered to ensure it would be able to accommodate the heavy upper stage and payload safely. When the weld of the hydrogen tank for a test article sprang a leak during a pressure test, a 4-month delay was introduced to strengthen the damaged structure and add the modification to the flight unit. Avionics qualification was another hurdle that resulted in delays and financial overruns. The first space-worthy unit of the Centaur-G-prime was not delivered to NASA until September 27, 1985.

Naturally, development of the Centaur-G for the USAF was severely affected by the issues encountered by its civilian counterpart, and further exacerbated by the self-imposed

---

[14] Historically, the space industry does not apply shift working to the manufacturing of space-worthy hardware. In fact, it enhances the risks of degrading the quality of assembly arising from human factor errors, such as incomplete handovers, typically arising when maintaining production across shifts.

requirement to make it compatible with both the Shuttle and expendable launchers such as the Titan 34D.[15] In fact, while the upper stage would be supported on both its front and aft ends within the Shuttle payload bay, on the Titan only its aft end could be secured. This difference substantially influenced the size and position of elements of primary importance. A "bolt-on kit" was proposed as a possible solution. In this way, the Centaur-G would be developed to be compatible with the Titan and if it was required to ride in the Shuttle the augmentation would provide the revisions to the structural interface. Another issue was that the Titan 34D was displaying lower-than-expected performance and it was being hoped the Centaur-G would be able to compensate for this shortfall. In fact, the USAF would have been almost better off to design two subtypes of the Centaur-G, one optimized for the Shuttle and the other for the expendable launcher. Indeed, in April 1985 General Dynamics suggested that the Air Force should terminate the Titan-based Centaur-G because the alterations would add risk, complexity, and might eventually prove to be impractical. A less-than-ideal relationship with NASA and concerns with their quality control processes were other factors impairing the design of the Centaur-G. Soon, however, all these issues would vanish because in the wake of the loss of *Challenger* on January 28, 1986, the Shuttle Centaur program was canceled.

Rarely is there satisfaction in the cancellation of a space program, but that of the Shuttle Centaur was warmly welcomed, particularly by the astronaut corps. STS-61F had been assigned the first such flight, with the upper stage carrying the Ulysses solar probe. Then on STS-61G another Centaur would have dispatched Galileo to Jupiter. Chief Astronaut John Young routinely described these two flights as the "Death Star" missions. This was a darkly humorous reference to peculiarities in the design of the Centaur that presented an extraordinary potential to kill the astronauts in charge of its deployment. One important issue was safety during an ascent abort. As STS-61F's commander Frederick H. Hauck put it, "If you've got a return-to-launch-site abort or a transatlantic abort and you've got a rocket in the cargo bay that is filled with liquid oxygen and liquid hydrogen, you've got to get rid of those propellants. You've got to dump them while you're flying through this contingency abort. To make sure it can dump safely, you need to have redundant parallel dump valves, helium systems that control the dump valves, and software to make sure those contingencies can be taken care of. And then when you land, you're sitting with the Shuttle Centaur in the cargo bay. If you haven't been able to dump all of it, you're venting gaseous hydrogen out of one side and gaseous oxygen out of the other side. That's just not a good idea."

The entire upper stage design was flawed by the fact that it was never meant to be human rated, and the human rating process was carried out in the face of continuous cost savings. "The whole concept of taking something that was never designed to be part of the human space program, that had his many potential failure modes, was not a good idea because you are always saying, 'Well, I don't want to solve the problems too exhaustively. I'd like to solve them just enough, so that I've solved them.' What does that mean? You don't want to spend any more money than you have to in order to solve the problem, so you're always trying to figure out, 'Am I compromising too much or not?' And the net result is you're always compromising."

---

[15] Although the USAF was ideally to use only the Space Shuttle for all of its payloads, uncertainty about how well it would perform led to the agreement to retain an expendable vehicle capability until the Shuttle had proved its worth.

To save on structural mass, the Centaur's tanks for both the Atlas and the Shuttle versions were made of very thin aluminum alloy panels. Lacking internal stiffening, both the hydrogen and oxygen tanks were pressurized to maintain their shape. As Hauck recalls, "If it's not pressurized, it is going to collapse by its own weight. If it were not pressurized, but suspended and you pushed on it with your finger, the tank walls would flex."

Another weight saving solution was to have a common bulkhead between the two propellant tanks, in a similar fashion to that of the second stage of the Saturn V. As Gary W. Johnson, who was Deputy Director of the Safety, Reliability and Quality Assurance (SR&QA) Directorate, recalls, "The big concern we had was… it would take very little delta pressure across that [wall] to cause it to break and really cause a problem [because if hydrogen and oxygen are mixed together the result is a violent explosion]. Instead of having a direct delta-P alarm system in there that would shut down the pressurization system or try to safe the systems if you were getting close to that pressure, they had a computer program that tended to try to manage the pressure in the oxygen tank and manage the pressure in the hydrogen tank, and between these two pressures that you're managing, then try to make sure you met the requirements for the delta-P. But there wasn't a direct measurement, and that was the real concern we had…The Safety Panel was insisting that we had to have something separate on that."

Owing to the volatile nature of liquid hydrogen and oxygen, the Centaur would be fueled on the launch pad just hours prior to lift-off. Johnson and his team had further worries concerning the propellant loading system. "Loading was to employ plumbing through the Shuttle and then into the Centaur. The plumbing lines had so little margin in them that if we suddenly had to stop propellant loading when loading the external tank, the hammer-type pressure you get – the pulse from suddenly shutting that off – stood a very good chance of rupturing the lines in the Orbiter, causing the loss of the entire vehicle on the pad. That was a big safety concern. We didn't have sufficient margin in those lines."

Also worrisome was the way in which the flight hardware was being approved for space despite its not having passed ground testing. This was an issue that profoundly concerned Johnson. "I was reviewing, going through some of the paperwork, and I noticed on some of the critical relay boxes that send commands from the computer, that they'd suffered failures in testing for vibration. I didn't see in the paperwork any real closeout that said that that hardware ought to be certified yet. Yet we had papers signed off by the people involved, saying things were certified and all ready to go. It looked to me, that we still had some open problems which we had yet to fix. I called in the engineer responsible for that system. I was quizzing him about how we'd gone through everything, and there were these problems. I said, 'What's the rationale for signing that this is ready to go?' I could see the young engineer was visibly shaken, and he confided that he'd been forced to sign off by his management."

Another safety issue that Hauck remembers bringing up at a meeting pertained to redundancy in the helium system which would pressurize the tanks and force feed the propellants into the combustion chambers of the twin engines. During a launch abort, this system would have expelled the propellants from the tanks to bring the Orbiter's weight down to a manageable level for a safe landing. "In early to mid-January 1986, we were working an issue to do with redundancy in the helium actuation system for the liquid oxygen and liquid hydrogen dump valves, and it was clear that the program was willing to compromise on the margins in the propulsive force being provided by the pressurized helium. We were very concerned. We went to a [safety review] board to argue that it was not a good idea to compromise on this feature. The board turned down the request." That meeting was quite revealing, so much so that soon afterward Hauck called in his crew and expressed his concern about how NASA was willing to sacrifice the safety of their mission. He reassured them that if anybody wanted off the mission, he would support their decision. The crew stayed together.

If you are wondering why these smart people were willing to risk their lives, fully aware of how dangerous it would be to fly with the Centaur, you need to understand the historical context.

As STS-61F's mission specialist John M. Lounge admits, "We assumed we could solve all these problems. We were still basically bulletproof. Until *Challenger* [was lost] we reckoned we were bulletproof and the details would be worked out." Hauck echoed this sentiment. "We'd never killed anybody in space up to that point. I mean, there was a certain amount of sense that it would not happen." Besides you must also consider that an astronaut would accept risk in order to fly in space, in particular for a high profile mission. As Lounge says of his thinking at that time, "It was a privilege to be assigned to an important mission, so our attitude was that we just had to work it out." And if the astronaut was a Navy aviator like Hauck, then pride was also a major factor: "I probably had an ego tied up with it so much that, you know, 'I can do this. Heck, I've flown off of carriers and I've flown in combat, and I've put myself at risk in more ways than this. I'm willing to do it.' So I didn't ever think of saying, 'Well, I'm not going to fly this mission.'"

While the *Challenger* investigation was underway, and as more and more details emerged of the poor safety culture within NASA, the Shuttle Centaur was still kept alive despite of the numerous safety issues. However, it was living on borrowed time. Well before the end of 1986, NASA Administrator and former astronaut Dick Truly decided to close the program for good. The Shuttle Centaur never flew in space, and missions were revised to use the less powerful Inertial Upper Stage.

As of May 2016, the sole surviving test article for the Centaur-G-prime is on display at the NASA Glenn Research Center in Cleveland, Ohio.

The idea of a cryogenic upper stage for use on the Shuttle was resumed for a brief period in the early years of the new millennium. In fact, Boeing's Advanced Shuttle Upper Stage (ASUS) concept was tested in 2000 at the NASA Marshall Space Flight Center. The plan was that the stage would be empty at the time of launch and then be filled during ascent by drawing liquid hydrogen and liquid oxygen directly from the Shuttle's external tank. The propellant transfer concept was a pressure-fed rapid chill and fill system which consisted of a spray bar located in the center of each tank and running its full length. At the start of the filling operations, the propellant would chill down the tank wall, and the vaporized propellant would be expelled through a typical vent system. When the wall had achieved an acceptable temperature, the vent valve would be closed to initiate the actual propellant loading, which would take 5 minutes to complete. This was a smart idea, because it would tap into the available reserve of hydrogen and oxygen that typically remained in the external tank after its separation from the Orbiter. In this way, there would be no reduction in the payload capacity of the Shuttle. Also, because the chill and no-vent fill process was being considered for on-orbit transfer of propellant, it was reasoned that the lessons learned from using the ASUS would assist the development of reduced gravity cryogenic propellant transfer technology. Although testing proved that the concept would work, in the wake of the *Columbia* accident it never received a chance to prove itself on a mission.

In the second semester of 2002, NASA Marshall Space Flight Center undertook a 6-month evaluation study of a High Energy Upper Stage (HEUS) concept. Proposed by Northrop Grumman, it again envisaged a liquid-fueled upper stage but unlike the Centaur or ASUS it would use storable propellants like monomethyl hydrazine and nitrogen tetroxide. These are simpler to handle than cryogens but, being hypergolic, these propellants will spontaneously ignite upon coming into contact. Two types of upper stage were considered: one with liquids and the other with storable gels. Apart from propulsion, their subsystems were based largely on heritage hardware in order to minimize cost, risk, and development schedule.

Unfortunately, this study did not progress past the concept stage. If it had reached the development phase and the assembly shop, the HEUS would have been an ideal candidate to deliver the James Webb Space Telescope (JWST) as the successor to the Hubble Space Telescope (HST) and the Space Interferometry Mission (SIM) which was to seek exoplanets.[16]

GATEWAY TO THE SOLAR SYSTEM

The demise of the Centaur upper stage for the Shuttle and the grounding of the fleet following the loss of *Challenger* had a severe detrimental effect on the deep space exploration program. White-dressed technicians preserved the Magellan, Galileo, and Ulysses

---

[16]The Space Interferometry Mission was canceled due to the complex and expensive technology needed to image alien words located tens or hundreds of light year away. The JWST is currently scheduled for launch in 2018 onboard an Ariane 5 launcher, marking a significant coup for the Europeans.

probes in space-worthy condition awaiting a revised launch schedule. In the meantime, new trajectories had to be computed for future launch opportunities based on planetary alignments that took into account the withdrawal of the powerful upper stage. In due course, all these probes would be safely dispatched using the solid-fuel Inertial Upper Stage.

The first was the Venus-bound Magellan. In any historical period, Venus, almost a twin of Earth, has captivated the imagination and attention of the amateur as well as the professional astronomer. To their frustration, however, the surface of the planet is hidden from view by an opaque atmosphere. It was not even possible to identify how rapidly the planet rotated on its axis. This was finally measured in 1961, when radio signals were beamed at the planet by the most powerful antennas and the reflections analyzed. Venus rotates in a retrograde manner, in the direction opposite to that of its orbital motion. In effect it spins up-side down. Furthermore, its rotation is extremely slow, its axial period of 241.5 terrestrial days being longer than the 224.7 days that it takes to travel around the Sun. Later it was found that the surface of the planet could be investigated by processing radar echoes to produce images. Then US and Soviet probes equipped with radar instruments were placed in orbit around the planet to map it. An almost global map was provided in 1978 by NASA's Pioneer 12 (also called Pioneer Venus Orbiter). This spanned 92% of the surface at resolutions in the range 30–84 miles. In 1983 the Soviet Venera 15 and 16 orbiters achieved the outstanding resolution of 1.25 miles, but their coverage was limited to the northern hemisphere above 30° latitude.

By 1988, NASA hoped to improve on this resolution and extend the map to cover the entire surface. In fact, in the wake of the successful optical mapping of Mars by Mariner 9 in 1972,[17] it started a feasibility study for an equivalent mission to Venus. This Venus Orbiting Imaging Radar (VOIR) was meant to use a Synthetic Aperture Radar (SAR) to map up to 70% of the planet at a resolution of better than 400 meters. The radar was also to collect altimetry data and operate passively to sense emissions from the surface. But VOIR did not make it past the drawing board and in 1982 the project was canceled. The following year NASA proposed the Venus Radar Mapper (VRM) as a simpler but more focused mission. All of the non-radar experiments had been removed. Only the core science objectives of the deceased VOIR were retained, namely to investigate the geological history of the surface and the geophysical state of the interior. By 1986 the rather unattractive acronym was replaced with the more fitting moniker of Magellan, to honor the Portuguese explorer Ferdinand Magellan whose expedition in the early 16th century was the first to circumnavigate the Earth to reveal the vast nature of its oceans. Likewise, it was hoped that the Magellan probe would provide a global understanding of our near twin.

To save on development costs, Magellan was mostly constructed with spare parts salvaged from other missions. For example, the 12-foot-diameter high-gain antenna, used alternatively for SAR mapping and data transmission to Earth, was a spare from the Voyager project, as were the primary structure and the small thrusters for attitude control. From the Galileo project, Magellan obtained the command and data handling system, the

---

[17] Mars has such a thin and transparent atmosphere that it is very easy to observe its surface with both ground-based telescopes and orbiting probes.

attitude control computer, and the power distribution units. The medium-gain antenna was a spare from the Mariner 9 mission.

The initial plan was for Magellan to be deployed by a Shuttle in April–May 1988 and pursue a Type-I trajectory that would reach Venus in just under 4 months. From launch to arrival, the spacecraft would have traveled less than 180° around the Sun. The *Challenger* accident in January 1986 grounded the Shuttle fleet until operations resumed in September 1988. The next available opportunity for a Type-I trajectory was October 1989. But this was the same period in which the Jupiter-bound Galileo probe was scheduled to launch. In order to avoid further delaying that mission, it was decided to launch Magellan in April–May 1989. The IUS could not to send Magellan on a Type-I trajectory. The relative positions of Earth and Venus at that time meant it had to fly a longer Type-IV trajectory where it traveled one and half times around the Sun on a cruise lasting 15 months. The lengthened journey time was compensated by a reduction in launch energy and Venus approach speed. On May 4, 1989, after just five revolutions of the Earth at an altitude of 160 nautical miles, Magellan was deftly deployed by STS-30 *Atlantis* to begin its journey to explore undiscovered lands. This was an important moment for NASA, because it marked the resumption of planetary exploration, which had been moribund for over a decade.

The Galileo mission was next. In the early 17th century the Italian astronomer Galileo Galilei spent countless hours observing the heavens using a crude telescope. One of the celestial objects he loved to watch was the majestic Jupiter which, to his astonishment, revealed itself not only as a planet but also to possess four prominent moons. This dealt the death blow to the Aristotelian-Ptolemaic view of the universe, supported by the Catholic Church, that all celestial objects revolved around the Earth. Indeed, using his observations of the Jovian system along with data gathered on the motions of other planets, Galileo strongly defended the Copernican concept in which all of the planets, including the Earth, revolve around the Sun. It was fitting that the Jupiter-bound probe carried into orbit by STS-34 *Atlantis* in the afternoon of October 18, 1989, was named in honor of this Tuscan astronomer. It would become the first spacecraft to enter into orbit around Jupiter to conduct long-term observations of the planet, its magnetosphere, and its moons. A small probe was to be released to make direct measurements of its atmosphere.

The cancellation of the Shuttle Centaur meant the Galileo mission had to use the less powerful IUS. As a result, it had to follow a very roundabout route that exploited a series of gravity assist fly-bys. It would not reach its destination until December 7, 1995, having flown by Venus once and Earth twice in what was dubbed the Venus-Earth-Earth Gravity Assist (or VEEGA) trajectory.[18]

The STS-41 flight concluded the brief period in which the Shuttle launched deep space missions. Lifting off in the morning of October 6, 1990, *Discovery* carried the Ulysses probe on a mission to study for the first time the polar regions of the Sun. The mission design required a close pass over the Jovian poles to steeply incline the plane of the spacecraft's orbit out of that in which the planets travel around the Sun. Although the probe was not very large, to compensate for the lack of a Centaur it was necessary to augment

---

[18] If the Centaur-G had been available, the voyage to Jupiter would have lasted just 2.5 years. On the other hand, the lengthy detour provided opportunities to study Venus using new sensors and the first close fly-bys of asteroids.

the IUS with a payload assist module based on that created as a perigee kick motor for geosynchronous satellites[19] as a third stage to reach Jupiter in 1992. This would be the first and only time that such a combination of upper stages developed for the Shuttle was used.

ADDITIONAL UPPER STAGES

The PAM-D and IUS were not the only satellite delivery systems available to Shuttle customers. On several occasions, large Hughes-built Syncom spacecraft were simply rolled out the payload bay in what was informally known as the Frisbee deployment mode. As the first satellite to be designed specifically for the Shuttle, it was sized to snugly fit the width of the payload bay. This wide body provided enough room to add perigee and apogee kick motors. In effect, the spacecraft had its own built-in upper stage. The fee NASA charged for carrying a satellite was scaled by the fraction of the payload bay that the package occupied. A wide but squat satellite that was carried on its side saved the client money.[20] The satellite was attached to a cradle in the payload bay at five contact points, four on the longerons and one on the keel.

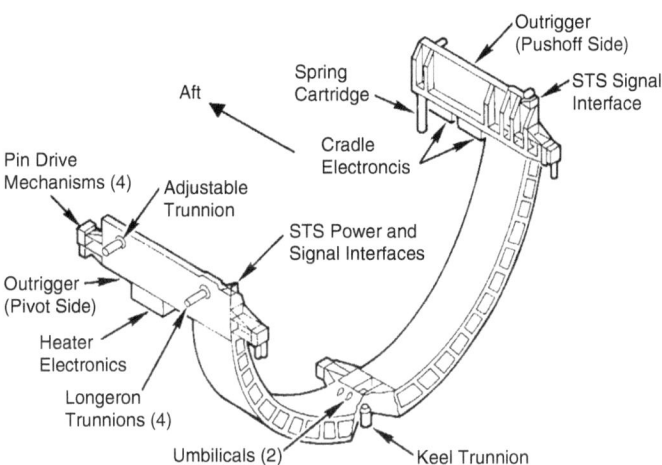

The main components of the cradle for the Frisbee deployment mode.

To prepare for a Frisbee deployment, the Shuttle would be oriented with its tail in the direction of travel and the payload bay facing down, in order that the spin axis of the

---

[19] This third stage was labeled a PAM-S.

[20] As will be discussed in greater detail in Chapter 12, the cost of deploying a satellite from the Shuttle was based on a formula that took into consideration the ratio of either the mass or the length of the cargo, whichever was the greater, in relation to the Shuttle's total capacity.

satellite would point in the correct direction for the perigee motor to thrust, as if it had just been released by an expendable launcher. Locking pins at four of the contact points would be retracted by electrical motors, each pin requiring about 5 minutes. A pyrotechnic device at the fifth contact would then be initiated to allow for the release of a spring that would push one side of the satellite upward as the other side pivoted. This provided simultaneously for a rotation and translation maneuver which imparted an initial separation velocity of 1.3 feet per second with a stabilizing spin of about two revolutions per minute. Thirty seconds later, the Syncom would be automatically activated. One effect was to start the onboard clock that would time the maneuvering events such as commanding the reaction control system to increase the spin rate to 33 rpm. Some 45 minutes later, and with the Shuttle at a safe distance and attitude, the perigee kick motor would ignite to achieve an elliptical orbit with an apogee of 9,000 miles and then be jettisoned. The liquid-fueled apogee motor would perform a series of maneuvers to raise the apogee to 22,236 miles and circularize the orbit.

A Syncom satellite rolling out the payload bay. In the background is the cradle covered in a gold-coated insulating Mylar blanket.

On October 23, 1992, *Columbia* deployed the Italian-built LAGEOS II satellite as the primary objective of the STS-52 mission. This Laser Geodynamics Satellite was a 900-pound, 24-inch sphere whose surface gave it the appearance of a large pitted golf ball because it was covered with 426 almost equally spaced three-dimensional prisms mostly made of a fused silica glass called Suprasil. From any ground station participating in the LAGEOS project, a laser beam illuminating the satellite would be reflected back by any one of the prisms. Measuring the round trip travel time and multiplying by the constant speed of light would precisely measure the distance that separated the spacecraft from the ground station. By tracking the LAGEOS satellites for a number of years, scientists would be able to characterize the movement of the ground stations. This would enable them to monitor the motion of the planet's crust, characterize the "wobble" in the axis of rotation, collect information on the Earth's size and shape, and more accurately determine the length of its day.

Such a simple spacecraft design, lacking even an onboard system, stemmed from several trade-offs. In fact, to serve as a stable orbital reference point its mass had to be large enough to minimize the effects of non-gravitational forces, yet light enough to be placed into a fairly high orbit. It had to be large enough to accommodate many retroreflectors, but small enough to minimize the influence of the solar pressure. Aluminum would have been too lightweight for the entire body, so it was decided to combine two aluminum hemispheres bolted around a brass core. This selection of materials had the bonus of reducing the interactions with the Earth's magnetic field and thus increasing its orbital stability. The first LAGEOS satellite was launched on May 4, 1976, from Vandenberg Air Force Base in California using a Delta launch vehicle.

With the premature retirement of the PAM-D, NASA was left without a suitable upper stage for deploying a mid-sized satellite such as LAGEOS II from the Shuttle. Instead, the Italian Space Agency (ASI) provided the Italian Research Interim Stage (IRIS) which, unsurprisingly, was a carbon copy of the deceased PAM-D. In fact, it consisted of a solid rocket motor mounted on top of a motorized spinning table that was enclosed within an airborne support equipment cradle equipped with sun shields and the necessary hardware and equipment for stage deployment. The satellite and its LAGEOS Apogee Stage (LAS) were mounted on the IRIS. After leaving *Columbia*, the satellite was propelled into the desired circular orbit at an altitude of 3,600 miles.

The Transfer Orbit Stage (TOS) was another upper stage which was used on the Shuttle only once. This resembled the IUS, but was smaller. Built by Martin Marietta Astronautics Group under the management of Orbital Sciences Corporation, it was a one-stage solid motor propulsion system with a gimbaled nozzle for thrust vectoring, a navigation and guidance system, a reaction control system to adjust attitude or local pointing, and a cradle to support the entire package in the payload bay and facilitate its deployment. The cradle resembled that of the IUS, in that it consisted of forward and aft elements. The forward cradle was a circular beam. Its upper section unlatched and rotated open to permit the aft cradle to pivot the TOS/spacecraft combination to a 45° angle. A pyrotechnic mechanism would then release the upper stage, and springs would gently push the mated pair out of the payload bay.

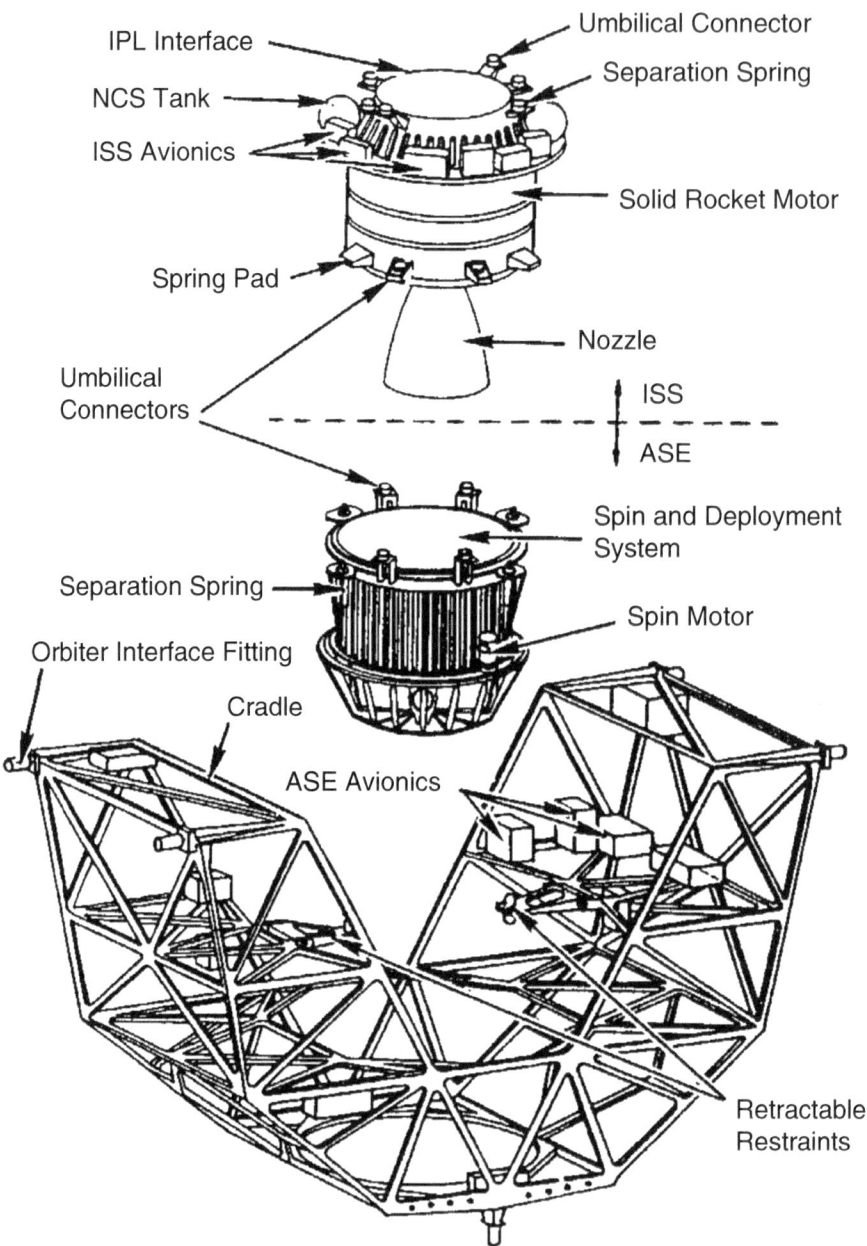

IPL Interface

NCS Tank

ISS Avionics

Umbilical Connector

Separation Spring

Solid Rocket Motor

Spring Pad

Nozzle

Umbilical
Connectors

ISS

ASE

Spin and Deployment
System

Separation Spring

Spin Motor

Orbiter Interface Fitting

Cradle

ASE Avionics

Retractable
Restraints

(Sunshields and Thermal Blankets Not Shown)

An exploded view of the Italian Research Interim Stage.

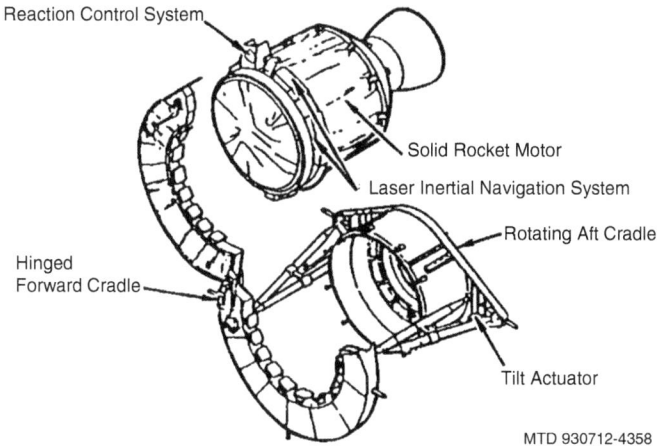

The main hardware of the Transfer Orbit Stage and its associated airborne support equipment.

The first mission for the TOS was in September 1992, when an expendable Titan launcher carried the Mars Observer spacecraft. It made its debut on the Shuttle a year later, when STS-51 *Discovery* deployed the Advanced Communication Technology Satellite (ACTS). This was a joint partnership by NASA and US industry aimed at maintaining America's predominance in the field of satellite communications in the face of intense competition from Europe and Japan. ACTS was a space-based test bed for advanced and high-risk technologies which NASA, industry, and government expected would dramatically enhance the capabilities of the satellite communications industry and reduce the cost of using such a system.

The deployment was successful, but a review of the video of the separation of the upper stage revealed the presence of extensive damage to the structure of the cradle. Sharp metallic debris, as well as non-metallic materials, were readily apparent. The immediate concern was that parts of the aft cradle might work loose and damage the Orbiter during re-entry and landing operations. However, analysis indicated that the payload bay thermal lining was capable of protecting the underlying structure in such an event. The post-flight inspection of the Orbiter mid-body and aft bulkhead found a total of 36 debris hit. Although most of them consisted of rips on the payload thermal blankets, there were three gouges on metal cable trays and one full penetration of the payload bay aft bulkhead. None of these, however, caused any harm to equipment in the aft section. Further investigation determined that incorrect wiring had caused two pyrotechnic cords to detonate, rather than just one.[21] The greater explosive force had damaged the cradle.

Despite its maker's ambitions, the TOS was never used again, either onboard the Shuttle or on an expendable launcher.

---

[21] The second cord was only a backup and was not meant to be used unless the primary one had failed.

3

EVA Operations

"THERE WAS NO REQUIREMENT FOR EVA"

Extravehicular activity (EVA), which is commonly known as "spacewalking," can undoubtedly be regarded as the quintessence of flying in space. In fact, the beauty of spacewalking and working in the void of space was best summarized by Edward H. White II on becoming the first American to step outside his spacecraft, when he said, "I feel like a million dollars." Training for and conducting a spacewalk is one of the most popular assignments. It is also vital to the program. Without spacewalks, how else could the Hubble Space Telescope have been serviced or the International Space Station assembled?

It is therefore rather perplexing that as legendary Apollo flight director Eugene F. Kranz recalls, "At the time that the Shuttle requirements were being levied upon the programs to build them, there was no requirement for EVA." James W. McBarron II, who had overseen spacesuit development at NASA since the Gemini program, adds, "The NASA perspective of a Shuttle was an airplane that could take off and land on runway, and the people inside…would not need suits. The original Shuttle program never had an extravehicular capability, in fact." It was certainly a bold approach, but as four-time spacewalker Jeffrey A. Hoffman notes, "They realized the Shuttle was still a new vehicle and that we should be putting all of our attention to just learning how to fly it safely, how to do the basic mission, how to launch satellites. If you put a spacewalk on top of that, people's attention is going to be diverted and that could be dangerous." Other people reckoned that spacewalking would simply raise additional and unnecessary safety issues. In fact, the unforgiving space environment becomes even more treacherous when the astronauts leave the safe habitat of their spacecraft and trust their lives to the protection of a spacesuit. Spacewalks were also considered an expensive capability, not only in terms of spacesuit and training development but also of the necessary added mass reducing the useful payload. In fact, not only would heavy, bulky spacesuits need to be stored in the crew compartment, there would also need to be an airlock which, regardless of how simply it was designed, would further reduce the payload.

© Springer International Publishing AG 2017
D. Sivolella, *The Space Shuttle Program*, Springer Praxis Books, DOI 10.1007/978-3-319-54946-0_3

But as the Orbiter design took place a number of Achilles' heels rapidly surfaced which, since they had the potential to seriously jeopardize crew safety, needed to be addressed.

The first weakness concerned the complex mechanisms to hermetically seal the large payload bay doors prior to re-entry. Although the doors did not form part of the primary structure, they had the all-important function of producing an aerodynamic fairing to prevent the hot plasma generated during re-entry from invading the payload bay with consequences all too easy to envision. As simply put by Robert L. Crippen, the pilot of STS-1, "They were absolutely critical. As far as re-entry was concerned they had to be closed." To make the requirement official and to get it accepted by the entire NASA community, Kranz incorporated it into the flight rules which governed how a mission was conducted. "We had a mission rule that said we've got to have these payload bay doors latched for entry. So unless all of these latches work, we're not going to have a safe entry." If the automatic closing mode didn't work, the only alternative was to close them manually. This meant going outside. And that made a strong case for a Space Shuttle-based EVA capability, at least on a contingency level. As Crippen continues, "So we started trying to come up with ways that if they didn't close normally, I could go out and get them closed. We devised winches with ropes and whatever, pulleys to pull them down, and then latches that we could go apply when the real latches didn't work. So we did quite a bit of simulation of that, and I felt fairly confident that if we ran into a problem, perhaps I could fix it. But it was not something I was looking forward to." Fortunately, the doors functioned satisfactorily on every Shuttle mission and this contingency measure was never put into action.

A second weak spot concerned two small doors on the underside of the Orbiter's aft fuselage section. Their function was to hermetically seal two cavities housing the structural attachments between the Orbiter and the external tank, and the disconnects through which the external tank fed the main engines during the ascent. While on the ground and during powered flight they were kept open. When the external tank was jettisoned they had to close to restore the continuity of the thermal protection on the Orbiter's belly. The doors were closed manually by the crew via a series of switches, but as Ronald L. Newman, a spacesuit engineer and manager who worked for the space agency, explains, "One of the big concerns was if the latches don't work, if the doors don't spring closed properly, re-entering could be fatal because you must have the tile protection there. So each door had backup mechanisms to help make sure it would close, but everybody was wondering, 'Well, sometime we may have debris in it, or maybe some other reason that it just doesn't close. We have to get astronauts out there to fix it.'"

Another key reason to develop EVA capability, at least on a contingency level, concerned the integrity of the delicate jigsaw puzzle formed by thousands of silica-based tiles on the Orbiter's belly. Even before the first Shuttle flight, the possibility of damage to the thermal protection system during ascent had prompted NASA engineers to devise ways for a crew to make repairs in space. Several studies were initiated with the objective of developing techniques for repairing a wide variety of possible damage. As Robert Crippen remembers, "The tiles were a big concern. Initially, when they put the tiles on, they weren't adhering to the vehicle like they should. In fact, *Columbia*, when they brought it from Edwards to the Kennedy Space Center the first time, it didn't arrive with as many tiles as it left California with. People started working very diligently to try to correct that problem, but at the same time people said, 'Well, if we've got a tile missing off the bottom of the spacecraft that is critical to being able to come back in, we ought to have a way to repair it.' So we started looking at techniques, and I recall we took advantage of a simulator that Martin [Marietta Corporation] had out in Denver [Colorado], where you could actually get some of the effect of crawling around on the bottom of the vehicle and what it would be like in zero-g. I rapidly came to the conclusion I was going to tear up more tiles than I could repair, and that the only realistic answer was for us to ensure the tiles stayed on."

As the launch of STS-1 approached, the effort to develop an array of tools and materials to enable an astronaut to restore the protection capabilities of the thermal protection was canceled, in part because of a variety of technical problems but also due to a renewed confidence in the tiles themselves.[1]

Contingency situations apart, the successful emergency repairs that the first crew of Skylab performed to save an otherwise doomed space station offered McBarron a strong case for advocating a Shuttle-based spacewalk capability. "We got approval to award contractors study proposals to show the benefits of EVA to a system like the Shuttle. We couldn't say it was for Shuttle, but like the Shuttle. These studies went out to all the satellite manufacturers and asked them how EVA could benefit the in-flight maintenance and servicing of satellites."[2]

The Gemini and Apollo programs had given NASA a lot of experience in the art of spacewalking. Sorties from either type of capsule required every astronaut to wear their complete ascent flight suit because, with no airlock, the capsules had to be fully depressurized prior to opening the hatch to commence EVA operations. This meant the habitable part of the capsule had to be able to cope with the vacuum of space and withstand the large thermal excursions as the spacecraft flew through the orbital day and night cycles. Consequently, development costs were driven up. For the Shuttle, this approach would have been highly impractical. With a regular crew numbering seven astronauts, can you imagine all of them donning spacesuits to allow just two of their number to work outside for several hours?[3] And what about the payload mass penalty of having seven heavy spacesuits?[4] Where would the suits have been stored inside the confined crew compartment? Plus, there was not enough funding to make the crew cabin and the sophisticated

---

[1] Unfortunately, on a Saturday morning 22 years later, when *Columbia* disintegrated over Texas as it was returning home, NASA realized that it would have been wiser to have continued the development of those studies for on-orbit repairs of the thermal protection system.

[2] As we will see in in Chapter 5, some satellite recovery and repair missions were flown that accounted for some of the most spectacular EVAs ever performed from the Shuttle.

[3] Normally astronauts would spacewalk in pairs. During STS-49 (May 7–16, 1992), which was *Endeavour*'s maiden voyage, a contingency situation developed in retrieving the Intelsat VI communications satellite and it was decided to send out three crew members.

[4] During ascent, Gemini and Apollo astronauts wore the same spacesuit they would use for spacewalking. Before the *Challenger* disaster, Space Shuttle astronauts wore just a blue overall. Subsequently ascent and re-entry were flown with orange partial pressure suits that provided some protection in case the crew had to bail out at low altitude. However, those suits would have been inadequate for EVA operations.

hardware it housed compatible with the vacuum of space and its unforgiving conditions. Although it was seen as an expensive item, it was decided to add an airlock to the Shuttle to allow two fully suited astronauts to be completely isolated from the rest of the crew to undertake an EVA without requiring the cabin to be depressurized. Although the airlock increased the overall weight, this was found to be acceptable because, as McBarron recalls, it was designed to "be self-sustaining, except for servicing things like oxygen resupply and water cooling while we were in the airlock from the vehicle."

Affixed to the rear wall of the middeck, the airlock was a cylinder 83 inches high and 63 inches wide. It incorporated two 39.4-inch diameter D-shaped openings with pressure-sealing hatches. It accommodated all of the EVA gear, checkout panels, and recharge stations. All hatches were mounted on the airlock itself. An inner hatch on the external wall isolated the airlock from the cabin. This was opened by an astronaut pulling it inwards some 6 inches and then pivoting it up to the right. It was closed by reversing the procedure. The outer hatch isolated the airlock from the unpressurized payload bay when closed, and gave access to the bay when open. During opening, it was first pulled into the airlock, then forward at the bottom and rotated down until it rested with the outer side facing the airlock's ceiling. Each hatch had six equally-spaced latches, a gearbox, and actuators, two pressure equalization valves, and hatch opening handles. Each hatch also had two pressure seals, one on the hatch cover and the other on the structural interface with the airlock. There were differential pressure gauges and relief valves. Later on, when the Shuttle carried out dockings with space stations, the airlock was relocated to a truss in the payload bay and augmented with a tunnel to the middeck and a docking mechanism on top.

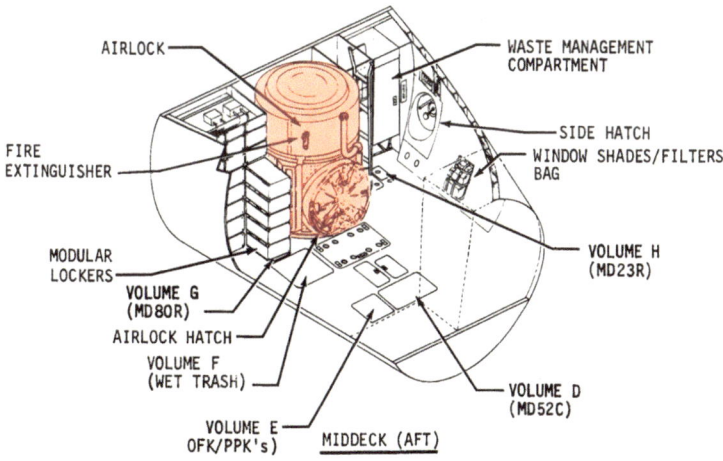

The location of the airlock on the middeck.

A view of the airlock relocated into the payload bay. The docking mechanism on top of the airlock is also visible. With the airlock in the payload bay, the astronauts were able to enjoy a much roomier cabin. This is *Atlantis* on display in a museum.

We have seen how a contingency spacewalk would be needed in the event of the payload bay doors failing to close fully. It is clear that the outer hatch of the airlock provided the only access route to the payload bay in flight. So if you are wondering what would have happened if that hatch failed to open, *Columbia*'s STS-80 flight is your answer. In addition to going down in the history books as the longest mission,[5] it was also the first and only one on which the airlock wouldn't open. The post-flight inspection found that a loose screw had jammed the hatch's gearbox, preventing the two spacewalks planned for the mission. As disappointing as it can be, especially for the astronauts assigned the EVAs, this was a critical condition because if the payload bay doors had not closed properly the contingency of sending out astronauts to do so manually would not have been possible. However, Story Musgrave, who served as a mission specialist on that flight, once told me, "We would have done something, we would have broken the handle, we would have broken the gears, we would have busted our way out that door. You can get access to the mechanism with the robotic arm, you could have shaken the mechanism because the latches were outside."

---

[5] In fact, *Columbia* logged 17 days 15 hours 33 minutes.

A WARDROBE FOR THE ASTRONAUTS

Having agreed to baseline a contingency EVA capability, it was now time to develop a spacesuit. NASA had already gained a long and valuable experience in designing and handling spacesuits. In fact, spacesuits worn during the Apollo program allowed a small group of spacefarers to walk on the surface of another celestial body. Given the available know-how and the chronic shortage of funding, the issue was whether the Apollo suit could be repurposed for the Shuttle, maybe with minor changes, or it would be better to fashion from scratch a brand new model. Perhaps not surprisingly, the decision hinged on the price. "I think I put together 13 different cost estimates for different ways of keeping the cost minimized for a new program," remembers James McBarron. "I included such things as using the Apollo backpack with a new suit, the Apollo suit with the old backpack, reuse of helmets and new designed helmet, reuse of bearings and suit hardware versus new hardware." After considering permutations, the space agency reckoned that a new space-suit model for use on the Shuttle should be developed in a way that salvaged as much as possible from the Apollo suits which were in storage. "It was resolved that we would develop a new [spacesuit] with reuse of some Apollo hardware such as the design of the helmet, neckring, gas connectors, and wrist disconnects. We actually stripped all that hard-ware off the Apollo suits (apart from those that flew and went to the Smithsonian) and provided them as GFE [Government Furnished Equipment] to the new contractor for the Shuttle spacesuit."

Despite the reuse of the Apollo hardware, the request for proposals NASA issued to industry contained a number of requirements that would bring the art of fabricating EVA outfits to a whole new level. Given that a flight rate of some 50 flights a year was envi-sioned, and that each flight would need at least two spacesuits for the EVA contingency capability, continuing with the disposable approach employed during the Gemini and Apollo programs would have been a useless drain on financial resources. A first essential requirement, therefore, concerned life expectancy, which was set to a minimum of 6 years and 100 sorties. In other words, the Shuttle suits would have to be reusable.

A second requirement had to do with making the spacesuit an integrated system. The Apollo suit was a masterpiece of engineering made of two separate components: the anthropomorphic pressure vessel that enclosed the astronaut, and a portable life support system carried as a backpack using straps and plugged into the suit by means of hoses.

For the Space Shuttle spacesuit or, more specifically the Extravehicular Mobility Unit (EMU), NASA demanded integration of the pressure vessel and the life support system to eliminate the straps, cables, and plugs that had proved troublesome on the Apollo suits. It also meant that only one contractor would be required, unlike Apollo where ILC and Hamilton Standard were the prime contractors for the pressure vessel and the life support system respectively. In 1977 Hamilton Standard was nominated as prime contractor responsible for the development and integration of the EMU. It would also be in charge of designing the life support system. ILC was subcontracted to supply the pressure suit.

The need for reusability led to the specification of generic sizing, or modularity. Gemini and Apollo astronauts had enjoyed the luxury of custom-built spacesuits that were tailored to their bodies. In fact, they received three such suits. In addition to the primary suit that

Scientist-astronaut Harrison H. Schmitt collects rock samples on the Moon during the Apollo 17 mission. The spacesuit's anthropomorphic vessel and portable life support system backpack are clearly visible.

they would wear in flight, they had a backup suit and a training suit. Given the high flight rate that was envisaged for the Shuttle and the large number of astronauts needed to fly so many missions, not only was the disposable concept out of question, the suit must be able to accommodate the largest possible range of male and female sizes. As Thomas V. Sanzone, who had worked at NASA on spacesuit hardware since Apollo, explains, "We had to fit between the fifth percentile female to the ninety-fifth percentile male. That's a whole bunch of folks."

Reusability, integration, and modularity combined to create the EMU that proved to be a real masterpiece of spacesuit engineering. For example, a consequence of the reusability requirement was a departure from using certain materials or mechanical elements which featured on the Apollo spacesuits, such as cables, swages, zippers for pressure seals, and Neoprene or rubber compounds for the soft parts. None of these would have been able to endure the wear and tear of frequent use, at least not without a costly refurbishment. Besides, during Apollo they all caused a considerable amount of trouble, such as pressure leakage and nicks, which the Shuttle program hoped to avoid.

The integration requirement split the EMU into two main components: the Lower Torso Assembly (LTA) and the Hard Upper Torso (HUT). These were the space suit assembly, the anthropomorphic pressure vessel within which the astronaut would be enclosed and sustained by the integrated life support system.

The LTA enclosed the waist, lower body, legs, and feet, and was comprised of a number of layers for protection against abrasion, thermal stress and micrometeoroids, and mechanical resistance to prevent the suit from ballooning when pressurized. The HUT enclosed the astronaut's upper body and had a joint on each shoulder for the right and left arm assemblies. The Apollo suit had required the astronaut to crawl into the pressure vessel and a second person to pull a long zipper running along the back. Shuttle astronauts would first don the LTA and then push up and crawl inside the HUT while that was hanging on the airlock wall by means of an adapter plate. Once settled, the astronaut would pull the lower torso up and connect the two sections at the mid-stomach using twelve equally spaced latches. All mechanical joints between the assemblies were dry-lubricated ball race bearings that afforded lots of flexibility and a broad range of movement for the arms and upper body.

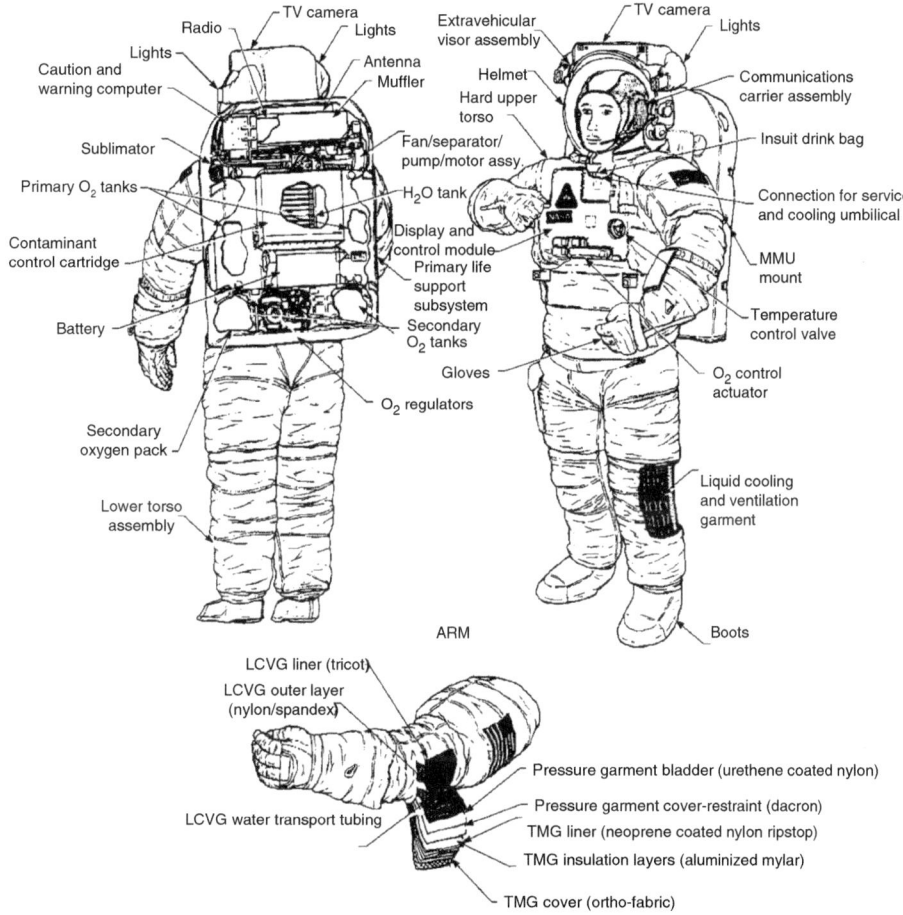

The principal components of the Space Shuttle Extravehicular Mobility Unit.

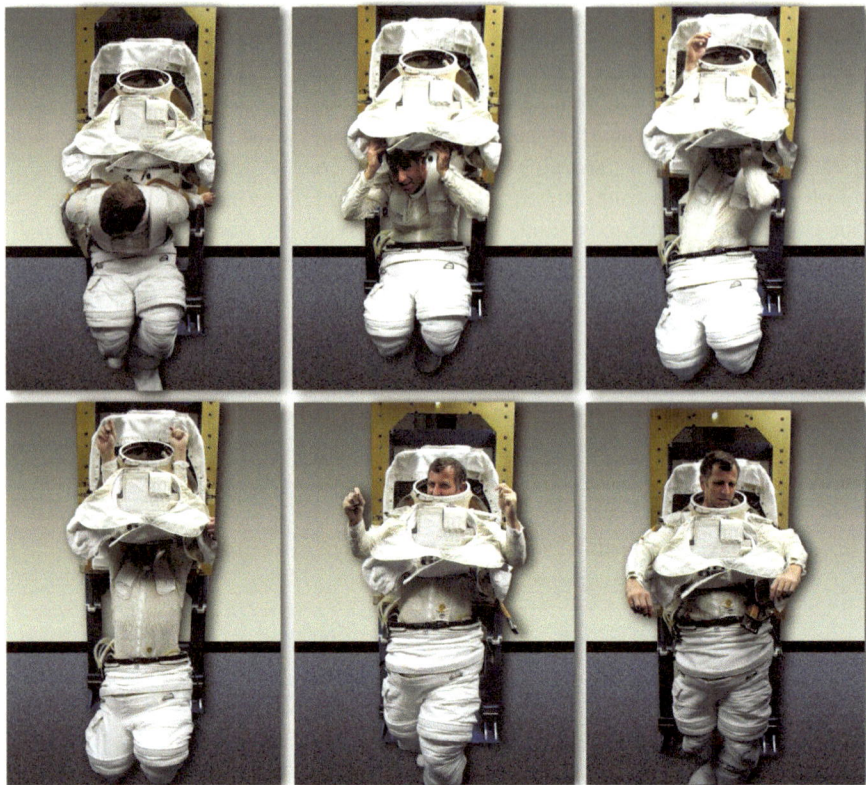

The procedure for donning the EMU. The arms of the suit have been removed to illustrate the range of shoulder and arm motion required to ingress the Hard Upper Torso.

The combination of upper torso, lower torso, and arm assemblies provided the necessary modularity. Sanzone continues, "We didn't really make suits…The best way to understand it is to think of all these parts being in bins. We would literally measure an astronaut like a tailor would, and then go to a bin and get the lower left arm that we thought would come closest to his size, and then the upper left arm, and we'd bolt those together. Then we had some ability to do some tweaking, to let it out a little or to pull it in a little. It was the same process with the lower torso." In other words, the EMU was a sort of a mix and match spacesuit.

The description of the upper torso assembly as being "hard" refers to its structure, which was markedly different from that of the complementary lower torso assembly. Its load-bearing structure was a fiberglass shell that provided mounting points for the primary life support system hardware on the back and the display and control module on the front. As Sanzone says, both the "primary life support system and the display and control module were bolted onto the suit, and were semi-permanent. From the perspective of an astronaut on-orbit, they were permanent."

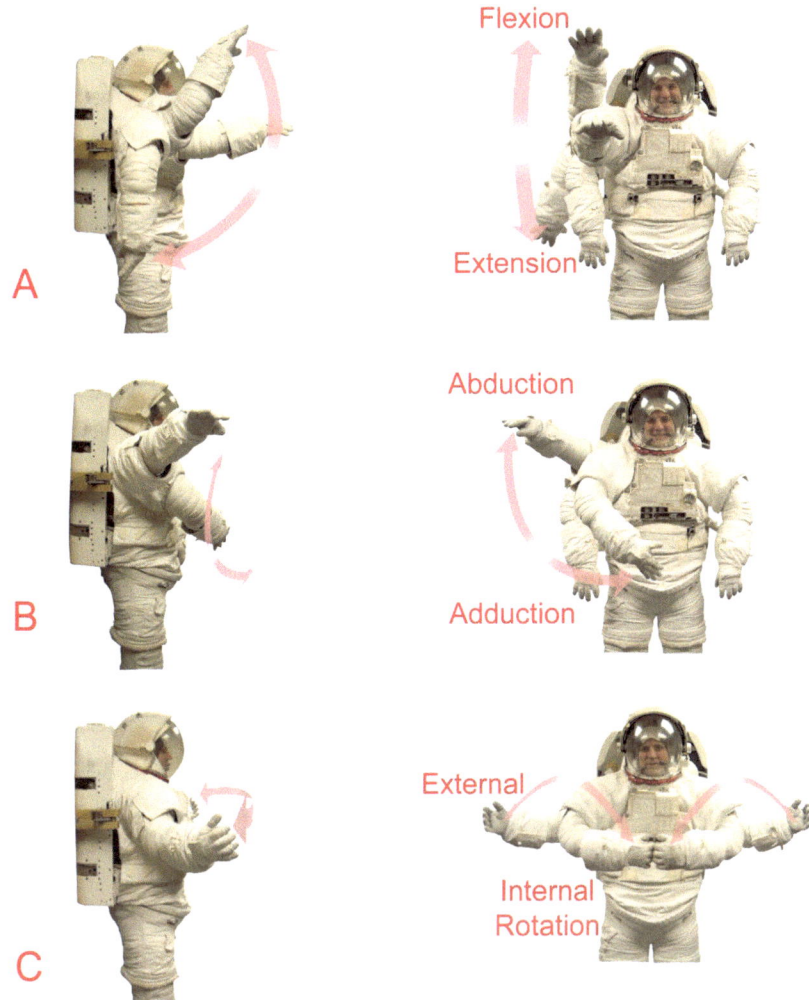

A subject demonstrating suited shoulder movements.

Passageways built into the fiberglass shell enabled cable bundles to run between the two hardware packages, so that all the electric cabling was accommodated within the spacesuit. As Sanzone recalls, there is a curious anecdote about the design of the Hard Upper Torso. "Hamilton designed the front-to-back dimension from the front of the display and control module box on the front of the suit to the back of the primary life support system to be 19.75 inches. They advertised as a special feature that they could get from the middeck where the suits were stored up to the flight deck through the opening of probably 20 inches." This was not a NASA requirement, but Hamilton was happy to advertise it, so

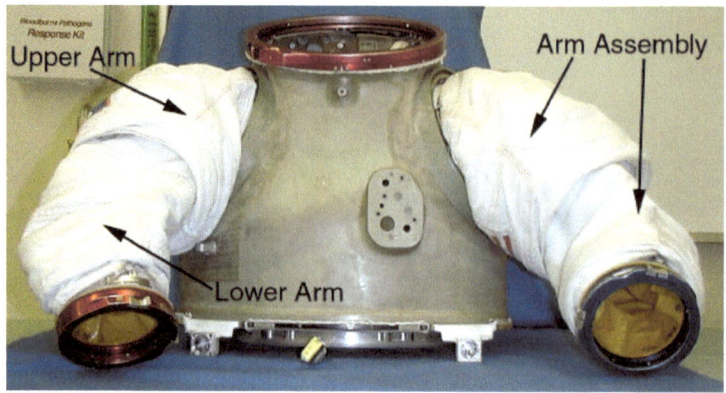

The hard fiberglass shell of the Hard Upper Torso is clearly visible along with the two arm assemblies.

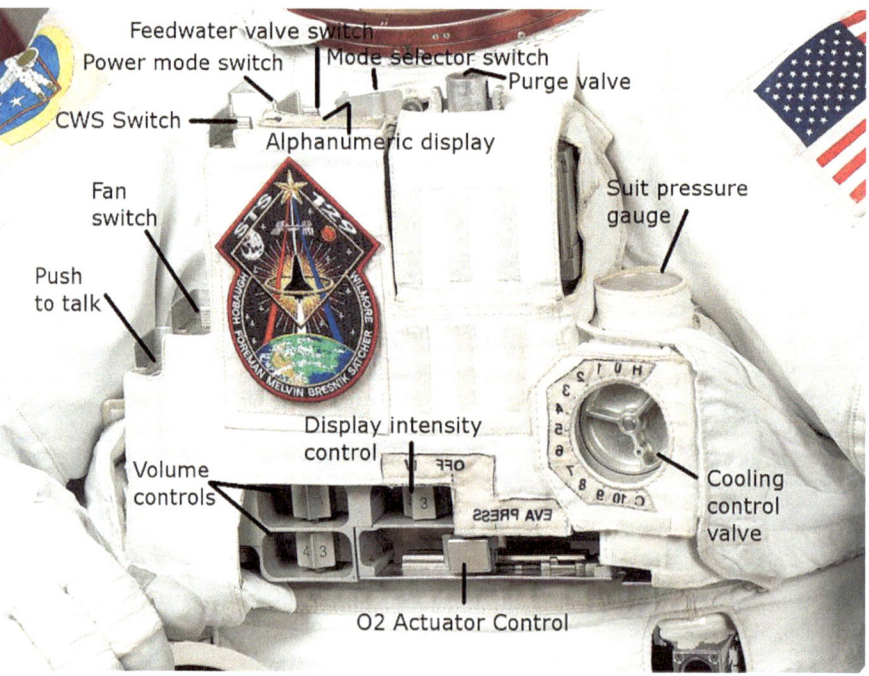

A close-up of the display and control module permanently fixed to the front of the Hard Upper Torso.

much so that Sanzone carried out some tests to confirm their claim. "For years people would come back to me and ask me, because I was the test subject, 'Can you really do this?' My answer then and now was, 'If the question is, could you physically get a person in a pressurized EMU from the middeck to the flight deck with someone pulling and someone pushing, with you having no control whatsoever because you had to go one arm forward and one arm back?' then it could physically be done. Would you want to consider doing it? No. I wouldn't recommend that."

Anchored to the rear of the upper torso was the sophisticated primary life support system whose role was to provide a safe living environment for the astronaut. To this effect, it addressed functions such as supplying breathing oxygen, suit pressurization, cooling, communications, and control and monitoring of the resources and operations of the EMU. Attached to the base of the primary life support system, the secondary oxygen system could back up the main oxygen system and provide a minimum of 30 minutes of emergency oxygen while maintaining suit pressurization and a degree of cooling.

Developing the spacesuit brought about its own set of challenges. For instance, as Harold Joseph McMann Jr, notes, "A life support system and a suit are very, very different. A life support system has a lot of metallic components. It is governed by drawings and very rigid specifications, and you can measure performance specs, and get tolerances down to ten-thousandths of an inch or so. On a suit, the soft goods in particular, the tolerances are much more liberal. They have to be. You sometimes get tremendous variance from lot to lot of material. You don't have drawings; you have patterns. So you have a different way to even describe what you are going to make." Hence the life support system and suit were two separate entities, each with its own life. "It would be nice if I could say a cycle on the life support system is the same as a cycle on the suit, but it isn't. So you have to define your certification requirements differently. The suit typically has components or materials in it that have a lot shorter life. The material ages, and maybe in 3 years, or 6 years, or 8 years, you've got to replace it. On the life support system, maybe you think that it's good for 15 years and you find out later that you can go to 20 years on it as maybe more handbook data gets gathered. For the logistics of suits, you size elements. You don't size elements for a backpack. It's an entirely different logistical problem. They are enormously different, yet they have to work together."

The Hard Upper Torso was something that demanded early attention, in particular after ground tests by astronaut Story Musgrave, who oversaw the development of the EMU. "It turned out to be very difficult for astronauts, particularly Story, to get into this hard upper torso," Sanzone says. "He was crawling through a tunnel. He had big shoulders. You had to see it to appreciate it. He would screw himself into this thing. It became obvious pretty quickly that we needed to make a significant change to the upper torso. The main change was that they actually shortened it. It was down like to your waist. They pulled it up maybe to your mid-stomach. That shortened this tunnel that you went through."

Further issues were being highlighted by some of the female astronauts who were participating in EMU testing. These mostly involved the smaller size of the average female body, as compared to that of a male body. Kathryn D. Sullivan, who on STS-41G became the first American woman to perform a spacewalk, remembers that two main factors were the "shoulder width compared to the placement of the shoulder bearings, and the depth of your chest cavity compared to the front-to-back dimension of the suit. The scye bearings, as they're called, that let the shoulders move are pretty large bulky complex assemblies. Physically, there is only so close you can put them on the hard upper torso before they'll start to impinge on each other and not work. Secondly, the front of that hard shell is the acreage to mount the display and control module on. If you pull those shoulder bearings too close together, you don't have any place to bolt the controls on." This imposed a limitation on how small the upper torso could be. Astronaut Anna L. Fisher continues the story, "They kept trying to get the upper torso small enough, but there's only so small you can get that hardware…They decided to just go with the 'medium' and the 'large' sizes. If you didn't fit in those, you could not be a candidate for a spacewalk. I didn't disagree with that decision, because the extra small HUT, the hardware that's in front, all your gauges and all the controls, you can't really make that smaller. Even if you make the torso small enough to fit, your useful arm reach, which is where you do most of your work when making a spacewalk, is going to be much less than one of the guys with their big orang-utan arms."

The constrained size of the upper torso also had an unexpected effect on training and astronaut qualification for spacewalking duties. Training for spacewalks is done in a large pool on the ground. The buoyancy provided by the water provides a good analog of the weightlessness of the space arena. Nevertheless, you are still affected by gravity and, as Sullivan explains, "If the suit doesn't fit you firmly front-to-back, you'll fall back in the suit so that your back is physically against the back of the suit, your fingertips pull back away from the gloves, and you won't really be able to do work. There is no reason to think small-frame people cannot perform well in zero gravity, when that phenomenon of falling to the back of the suit wouldn't happen." However, as Sullivan continues, "It can make it impossible or extremely hard for you to demonstrate top competency at EVA tasks in training if every time you reorient your body, your hands may come out of the gloves. It means you can't grip and work effectively. That's where some of the complexity of the argument comes in." Bottom line, it can be difficult for the training team to assess if an astronaut is indeed suitable to perform a spacewalk. He or she might not perform as well as expected due to the lack of a tight fit in the suit, but what if they are not really cut out for extravehicular activities? If you think that performance is biased by the combined effects of suit size and gravity, you might actually select the wrong person. Sullivan further explains, "If I'm training six people as candidates for that task, and I can see five of them perform it superbly underwater, and I never get to see one of them perform it well…how can you expect me responsibly to pick the person that I haven't seen? I wouldn't. You have to live up to your responsibility to the program."

Sullivan also experienced additional fitting difficulties. "The torsos fit me well. I was almost getting bruised getting in and out of the suit. If I inflated my chest really fully, I'd feel the hard points on my ribcage, on the sides, and could feel it front-to-back. So it is a good incentive to keep your breathing nice and even…The placement of the mechanical elbow should be lined up with your anatomical elbow. Mine was always about an inch and a half or two lower. When I wanted to bend the suit arm, I wasn't actually pivoting it. It was like bending a balloon. I had to force the lever, so it took some extra strength to bend that. I adapted to that. My knees also never actually lined up with the suit knees. The joint of the suit was below my anatomical knee. So it wasn't a nice easy bend here. The suit would actually hit my calf, and I would have to push around the corner to make the knee bend."

The life support system's oxygen supply also had its share of troubles. In fact, on one occasion it almost claimed the life of Robert Mayfield, a Hamilton Sundstrand technician, as he was preparing to run a test to get the EMU ready for its first manned chamber test run. As Sullivan explains, "One step of the procedure is you take the mode controller, just like you'd do in flight, and move it from the IV [intravehicular] position where all the fans and pumps are running and the suit is pressurized, you move it to the EV [extravehicular] position, then you go through the door. The main thing that happens from IV to EV, is a valve is opened which enables the emergency oxygen bottle. The isolation valve that shuts it off from the system is opened so that that 6,000-psi oxygen flow is sitting on a regulator, ready to flow into the suit if the suit's pressure drops. If ever you rip the suit, you don't have to do anything to start flooding the suit with emergency oxygen. As soon as the pressure gets low enough, the bottle will start to flow. When they moved the switch of the mode controller from IV to EV, the suit exploded. The whole torso just exploded in a huge big oxygen ball fire that needlessly to say severely burned the suit tech that was working on it."

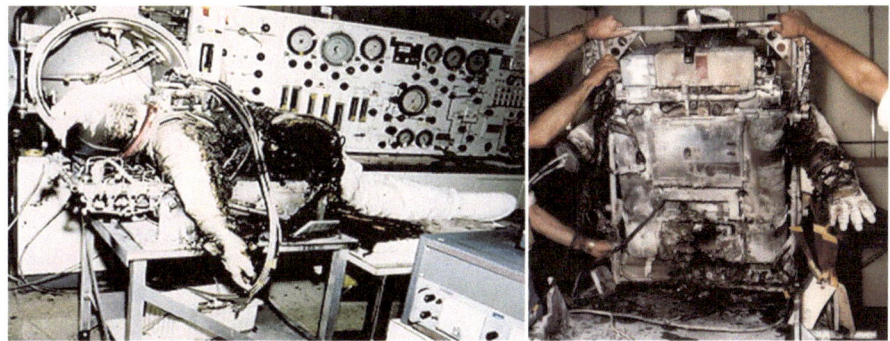

This is what $2 million worth of EMU damage looks like!

The investigation was unable to duplicate the failure, but the probable reason was localized heating caused by shockwaves that were made by the high pressure oxygen flowing from the secondary oxygen storage tanks into the regulator assembly whose task was to reduce the oxygen supply pressure to a breathable 3.5 psi. As Sullivan says, this revealed "subtle but significant factors in the shape of the channel that high pressure oxygen flowed through, and the likelihood that there was a little bit of residual organic matter. The suit had not been really fully properly cleaned. You slam in oxygen, the rising pressure heats things up, and there's just enough organic stuff there to be volatile."

After investigating for 5 weeks, the team made several recommendations such as to redesign the high pressure oxygen valves and regulators to avoid stagnation zones and minimize the chances of internal contamination; to replace silicone-rubber seals with a material with greater resistance to ignition; to make more detailed inspections during manufacturing; and to replace the flammable aluminum alloy of the pressure regulator with the highly heat-resistant Monel alloy.[6]

---

[6] Monel is a nickel-copper alloy whose high strength properties at high temperatures make it ideal for use in applications such as jet engine turbines and combustion chambers.

A TALE OF TWO SPACEWALKS

Each of the first four orbital flight tests included in their inventory two spacesuits for use in an emergency, such as the payload bay doors not latching properly. Although this contingency never developed, the EMU received its first orbital trial run during STS-4, in preparation for the first spacewalk of the Shuttle program scheduled for the next mission. In fact, due to the need to have at least two crew members outside for mutual support (called the buddy system), clearly the first spacewalk had to wait for the first opportunity on which there was a crew of at least four people. And this was STS-5.

STS-4's commander Thomas K. Mattingly was to don the EMU inside the cabin to address two fundamental questions. The EMU had been extensively tested on the ground, and was considered ready for use when *Columbia* flew its first mission, but there was no substitute for testing hardware in the environment for which it had been designed. For the EMU, the first question was whether an astronaut would be able to latch together the upper and lower assemblies of the suit, possibly by himself or with minimum assistance from a crewmate.

As spacesuit engineer Ronald L. Newman says, "After a person climbs into the upper torso…you have got to attach the pants to the upper part of the suit. It's a ring with around 12 latches…and each of those latches has to close."

The second question was whether Mattingly would be able to fit into his spacesuit at all. It is a known phenomenon that the weightlessness experienced during orbital flight relaxes tendons and ligaments. Because the body is no longer compressed by gravity, body fluids seep into the vertebrae. As a result, the body undergoes a stretch of an inch or so. Apart from causing backache as the skin and nerves are offset, this might cause some difficulty fitting into a spacesuit. During training, astronauts like to have a tightly fitting spacesuit which gives them more mobility. If it were loose, they would slip about inside the suit as they moved, with the result that their hands would move in and out of the gloves, impairing their dexterity. Newman remembers, "There was a design feature early on in the Shuttle suit where there was a bracket and if you pushed a button the suit would get one inch longer." But because this was a complex mechanism, difficult to manufacture, and heavy, it was soon discarded for a simpler solution. It was decided that the astronauts should use a tightly fitting suit in training and then add anywhere between one to two inches to the length just prior to launch. Thus they would still be able to achieve a tight fit. "When Ken Mattingly put the suit on, it was a nervous time for us, because no one had ever done this before. It was a big deal for him to put the suit on. He put the suit on. He was able to close the suit. Over the com loop he said, 'They were supposed to make the suit longer, but it feels just like it did on Earth.' It was a problem averted." The EMU was now ready for its first real outing from the Shuttle.

On the fourth day of STS-5, Joe Allen and Bill Lenoir were eager to write another chapter in the history of space flight, by making the first spacewalks of the Shuttle era. They both donned their white EMUs and "just as the spacewalk was to begin my spacesuit failed," Allen says. "It was an electrical failure in the suit. When one is in a spacesuit and you power it up, you hear a very high-pitched hum there someplace in your ear, just a high-frequency hum. When I powered mine on, the hum started but it didn't sound like it was healthy. It sounded more like an angry mosquito. I had never heard that before. I proceeded to make various electrical checks of the suit systems, and none passed the check." Allen clearly was not going to make history that day. Lenoir's suit was also suffering minor malfunctions. As he recalls, "The spacesuits are supposed to hold 4.3 psi pressure, pure oxygen. Mine was holding 4.1 and it was rock solid, which is safe." A proposal that Lenoir should go outside on his own was rapidly dismissed for fear that his suit might develop a serious malfunction when he was without a buddy. But he did have the opportunity of spending some quality time in the airlock. "What we did, was I got into the airlock and closed the door. We did a partial decompression, trying to see what was going on with my suit. So I got in near-vacuum, although I was never outside, and the suit held lock-tight."

On April 7, 1983, it was show time for the crew of STS-6. NASA was still very eager to try a spacewalk outside the Orbiter, and the gremlins in Allen's and Lenoir's spacesuits had been attended to. In particular, the inability of Lenoir's spacesuit to achieve the required 4.3 psi was traced to two tiny plastic inserts within the pressure regulator. Despite manufacturing documentation stating these were where they were meant to be, there were missing. Allen's troubles, on the other hand, were caused by an electric motor running intermittently due to a magnetic sensor damaged by excess moisture. In turn, the sensor prevented the oxygen circulation fan of the suit from working as intended. As Newman remembers, "The sensor would sense either a fan operating too slow or too fast. If the fan was too slow it wouldn't give you enough circulation to clear out the CO_2 from your breathing gas. The logic wouldn't let you operate; it would shut off. If your fan was going too fast, it might be an indication of an electrical problem and so it would shut off. The fan speed had to be a certain rate for the system to keep operating. The sensor itself didn't work. It wasn't sending the right signal to keep the system operating. So that suit didn't work."

Losing the chance to carry out an extravehicular activity is undoubtedly always a bitter pill to swallow, particularly when it could feature prominently in the history books. However, as wryly noted by Allen, "The good news was we were not outside the ship when it failed. It would have been considerably more traumatic had it failed outside. It would not have been fatal to me, but it would have gotten my attention, for sure. I would've had to scramble to get back in, button myself up, and get out of the suit before other parts of it started to fail." After the post-flight investigation, NASA implemented more rigorous spacesuit testing to prevent a recurrence of these faults, and indeed others.

On STS-6, Story Musgrave and Donald H. Peterson were to make a spacewalk. As a precaution, *Challenger* had been provided with a third upper torso assembly. In preparing, the two men first put on Liquid Cooling/Ventilation Garments (LCVG) similar to "long john" underwear but made of a dense mesh of flexible tubing which circulated water from the primary life support system to pick up the astronaut's body heat and maintain him at a comfortable temperature. The garment also incorporated a network of ducting to avoid stagnation of carbon dioxide-laden air at the astronaut's extremities and hence complete the suit ventilation loop. Both astronauts then moved into the airlock to put on their EMUs. They first donned the lower torso, followed by a shuffling crawl up inside the upper torso that was hanging on the wall. As Peterson explains. "You make sure everything's laid out properly, and everything that you can check is working properly. The whole thing is set up so that you test as you go along. In other words, you don't just put the suit on and open the hatch. You make sure that everything's working before you go that final step."

The EMU was designed to operate at a pressure of 4.3 psi, which would afford a good degree of hand dexterity while reducing the amount of effort required to move the joints of the arms. In fact, if the suit were to be pressurized with nobody inside, it would assume the so-called neutral position. As Peterson says, "If you move it away from that position, it tries to come back because the arms and all over are very rigid and they are under pressure." However, working in the suit was not for the faint of heart. "EVA would be fun if the suits weren't so hard to work in. The suits are fairly uncomfortable…So anytime you're doing anything, you're typically fighting against the suit itself. The gloves are the same way. If you look at a lot of photographs from spacewalks, you see…they don't grab something, because to do that, you must fight the glove. They wedge things between their fingers, because that way you don't have to exert pressure."

As the Orbiter cabin was pressurized at 14.7 psi, prior to starting a spacewalk the two men had to expel nitrogen from the soft tissues and fluids of their bodies. In fact, once sealed into the spacesuit at a much lower pressure than that of the cabin, their bodies would quickly transfer the nitrogen into the blood stream, leading to all kinds of unpleasant phenomena such as blood vessel clogging in the heart or brain, pain in the joints, and skin rashes. In the worst case, it can cause paralysis and death. In other words, they would suffer decompression sickness, a phenomenon known to deep-sea divers as "the bends" because the nitrogen which accumulates in the joints forces the unfortunate person to bend in pain. Although it is possible to design a spacesuit for a higher pressure and thus mitigate the danger of decompression sickness (for example the Orlan spacesuit utilized by Russian spacewalkers operates at 5.8 psi) it is easy to appreciate the downside of the increased physical effort required to operate it, which in turn translates into fatigue and exhaustion.

To neutralize the threat of pain and possible death from decompression sickness, Musgrave and Peterson spent 4 hours in the airlock leisurely breathing pure oxygen. This was the so-called pre-breathe protocol. As Peterson explains, "You put the suit on, and because the suit's fastened to the wall of the airlock you are essentially just hanging on the wall there. All you can do is just wait. Now, what happens is, as you breathe oxygen, the nitrogen, the gases that are inside your body, try to equalize with the gases that are on the outside. As you breathe, the nitrogen is being displaced by oxygen and, little by little, the nitrogen all comes out of your body – not all of it, but most of it comes out of your body. It literally goes out into the suit, but because you are hooked to the Shuttle system it gets flushed into the Orbiter." Because a spacesuit is not absolutely airtight, during this time the pressure inside the airlock was kept at a slightly lower level than that of the suit to prevent the nitrogen that was flushed into the airlock from being returned into the spacesuit and into the astronauts' bodies.

A few years later, NASA attempted to design a spacesuit to operate at double the EMU pressure to eliminate the pre-breathe protocol. Veteran spacewalker Jeffrey A. Hoffman participated in the effort to develop this Zero Pre-breathe Suit (ZPS), "Four hours [idle] in space is unacceptable…If you could develop a spacesuit to work at 8 psi, you wouldn't have to pre-breathe, because if the change in pressure is less than a factor of two, experience shows us that people don't develop the bends. So that was the motivation behind the 8-psi suit." As he explains, "We had two different kinds of suits, one developed at JSC and the other at Ames [Research Center in California]. The Ames suit was a hard suit, looked very much like a robot with a lot of articulated robotic type circular joints. It was very easy to move around in, but it was very bulky and heavy. We had to test these suits by doing a lot of construction jobs." In the large swimming pool used for EVA training, they replicated the same construction work performed by the crew of STS-61B when they tested different possible methods of assembling large structures on-orbit."[7] They tested both types of 8-psi suit against the regular EMU. "In the end, our recommendation was that neither of the suits were adequate. This was on top of the fact that nobody had been able to develop gloves to work at 8 psi." Once again, the flaw was loss of dexterity and the increased physical exertion to accomplish a job. It would have been possible to eliminate the 4-hour pre-breathe protocol, but a spacewalk may well have had to be curtailed by the astronauts becoming fatigued. Shorter spacewalks would translate into undertaking more EVAs to accomplish the same number of tasks as was possible during a single outing with a regular EMU.

Throughout the Shuttle program, every spacewalk was preceded by a lowering of the cabin pressure to 10.2 psi prior to the pre-breathe protocol. To save time, the pre-breathe protocol was usually performed during the sleep period ahead of the EVA. If you think that hanging on a wall for all that time is uncomfortable, Peterson's words will make you change your mind. "[That was] probably the best sleep I had on-orbit, because you've got fresh oxygen coming in over your head, and it kind of makes a nice whishing sound, and there's no other noise. We turned the radio receivers way down so we were not bothered by people talking. So we got some really good sleep before we went outside."

Once sufficient nitrogen had left their bodies, Musgrave and Peterson were ready to make history. After the hermetic closure of its hatch, the airlock was depressurized to 4.3 psi to allow the spacewalkers to verify that their spacesuits would maintain the pressure. Then the remaining air was bled out and the astronauts were given the go-ahead to open the outer hatch and venture into the payload bay.

As Story Musgrave left the airlock he became the first spacewalker of the Shuttle era. All in all, it was only fitting that he would be the first, since he had played a key role for many years in developing the EMU and EVA operations. In fact, as Peterson recalls, "Story knew everything about the suit there was to know. He had spent 400 hours in the suit [testing it] in the water tank, so he didn't really have to be trained." Although the Orbiter had the capability to maneuver to recapture a spacewalker who became adrift, both men anchored themselves in the payload bay using safety tethers. As Peterson explains, "You have tethers. You have a safety tether that is a little short tether. Then you have the safety reel that you use when you're outside, because you need to be able to move around. So what you do is you hook the safety tether inside the airlock, and you go out through the hatch. There's a line that you hook your reel to. Now you're hooked on the slide wire, and you can run up and down the handrail and drag this tether up and down with you. Then you unhook the safety tether that is hooked inside the airlock. So you're always tethered to something."

---

[7] Refer to Chapter 4 for more details of the fascinating STS-61B mission.

Musgrave and Peterson were tasked to verify the operational adequacy of the full EVA system, including the EMU, airlock, payload bay provisions for spacewalking, procedures, timelines, and the fidelity of training. As regards EMU testing, Peterson recalls, "We used different tools and filmed everything. We did certain motions with our arms and legs and head and all that to make sure that the suit was okay. But the basic test was just to get the suit out in the environment and see if the suit was going to work properly…We were just testing equipment and the deal was, if something went wrong, you'd just stop and come back inside…The suits worked differently, a little differently, in the real environment than we had seen them work on the ground, but basically they did exactly what they were supposed to, and they worked the way we expected them to."

The availability of the airborne support equipment that had been used to deploy the IUS/TDRS-1 combination earlier in the mission[8] offered a useful playground for the two astronauts, in particular to assess contingency EVA operations. As Peterson says, "It had to be tilted back down before we could close the payload bay doors and come home. So instead of driving it with the electric motors, they said, 'Let's go back and see if we can crank it down with a wrench, to simulate a failure. We'll see if we can do that.' So that was one of my jobs. We had foot restraints, but it took so long to set them up and move them around that we didn't want to do that. So I just held on with one hand to a piece of sheet metal, which is not the best way to hold on, and cranked the wrench with my other hand while my legs floated out behind me."

All of sudden, Peterson's EMU declared an emergency when its internal sensors detected an oxygen leak. Together with Musgrave, he performed a number of cross-checks. Then some 20 seconds later the alarm ceased. So the two men resumed work as if nothing had happened. "Everything seemed okay," Peterson recalls. "So we just finished what we were doing. I mean, everything seemed all right." Nevertheless, the cause of the anomaly had to be identified. The only explanation that Mission Control could offer was that as Peterson was working so hard that his increased demand for oxygen had fooled the sensor into believing there was a leak. However, Peterson did not believe that was the true problem. "I talked to some of the doctors, and they said, 'We don't think that's right.' Well, my heart rate was very high. Working in the suit is very hard. My heart rate was 192, okay, when I was cranking that wrench, so I was working very hard at the time. But a guy my size can't work hard enough to breathe enough oxygen to set off the alarm that way."

The real cause only came to light 2 years later, when astronaut Shannon W. Lucid was performing a test run of her spacesuit. As she was walking on the treadmill, the alarm on her suit went off to warn of an oxygen leak. That a leak was in progress was also apparent from the rising pressure inside the vacuum chamber. When Shannon stopped and stood still for a while, the alarm (just as in Peterson's case) cut off. For a technician monitoring the test, this was a Eureka moment. As Peterson explains, "He said, 'I have seen this same thing before. I don't remember the details, but I've seen this same phenomenon before.' They went back to the video of my flight and looked at it. The guy said – and this is kind of interesting – he said when Shannon Lucid was walking, since she's a woman, her hips swivel, and her suit was actually rotating, and we had never seen that with a guy, because

---

[8] Refer to Chapter 2.

guys don't walk that way. But he said, 'That's the same thing that happened to Peterson's suit 2 years ago.' As I cranked, my legs were flailing like a swimmer, in reaction to the loads I placed on the wrench. The waist ring was rotating back and forth, the seal in the waist ring popped out, and the suit leaked badly enough to set off the alarm." The waist seal was redesigned and the problem never recurred.

The 4.5-hour EVA by Musgrave and Peterson was a fairly unpretentious exercise to get NASA back into the spacewalk business. By the time the Shuttle retired, three decades later, the tally was an astonishing 160 spacewalks. No other space program by any nation can claim such an achievement, and never will for a long time. These spacewalks were performed to rescue and repair satellites, service the Hubble Space Telescope, and assemble and maintain the vast International Space Station. Hundreds of hours spent in EVA training and in space provided a wealth of valuable know-how that is being applied to the development of new spacesuits and future activities which will occur not only in open space but also on the surfaces of the Moon, Mars and the asteroids.

EMU IMPROVEMENTS

Spacewalking from the Shuttle and the International Space Station owes its success mainly to the EMU. The fact that this has remained basically unchanged attests to its excellent design and margin of performance. As Harold Joseph McMann Jr, says, "The Shuttle EMU was supposed to be good for three EVAs before it required some ground servicing. We tested it, and realized it is good for twenty-five EVAs between servicings. That's a damn good thing…How can we do it? Answer: Margin." In fact, the current spacesuits used on the ISS are nothing less than enhanced EMUs which offer the capability of resizing the arm and leg segments on-orbit by using aluminum sizing rings and adjustable-length restraint lines. This means that the spacesuit can be tailored to any space station resident in space, as opposed to on the ground prior to a Shuttle mission. It is easy to see that for ISS astronauts this was not a viable option, as it would have required sending a tailored spacesuit every time a spacewalk was due. Apart from the cessation of the Shuttle program, that approach would have been very expensive and impractical. In fact, on several occasions contingency spacewalks have been performed to repair a given critical system on the ISS. In such a situation, where an EVA must be made as soon as possible, it is essential to have spacesuits promptly available for use.

A significant upgrade to the EMU during the Shuttle era was the addition of the Simplified Aid For EVA Rescue (SAFER). It was a small, self-contained propulsive system designed to provide a self-rescue mode for spacewalkers who had come adrift in situations where the Orbiter was not available to provide rescue capability, such as when docked to the ISS or the Hubble Space Telescope. The unit was installed on the base of the existing backpack. It is a standard feature of the enhanced EMUs now in use on the ISS. The system has 24 gaseous nitrogen thrusters and control authority to enable a drifting astronaut to stabilize and return to safety. It has propellant for some 13 minutes of use. Control is performed by means of a single hand controller that is extended from the unit when required.

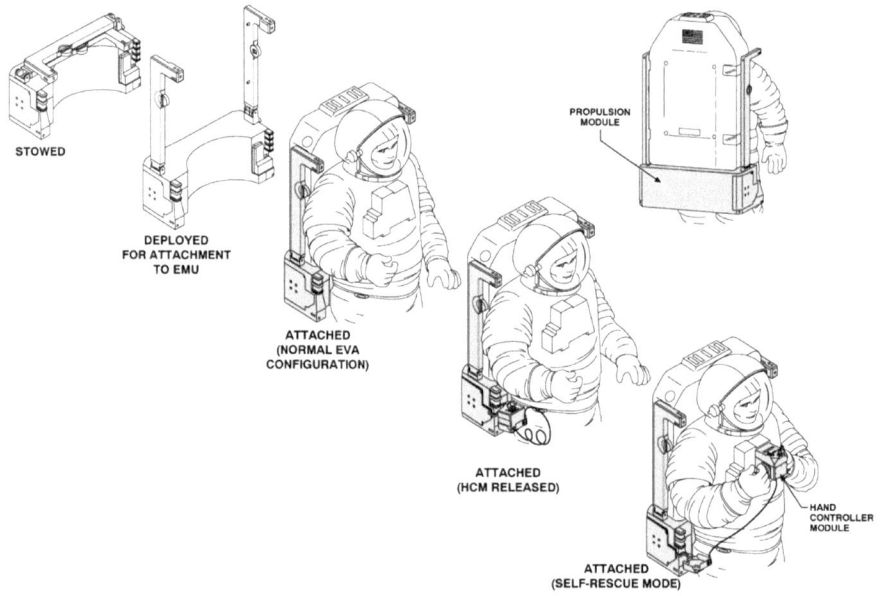

STOWED

DEPLOYED
FOR ATTACHMENT
TO EMU

ATTACHED
(NORMAL EVA
CONFIGURATION)

ATTACHED
(HCM RELEASED)

ATTACHED
(SELF-RESCUE MODE)

PROPULSION
MODULE

HAND
CONTROLLER
MODULE

The attachment of SAFER to the EMU.

One element that received periodic attention were the spacesuit gloves. Being the active and direct interface between the astronaut and the work to be performed, a lot of attention was devoted to enhancing dexterity and comfort and reducing fatigue. As Thomas Sanzone explains, "The glove technology improved pretty substantially, and continued to improve throughout the whole Shuttle program. When you think about what astronauts were able to do with Hubble, it would have been virtually impossible with Apollo era gloves, and even the early Shuttle gloves. Those were just too bulky to be able to do those kinds of things. So there was a lot of technology development put into the gloves by the folks at ILC."[9]

Astronaut Jerry L. Ross performed the first two spacewalks of his career during the STS-61B mission in November 1985. He recalls, "The design of the gloves was such that most people were getting numb thumbs and maybe other fingers, but the problem was that it had too much bladder…It was kind of like every time you tried to grip something, the nerves that go right through this part of the thumb were getting crushed across [what] felt like a pencil. If you could put a pencil in there, and just continually crush across it for 6 hours at a time, that is kind of what it felt like. And that was irritating the sheaths of nerve bundles and causing numbness in the fingers. Now that, plus the fact that we couldn't get good fits on a lot of people's hands, was convincing the program to do something about it." The result was a glove that would almost perfectly match the profile of the hand. "They basically made plaster molds of the hand, and then tried to fabricate a glove that would fit it like a glove. They made the joints on the fingers out of convolutes, so they would be

---

[9] ILC was the manufacturer of the EMU pressure vessel.

more flexible and less fatiguing to bend the finger joints....They had a thing that they called a segmented palm. This was a device that kind of tried to keep the palm from ballooning out too much, and would keep the concave part of the glove as you'd like to have it. We had adjustable capability for where you put the palm bar. The palm bar is a metallic bar that goes across the palm to try to keep the hand from getting too much of a balloon. Anytime you pressurize something, it's going to try to go into a cylinder or a sphere. When you want the palm of the glove contoured to the concave nature of the palm, it doesn't do that very well when you pressurize the glove. So a palm bar tries to keep the glove closer to the surface of the skin, on the inside of the palm, and it has a strap that you can cinch down to make it as tight as you can stand it and still be able to function inside the glove, to keep that conformal nature. It had the capability to be moved, as opposed to some other gloves which were fixed in place. It also had a flat bar, as opposed to a bar with a circular cross-section that wasn't as good."

Another key change was the addition of fingertip heaters. Astronauts had always complained that their hands were getting very cold during orbital darkness, or when working in areas of the payload bay in the shadow. As Sanzone notes, "Ultimately, you're getting down to the fingertips being the interface point that the astronaut is dealing with...So the gloves actually have the least amount of insulation of anything in the suit. The suit itself, the arms and the legs, have a lot of insulation. The body produces heat, so we've never in our suits and life support systems had to heat up an astronaut, even when he's in an environment of minus 300°F, because he's inside the 'thermos bottle.' His body is producing this heat."

The problem, became particularly worrisome when Story Musgrave ran a vacuum chamber test in preparation for the first Hubble Space Telescope servicing mission. They had to simulate the exact thermal environment which he would experience. As Sanzone says, "We had him in a very cold environment, which was planned, because that is what they were going to experience [in space]. He ended up getting frostbitten on his fingertips pretty badly. That scared a lot of folks. Not just us, but oh my gosh, because this was going to be the environment that we're working in. So for the first time we actually developed glove fingertip heaters. We put little heaters in the tips of the fingers, and if the astronaut was in that kind of a cold environment, we could ask him to, or he could on his own, activate these heaters. Throw a switch and actually turn these things on. It would provide some heat to his fingertips."

HUMAN SATELLITE

Moonraker, the 11th release of the James Bond movie series, requires the fearless British spy to save the entire planet from a threat coming directly from outside Earth. Rather than fighting green aliens, he has to stop the villain, Hugo Drax, from erasing humankind in order to establish a new master race. Drax is the founder and owner of Drax Industries, a large private manufacturing company based in California, where a fleet of winged reusable spacecraft known as Moonrakers are constructed for sale to governments worldwide. Moonrakers are nothing less than Space Shuttle Orbiters, but as *Columbia*'s first flight was still a few years in the future, the movie producers took care to avoid any reference to the real Shuttle program. Disgusted with how the human race is ruining the environment, Drax has secretly built a space station in low-Earth orbit from which he plans to bombard the planet with bombs carrying a nerve gas that is fatal to humans but harmless to animals. His space station is also destined to house a small group of perfect Adam and Eve couples so that, once the lethal gas has cleared, he will be able to give birth to a new master race. In one of the most epic scenes towards the end of the movie, Drax's defense forces engage in a zero-g battle with soldiers sent by the American military. Part of the confrontation occurs in open space, with astronauts from both sides swiftly maneuvering in mortal combat using a personal self-propelled backpack. Obviously, Bond and the American military regain full control of the space station, Drax's plans are foiled, and his bombs are zapped by laser cannons before they can strike their targets.

Moonraker went on worldwide release in cinemas in 1979, some 2 years ahead of *Columbia*'s maiden launch. On February 7, 1984, one aspect of the movie became reality when astronauts flew self-propelled backpacks (without laser cannons) out of the payload bay of *Challenger* during the STS-41B mission. Earlier, the five-person crew led by Apollo and Shuttle veteran astronaut Vance D. Brand had deployed two communications satellites, both of which then failed to reach their target orbits due to malfunctions in their PAM upper stages. An inflatable balloon was also deployed to serve as a target to rehearse rendezvous and proximity operations in readiness for the rescue of the Solar Max satellite scheduled for later in the year.[10] Regretfully, the balloon ripped as it was inflating, precluding the planned rendezvous training. The crew managed to track the shredded balloon out to 30,000 feet, obtaining good short-range sensor data. The fifth flight day brought a fourth satellite deployment, one that had to be successful because lives were at stake.

As astronaut Bruce McCandless II emerged from *Challenger*'s airlock into the payload bay followed closely by colleague Robert L. Stewart, he moved toward the port side of the Orbiter to initiate the most ambition EVA scheduled since the first spacewalk performed by Alexei Leonov just two decades earlier. The fragility of the tiles that covered the Orbiter's belly to protect it from the intense heat of re-entry was already known from the onset of the Shuttle program. The fear of losing an Orbiter during re-entry prompted NASA to devise a means to inspect the hidden surfaces to check for flaws in the thermal shield and test methods for making temporary repairs. Since the Orbiter's robotic arm was too short to place an astronaut in position to view the belly, and in particular its aft section, Martin Marietta of Denver, Colorado, was tasked with developing a self-propelled platform that an astronaut could use to freely maneuver around the Orbiter.

Experiments for such a platform dated back to when Gemini IX astronaut Eugene A. Cernan exited his capsule in 1966 with the intention of retrieving the Astronaut Maneuvering Unit (AMU), a rocket pack that had been stored in the aft section of his spacecraft. The concept of the AMU was to allow an astronaut to move freely around in space to carry out a variety of tasks. In other words, astronauts would become real 'Buck Rogers' characters. But at that time NASA did not recognize the difficulties of spacewalking, and training was rudimentary and inadequate. As Cernan struggled to stabilize himself in weightlessness his suit overheated and his helmet visor fogged. With the AMU task canceled, he was exhausted by the time he managed to re-enter the spacecraft. The objective was not repeated until when Skylab became available. In fact, the members of the

---

[10] Refer to Chapter 5 for details.

second and third crews that visited America's first orbital outpost spent some time with the M509 experiment, which included both a backpack maneuvering unit known as the Automatically Stabilized Maneuvering Unit (ASMU) and a hand-held maneuvering unit similar to that used by Edward White on the first American spacewalk in 1965. Both units were tested inside the cavernous laboratory environment, logging some 14 hours of trials. Several hours of trials were performed with the astronaut wearing a full space suit.

Alan L. Bean, Skylab 3 mission commander, flying the M509 inside the Skylab space station.

The lessons learned from those experiments allowed Martin Marietta to develop the Manned Maneuvering Unit (MMU). It was a sort of large armchair into which an astronaut, fully suited with an EMU, could sit and become an autonomous spacecraft.

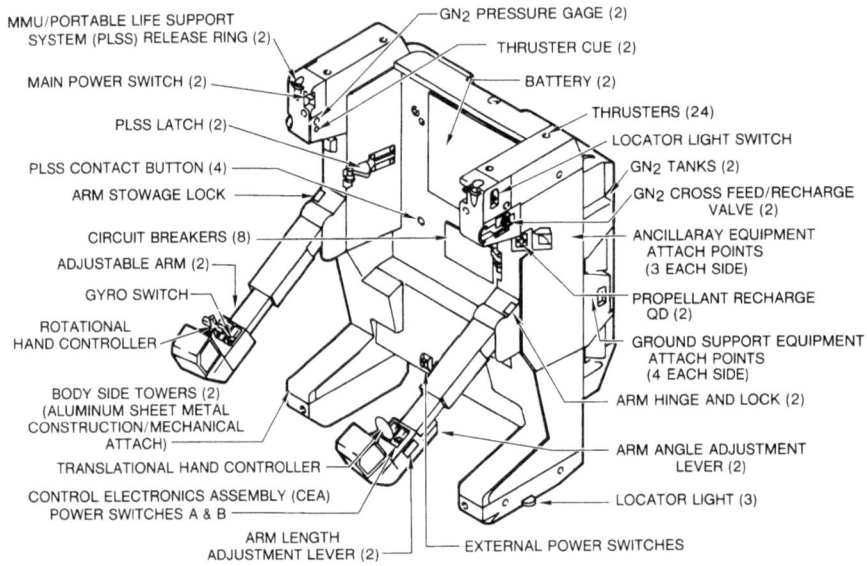

The Manned Maneuvering Unit.

The MMU consisted of a white coated aluminum frame that consisted of a central structure which connected two side towers and arms. The center structure contained a pair of Kevlar filament wrapped aluminum tanks for gaseous nitrogen at 3,000 psi, along with associated lines and fittings to feed two groups of 12 thrusters that were distributed along the two lateral towers in order to provide six degrees of control in attitude and translation. Each tank fed a specific group, but cross feeding was a built-in feature to interconnect both gaseous nitrogen tanks to one system of 12 thrusters. The selection of nitrogen as propellant was because it is chemically inert, preventing contamination of delicate payloads in proximity to the astronaut using an MMU. A hand controller on the right arm provided attitude control in pitch, roll and yaw while a hand controller on the left arm provided for translations. Both controllers could be operated at the same time, for a full range of movement within the operating logic of 729 command combinations within the control logic. One feature was the automatic attitude hold mode that would enable the astronaut to maintain position in relation to another object, such as when performing an Orbiter's tile repair. This mode would be commanded by pressing a particular button on the right-hand controller.

The center structure also housed the control electronic assembly with three gyros (one for each rotational axis), control logic, thruster selection logic and motor-driven thruster

Astronaut Robert Stewart wearing his EMU and the MMU on Earth. Note the large size of the MMU compared to a human body.

valve drive amplifiers. It also had a pair of 16.9 volt direct current silver-zinc batteries and an electrical power distribution system with associated circuit breakers, switches, and relays. Two heater systems for each battery made sure they would stay at the required temperature for optimum performance. Care was taken in designing a fail-safe electrical and propulsive system, so that in the event of a single operational failure the astronaut would still be able to complete a critical task and return safely to the Orbiter using the remaining viable system, possibly with energy or propellant to spare. In the unlikely event of a second failure, the EVA flight crew member would await rescue by the Orbiter. This would be possible even if the MMU had exhausted its propellant in accelerating the space-walker directly away from the Orbiter.

STS-41B carried a pair of MMUs, one in a Flight Support Station (FSS) on either side of the payload bay. These were installed near the forward bulkhead in order to minimize the time and effort in moving from the airlock's outer hatch to the MMU. The FSS not only provided for storage during launch, re-entry, and non-MMU orbital activities, but also allowed the Orbiter to recharge the MMU batteries and run their heaters when not in use. It could also replenish the nitrogen propellant tanks from the Orbiter's environmental control and life support system.

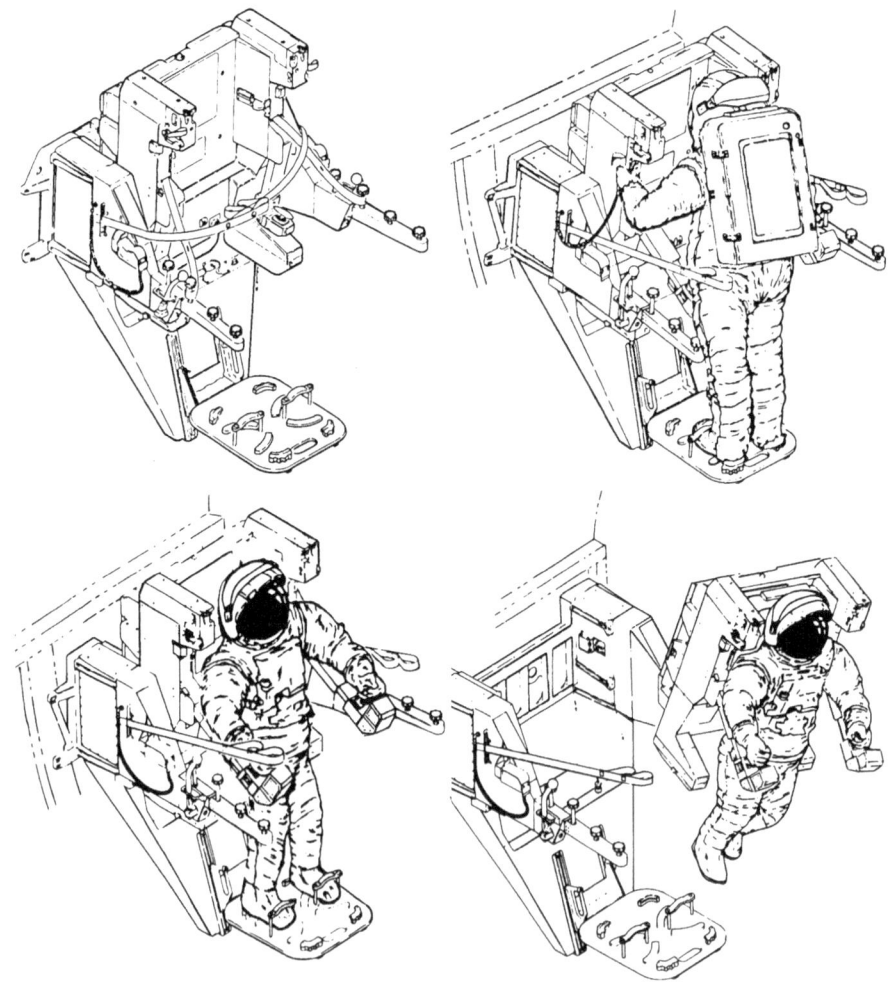

FSS and MMU operations.

On approaching the FSS on the port side, McCandless climbed onto an adjustable foot restraint that allowed an astronaut to regulate its height for donning and doffing the MMU. After checking his MMU, he rotated to face its back to the MMU, then as he reversed, mechanical latches located on the inside of each tower, above the arms, sprung into receptacles on the EMU backpack to hold him firmly in place. He and the MMU were now one. For good measure, McCandless buckled a belt crossing his lap. Then on disconnecting the safety tether which anchored him to the Orbiter, he stood up on the foot restraint and checked the functions of both hand controllers, including the automatic attitude hold mode. Satisfied with the results, he finally stepped off the foot restraint and carried out a series of translation and rotation exercise to acquaint himself with the MMU. Then history was made.

"One small step for Neil, a heck of big step for me," McCandless exclaimed as he hovered over the payload bay, slowly moving away from *Challenger* to a distance of 105 feet. It was the first time that an astronaut spacewalked without any connection to the mother ship.

After returning, McCandless flew off again to double the previous distance. Back in the payload bay, he set out to test the Trunnion Pin Attachment Device (TPAD), a critical apparatus developed for the rescue of Solar Max. In the middle of the payload bay was the German-build SPAS[11] platform. On it was a receptacle identical to that on Solar Max. McCandless performed a series of docking exercises.

Meanwhile, Stewart tested the Manipulator Foot Restraint (MFR), a humble EVA tool that would subsequently prove its utility on many a spacewalk. It was a platform to provide unrestrained access to EVA worksites within reach of the Orbiter's robotic arm. In fact, it could be installed at the end of the robotic arm as well as at other work sites in the payload bay to allow an astronaut to stay in place, releasing his hands for work instead of maintaining position. It also had provisions for tethering tools as well as the safety tether of the EMU.

On completing the TPAD tests, McCandless returned to the FSS to replenish the nitrogen tanks of the MMU. Then the duo traded places, with Stewart flying in the MMU and McCandless continuing the evaluation of robotic arm and foot restraint operations.

Two days later, a second spacewalk was conducted, this time using the starboard MMU. It was intended to be a dress rehearsal of the first spacewalk of the Solar Max mission, when that satellite would be captured and stowed in the payload bay. In this case the robotic arm was to lift the SPAS platform and maneuver it in a slow turning motion about its vertical axis to simulate a tumbling satellite. The astronauts were to rehearse their approach path, match the rotation rate, and then dock using the TPAD. But while the arm was maneuvering to grapple the SPAS, its wrist yaw joint failed. After some troubleshooting, the crew determined that it was a hard failure rather than a software or sensor glitch. The arm was then returned to its stowed position. So the Solar Max rendezvous and docking exercise was done with the SPAS sitting in the payload bay. Afterward, Stewart had time to conduct an engineering demonstration of on-orbit refueling to simulate the transfer

---

[11] SPAS stands for Shuttle Pallet Satellite, and it flew a number of times as a free-flying platform in different configurations housing hardware for experiments and/or astronomical observations that needed to be performed away from the Orbiter in an undisturbed environment.

Riding the MMU, Bruce McCandless recedes from *Challenger.* He appears at an angle due to the attitude of the Orbiter when Robert "Hoot" Gibson took the picture.

of hydrazine to a mockup of the fueling port of the Landsat 4 satellite, which NASA was hoping to restore to full status on a later Shuttle mission.

Despite the problem with the robotic arm, the EVAs of STS-41B were a welcome success for NASA. The EMU and MMU worked well together, and the performance of the MMU was nominal. The only nuisance was a "chatter" during +X translations. This proved to be due to an 0.6-inch offset between the geometric center-of-thrust of the MMU and its center of mass. It imparted a positive pitch motion while translating in the +X direction. If attitude hold was on during translation, then this control logic would maintain a very low pitch rate by cutting off two of the four thrusters that were being fired to create the +X translation. The control electronics assembly would then cycle between commanding four and two thrusters extremely rapidly, producing the observed chatter.

McCandless and Stewart proved the MMU would be a valuable tool in the hands of a spacewalker. By the time the MMU project had matured sufficiently to become a space-worthy unit, other uses had been recognized in addition to just being an aid for carrying out Orbiter repairs. These tasks envisaged payload deployment or retrieval, adjustment of instruments, servicing of free-flying payloads, cleaning of sensors and lenses, assembly of large structures, rescuing astronauts from another Orbiter, and so on. Unfortunately, the MMU would never live up to such expectations. As we shall see shortly, only two more flights would use it.

4

Learning to Build a Space Station

INTRODUCTION

On January 25, 1984, President Ronald W. Reagan, recently installed in office, stood before a joint session of the US Congress and delivered a State of the Union Speech in which he reported upon the condition of the country and traced out his agenda and priorities for the year. Halfway through the 43-minute speech, Reagan reminded his audience of America's next frontier: Space. He was emphatic, "Nowhere do we so effectively demonstrate our technological leadership and ability to make life better on Earth," He added, "America has always been greatest when we dared to be great. We can reach for greatness again. We can follow our dreams to distant stars, living and working in space for peaceful, economic and scientific gain." Everyone in attendance in the House of Representative chamber of the United States Capitol was now primed to receive the next big announcement. "Tonight, I am directing NASA to develop a permanently manned space station and to do it within a decade."

This wording was clearly intended to evoke a memory of when, on May 25, 1961, President John F. Kennedy challenged the nation to place a man on the Moon before the decade was out. Reagan was directing NASA to pursue a dream which had been on the agenda even before Apollo astronauts left bootprints on the lunar surface. This was what Administrator James M. Beggs had been waiting for, having invested years in defining a feasible concept for a permanently manned space station and persuading the White House of its benefits, "I told him what it would do. I gave him a number of presentations on the potential of the space stations, what we could learn, the potential of commercial activities, the potential for long-term research in space, microgravity research…So I pointed out to him that we could do long-term research. Research is a thing where you set up goals and objectives for an experiment, but often you find in doing the experiment that other results show up which you must pursue if you want to take advantage of them. And that's what we will have… I pointed out, which is an argument that he liked, that you'll be able to see it with the naked eye, which you are. It is an evening and morning star. He liked that.

© Springer International Publishing AG 2017
D. Sivolella, *The Space Shuttle Program*, Springer Praxis Books,
DOI 10.1007/978-3-319-54946-0_4

And I said, 'With a modest increase in my budget, if you give me just a 2% increase in real terms, I think we can do this without any additional increase in budget.'"

Selling the space station to Reagan had not been difficult because he was fond of the space program. He recognized its benefits as a crucial strategic asset for national security and a catalyzer for America's burgeoning economy. And a space station that promised to deliver advanced technology and science would be a shining reminder of the superiority of the Free World over the ideology that ruled on the other side of the Iron Curtain.

As Reagan continued in his State of the Union speech, he broadened the appeal of the space station project. "We want our friends to help us meet these challenges and share in their benefits. NASA will invite other countries to participate so that we can strengthen peace, build prosperity and expand freedom for all who share our goals."

In following up, Reagan granted Beggs authority to initiate preliminary talks with prospective partners in this new venture, "The president said, 'I will give you a letter of introduction to all of the presidents and prime ministers around the world.' Let me tell you, if you get an introduction by the President of the United States, you get in. The vice-president was instrumental and gave me an airplane to fly around in. He got me an Air Force airplane. We took a trip around the world."

Canada, Japan and Europe (with the notable exception of the UK) all expressed a robust enthusiasm and willingness to participate and contribute hardware that would capitalize on their individual capabilities. For instance, the European Space Agency (ESA) which had gained valuable skills and knowledge from designing and building Spacelab as a laboratory for the Shuttle, offered to provide a permanent laboratory named Columbus for the space station.[1] Likewise, the Canadian aerospace industry had developed the robotic arm for the Shuttle, and it was only natural to offer a more advanced tool to ease the construction and maintenance of the space station. Despite not having any experience in manned space flight, the Japanese aerospace industry was determined to make a major contribution, and proposed an additional laboratory with an independent logistical support system.

Although the space station was intended to be open to business by the end of the 1980s, the project became mired in controversy in terms of its configuration and cost. By the early 1990s the budget touted when the project was conceived had been fully spent without a single piece of flight hardware being built. It was only after the fall of the Soviet Union and the inclusion of Russia as an international partner that the plan really got underway. Even so, the first module, provided by Russia, was not launched until November 1998. NASA added its first module a month later. The final element, an experiment in astrophysics to study antimatter, was installed in May 2011, shortly before the Shuttle was retired.

Fulfilling Reagan's mandate took much longer than expected, but the engineering challenges were pursued against a background of political games and ever increasing budgetary constraints. What was envisaged as a large multi-role outpost on the space frontier was repeatedly constrained by the projected available funding and reduced in terms of both size and capability. What was left, and it was a great deal, became the International Space Station.

---

[1] Years later, NASA would also pen a contract with the Italian Space Agency (ASI) for three pressurized logistic modules to be employed by the Shuttle to ferry cargo to and from the ISS.

As NASA was struggling to persuade Congress and the Office of Management and Budget to authorize adequate funding, the space agency recognized that it lacked the skills to build a space station efficiently. In fact, the short-lived Skylab had been launched fully assembled and ready for use. This new and much larger station was to be assembled in space from elements launched by Shuttle. This would require many hours of spacewalking and robotic work, of which it had little practical experience. In the years to come, several missions were assigned primary or secondary objectives to rehearse and evaluate techniques for orbital construction work or to validate possible engineering models.

HEAVY PAYLOAD LIFTING

The first opportunity to undertake such testing arose unexpectedly when TDRS-1 left *Challenger*'s payload bay in the early hours on April 5, 1983.[2] Following a perfect burn by the inertial upper stage's first stage, the second stage rocket motor shut down prematurely. For almost 3 hours, America's first TDRS appeared to be lost, deaf to all communications. At 9.00 a.m. EST, the Goldstone tracking station in California received a faint indication that the satellite had separated from the second stage, but it was in an elliptical orbit ranging between 21,950 and 13,540 miles at an inclination of 2.4°. Everything about this orbit was wrong, because the TDRS constellation was designed to operate in a circular equatorial orbit at 22,236 miles. In addition, it was spinning out of control at an alarming 30 rpm.

But there was hope of recovering the satellite. Working with its manufacturer, engineers at the NASA Goddard Space Flight Center wrote a complicated procedure which, by means of 39 burns of the spacecraft's attitude control system jet thrusters, delivered it to its original intended place in space and it was able to start relaying for a variety of NASA missions, in particular the Shuttle. The next TDRS was scheduled for launch on STS-8, to have the space communications network up and running for the long-awaited first flight of Spacelab. However, the inability to identify the root cause for the upper stage's shutdown obliged NASA to delete TDRS-2 from STS-8. This decision entailed a significant financial loss, but it would have been even more detrimental to launch another faulty IUS and strand a second TDRS in an incorrect orbit. The STS-8 flight was revised to deploy INSAT-1B, an Indian communications satellite.

The small size of INSAT-1B left a large part of *Challenger*'s payload bay empty. The sudden availability of premium capacity gave NASA the unexpected opportunity to install an awkward-looking contraption known as the Payload Flight Test Article (PFTA). This structure consisted of a 19-foot-long hollow aluminum beam capped at each end by an aluminum screen about 16 feet in diameter, an arrangement intended to give the illusion of a "full volume" cylindrical payload. Most of this 7,460-pound mass was placed at the aft end and consisted of lead ballast. Due to its mass and size, the PFTA was carried into space by five attachment points, two on each longeron and one on the keel of the payload bay. It was the first demonstration of this attachment configuration. As its name implies, the PFTA was completely passive, meaning that it did not have any autonomous guidance system or any on board systems at all. Its simplicity, however, was deceptive. In fact, the

---

[2] Refer to Chapter 2 for more details.

PFTA had been designed to test the operational envelope of the robotic arm when handling heavy payloads, such as those of the future space station. To date, only small payloads had been moved in, out, and around an Orbiter. Future missions would require the arm to manipulate heavier and bulkier payloads.

An example was the Long Duration Exposure Facility (LDEF), a 20,000-pound passive satellite carrying experiments to test the endurance of materials to the harsh environment of space. It was to be deployed by STS-41C. On that same mission, the large Solar Max satellite was to be retrieved to accomplish the first rescue and repair of a spacecraft on-orbit.[3] The Hubble Space Telescope, whose mass and size were comparable to LDEF, was also on the list. Furthermore, if the Shuttle had to be used to assemble large structures in space, such as the much-touted space station, this task would involve maneuvering heavy and cumbersome payloads.

It was evident that NASA needed to understand not only the performance of the robotic arm when handling payloads with large inertia, but also the behavior of the Orbiter's attitude control system in such conditions. STS-8 and PFTA were meant to provide those answers.

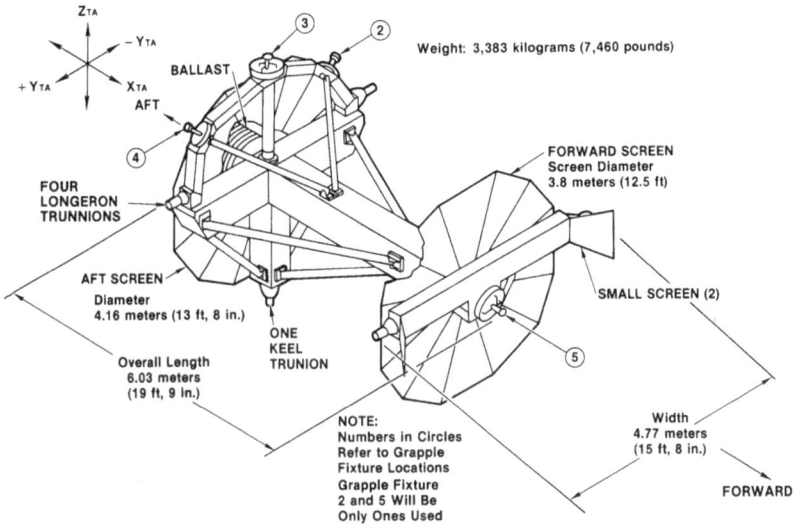

The simple structure of the PFTA.

After INSAT-1B was deployed, flight days three to five were spent maneuvering the PFTA and performing a series of tests of the arm/Orbiter combination. Tests were made by grappling the PFTA at two different points in order to simulate the berthing and unberthing

---

[3] See Chapter 5 for a full account of the first satellite rescue and repair mission.

of heavy payloads possessing different moments of inertia.[4] The arm was operated both in the augmented mode and in single joint drive mode; the latter representing the most degraded mode of operation.[5] As Richard H. Truly and Dale A. Gardner were operating the robotic arm, data was collected for post-flight analysis to improve knowledge of *Challenger*'s flight attitude control system and structural responses. The data gained from the arm revealed its dynamic characteristics, both in terms of natural frequencies and damping properties, and its ability to follow a pre-programmed automatic sequence path, then halt at a specified position and attitude. It was essential to understand the dynamics of the arm, as STS-8 flight director Randy Stone explained at a pre-mission press event, "The purpose of manipulating a heavy article like the PFTA is to understand the arm dynamics when you start moving this payload and when you stop the arm. When you put the brakes on the arm, it actually moves on a little bit further than where you stop it, and then it moves back. With the data from this test article, we will be able to extrapolate to payloads of 20,000 pounds or more." The results from the in-flight trials were promptly fed into the robotic arm simulators, in order to improve the training of astronauts assigned as arm operators.

STS-8 also demonstrated that payloads could be safely maneuvered without the grapple fixture on the payload being directly visible to the arm operator. When the arm was used on previous missions, the operator had a clear view out of the aft flight deck windows of the payload's grappling point. On STS-8 the operator did not have this luxury. Instead, he had to watch the action on screens fed by CCTV cameras on the arm and in the payload bay. This was an important test, since the deployment of large payloads and the construction of the space station would hinge upon being able to operate the arm out of direct line of sight of the operator. STS-8 gave NASA great confidence in the arm. The success in assembling the International Space Station can be traced straight back to the PFTA tests.

Considering how vital the robotic arm proved to be during 30 years of Shuttle operations, and the iconic status it attained, it is startling to think that its development and installation on the Orbiter was met with open hostility. Even from the astronauts themselves. Eugene Kranz, the mastermind behind Mission Control in the Mercury, Gemini and Apollo eras, appointed Jerry C. Bostick to fix the situation. As Bostick says, "Late one afternoon after normal working hours, Dr. Kraft, who was the [JSC] director…strolled into my office and said, 'Jerry Carr[6] just left my office, screaming and hollering about how terrible the RMS – the mechanical arm that's going to fly on the Shuttle – how terrible it is. It'll never work. It's going to be like a wet noodle, he called it. It's unsafe. Even though the Canadians are building it and they're going to give it to us, we shouldn't accept it. Or we ought to

---

[4] As a layman's definition, the moment of inertia of an object indicates how difficult it is to start or stop its motion. It depends on the geometry of the object and the distribution of mass about the center of gravity.

[5] For more details on the robotic arm's modes of operation, refer to Chapter 2 of my previous book *To Orbit and Back Again: How the Space Shuttle Flew in Space*.

[6] Gerald P. Carr was a veteran of the Skylab program, having spent 84 days on the final manned mission to the orbiting outpost.

spend a lot of money and fix it, because it's terrible. Won't work. Useless. Tomorrow morning, I want you to talk to Aaron Cohen[7]…and see if you can't fix it.'"

That is how Bostick established the Office of Payload Deployment and Retrieval Systems with the robotic arm as the main piece of hardware.[8] But before addressing the technical issues, he had to get some unbiased people to work the bugs out of the system. "There were a lot of people, it turned out, who shared Carr's opinion. In fact, one of them was Glynn Lunney.[9] I went to see him soon after I talked to Cohen and I asked, 'What do you think about the RMS?' He said, 'It's a piece of junk. Get rid of it. It won't ever work. We're working on some other ways to deploy payloads, [for instance] rotating arms. This thing with a cherry-picker sticking out of the back of the cockpit, every time you fire the jets on the Orbiter, it's going to vibrate. It's probably going to break off. Just forget it. Career-wise, this is a dead-end. Do not go do that, because you're going to fail.'" Despite the adverse environment and the prospect of working on a career-killer assignment, Bostick managed to gain the collaboration of astronauts who were willing to test the robotic arm in an objective manner.

As Bostick says, "Bill Lenoir, Norm Thagard, Sally Ride, and Judy Resnik for a while, were very good, very objective. They took the attitude that there's a job to be done here, and, yes, the Astronaut Office is against the RMS, they don't like it or are afraid of it or whatever, but they kept an open mind and were very helpful in deriving reasonable requirements and working with the Canadians and the Orbiter people on how we were going to integrate this whole thing together."

Sally Ride remembers, "I was one of a couple of astronauts that became heavily involved in the simulator work to verify the simulators accurately modeled the arm, develop procedures for using the arm on-orbit, and develop malfunction procedures so astronauts would know what to do if something went wrong."

Operating the Orbiter in close proximity to a deployed or an about-to-be-retrieved payload was also a matter of concern. It is easy to appreciate that the flimsy robotic arm could have imparted significant oscillations to an attached payload if the Orbiter were to maneuver too rapidly. On the basis that the arm would be wholly or partially extended, Bostick and his team came up with combinations for how attitude control thrusters could be fired in sequences which would not result in damaging dynamics. Procedures were also written on how quickly a payload could be released, or when a free-flying platform could be grappled, particularly if it was spinning, or even worse, tumbling, without creating torque sufficient either to break the arm or snap it off the Orbiter. As Sally Ride recalls, "There weren't any checklists when we started. We developed them all."

As the arm had been designed as a plug-and-play piece of hardware, Bostick had also to consider where the interface with the Orbiter would be. The ideal location was on one

---

[7] Aaron Cohen was the Orbiter project manager.

[8] Other components of the Payload Deployment and Retrieval Systems included attachment fittings, visual aids, and software.

[9] Glynn S. Lunney was a flight director who at that time was managing the Shuttle Payload Integration and Development Program.

of the payload bay longerons, "Attaching it to a longeron on the Orbiter was another engineering challenge, because the longeron, the side beam of the payload bay, was already built. So it had a certain strength. And there was a limited volume because when you close the payload bay doors the arm must fit into that little space. So it was an engineering challenge to build the base or the shoulder of the arm there strong enough and small enough that it would fit."

Considering that the arm had shoulder, elbow and wrist joints, the positioning of its end effector[10] at a given point in the operating envelope posed its own challenges. As Bostick illustrates, "You've seen cherry pickers out on construction jobs. Most of them are limited to movement of one joint only at a time, so you move one joint and then another. When you do that, you get some strange movements. Sometimes you think you're going to go up, and when you rotate this joint up, the end actually goes down." The designers wanted to operate the arm one joint at a time but Bostick was adamant that the astronauts input where they wanted the end effector to go, then the software would position the arm to suit. This would allow the astronauts to operate the arm in an accurate and speedy manner, because they would not need to consider joint positions and geometry. Wisely, the ability to move each joint separately was retained to provide a downgraded mode if a failure disabled the fully software-driven mode.

Even with the convenience of the software-driven mode, the astronauts would still need visual cues in order to maneuver the arm safely. As Sally Ride says, "The astronaut controlling the arm looks at it out the window and also monitors its motion using several cameras. Often critical parts of the view are blocked, or the arm is a long ways from the window, or the work is delicate. In those cases, the astronaut needs reference points to help guide the direction he or she moves the arm. How do you know that you're lifting a satellite cleanly out of the payload bay, not bumping it into the structure?"

These investigations led to the development of end effector targets, visual cues, and a camera system on both the arm and in the payload bay to give the operator the best possible situational awareness.

The success of assignments such as servicing the Hubble Space Telescope and assembling the International Space Station are testament to the efforts of Bostick and his team to overcome the challenges of introducing this key piece of engineering.

SPACE STATION HARDWARE TESTING

The International Space Station generates a total of 75 kW of electrical power from eight solar arrays so large that the reflected sunlight makes the outpost easily visible in the twilight with the naked eye.[11] In fact, they are the largest parts of the station. Their flimsy and delicate design is another marvel of space hardware, the success of which can be traced back to the OAST-1 experiment.

---

[10] The end effector is the robotic arm's grappling hand.

[11] Obviously you need to observe when the station is overflying your location and the sky must be clear.

After suffering a launch pad abort,[12] *Discovery* was rolled back to the Vehicle Assembly Building for an engine refurbishment and the replacement of engine #2. When rescheduled to launch on August 29, 1984, a timing issue between the flight software and the Master Event Controller[13] prompted a third launch scrub. Although the crew of six finally managed to lift off the following day, this was not until after an additional delay of 6 minutes 50 seconds caused by a problem with the Ground Launch Sequencer[14] and two private aircraft that encroached on the restricted area for launch operations. The mission of almost 7 days allowed a variety of objectives, such as the first deployment of three satellites on a single flight.[15]

A substantial amount of time was devoted to testing various configurations of the NASA's Office of Aeronautics and Space Technology (OAST-1) payload that was to investigate solar energy and large space structures technologies, both of which were vital for a space station. The central element was the Solar Array Experiment (SAE), a solar array wing 103 feet long and 13 feet wide, the primary structure of which was a thin blanket of a plastic called Kapton with 84 panels that unfolded accordion-style as the structure was drawn out of its box.

As STS-41D mission specialist Steven A. Hawley says, "It was actually similar to the solar arrays we have on the ISS, and that was kind of the point of the experiment back then. One of the things we were supposed to do was to have this thing extended and then input pulses into the Orbiter's reaction control system and watch how the dynamics were generated in the solar array, how it damped over time, and see if the predictions matched what we saw. They had cameras set up to measure actually how much the tip of the mast deflected, to be compared with what they had predicted." He recalls the experiment as truly captivating, "The thing I remember most about it is at sunrise or sunset, the thing that either the first bit of sunlight or the last bit of sunlight would hit was the solar array, and that would make it almost look like it was lighting up with its own source of internal illumination. Everything else would be dark and the solar array would be glowing gold. It was really pretty."

On the truss section spanning the width of the payload bay which held the SAE, were the Dynamic Augmentation Experiment (DAE) and the Solar Cell Calibration Facility (SCCF). DAE gathered information on the structural vibrations of the solar array in order

---

[12] During a launch pad abort the ignition sequence of the three main engines, which started 6 seconds prior to the scheduled time of lift-off, was curtailed due to the engine health monitoring system detecting a violation of one or more safety criteria. STS-41D was the first of five launch pad aborts experienced by the Shuttle program. The last one occurred on STS-68, when the main engines were shut down at just 1.9 seconds prior to when the solid rocket boosters would have been commanded to ignite.

[13] The Master Event Controller is the internal clock of a spacecraft, setting the time for all major flight events such as engine ignition and shut down.

[14] The Ground Launch Sequencer is the clock that regulates all events in the countdown sequence leading to lift-off.

[15] These were the SYNCOM-IV, Telstar 3-C and SBS-D communications satellites. They were all boosted to their operational orbits using PAM upper stages.

to validate an on-orbit method of defining and evaluating the dynamic characteristics of large space system structures. SCCF evaluated and validated solar cell calibration techniques. Overall, OAST-1 gave NASA a critical understanding of the performance of a large, low-cost, lightweight, deployable/retractable solar arrays.

Unfolding the Solar Array Experiment of OAST-1.

STS-48 conducted further testing and validation of hardware for a space station, with the crew working on the Middeck Zero-gravity Dynamic Experiment (MODE), an MIT-sponsored study to improve understanding of the design and construction of large structures in space.

The accurate testing of such structures on Earth is not possible, owing to the ever-present force of gravity that alters the dynamic response. One way to mask gravity is to hang the structure on a complicated suspension system but this cannot eliminate the gravitational influence entirely. For a more realistic test, MIT developed MODE. This used special electronically instrumented hardware that occupied three and a half lockers on *Discovery*'s middeck. The experiment consisted of four fluid test articles, a partially assembled structural test article, optical data storage disks, and a "shaker" that was to be mounted on the experiment support module.

As mission commander John O. Creighton says, the structural test article was "a truss structure that was very similar to what they built the International Space Station out of. There is about 7 feet of height between the floor and the ceiling in the living quarters [of the Orbiter] and we erected this thing. It was almost being a kid with a Tinker Toy set, putting all those pieces together." The test article was equipped with four strain gauges and eleven accelerometers, and was vibrated by an actuator. "Then we had a little shaker with a bunch of string gauges on it to vibrate this structure to see how it reacted in space. What they were trying to do, was to verify the computer models down on the ground, to make sure that the vibration and the characteristics of this truss would react in space the way they were predicting in computers. That was successful." Using the structural test article, the astronauts tested both deployable and erectable types of truss[16] arranged in different configurations. The first configuration comprised an in-line combination of truss sections, with an erectable module flanked by deployable modules mounted on either end. The second configuration replaced the erectable section with a rotary joint similar to the "alpha joint" that controls the orientation of the solar arrays of the ISS. The third configuration had an L-shaped combination of a deployable truss, rotary joint, and erectable module (all mounted in-line) and another deployable section at 90° to the end of the erectable truss. On the fourth configuration, a flexible appendage was attached to the elbow of the L-shaped arrangement in order to simulate a solar panel or a solar dynamic module.

The fluid test articles were to simulate the dynamics of liquid propellants in tanks. Creighton explains, "We'd been flying spacecraft for years, and we understood how propellants in tanks worked when they were full and how they worked when empty, but there was not a whole lot of real good evidence on how these fuel tanks operated when partially empty. So we carried up some small trials and vibrated them on this special device. Again, it was an experiment designed by MIT that we took up there to study how fluids would migrate inside clear Plexiglas tanks. We weren't using real rocket fuel – if it had gotten leaked, that stuff is potentially lethal – so we used fluids with similar characteristics to them, to see how they behaved in a sphere type of an experiment."

As stated in the post-flight mission report, MODE was very successful. All four fluid test articles were studied and all four structural test article configurations were assembled and shaken. Over a period of 2.5 days, the crew completed six planned fluid test protocols, and eight planned and an additional five unplanned structural test protocols. The live video downlink of these activities provided valuable and timely confirmation of the predicted behaviors.

---

[16]The deployable structures were stored folded and were unhinged and snapped into place for the tests. The erectable structure was a collection of individual truss elements that screwed into round joints or "nodes."

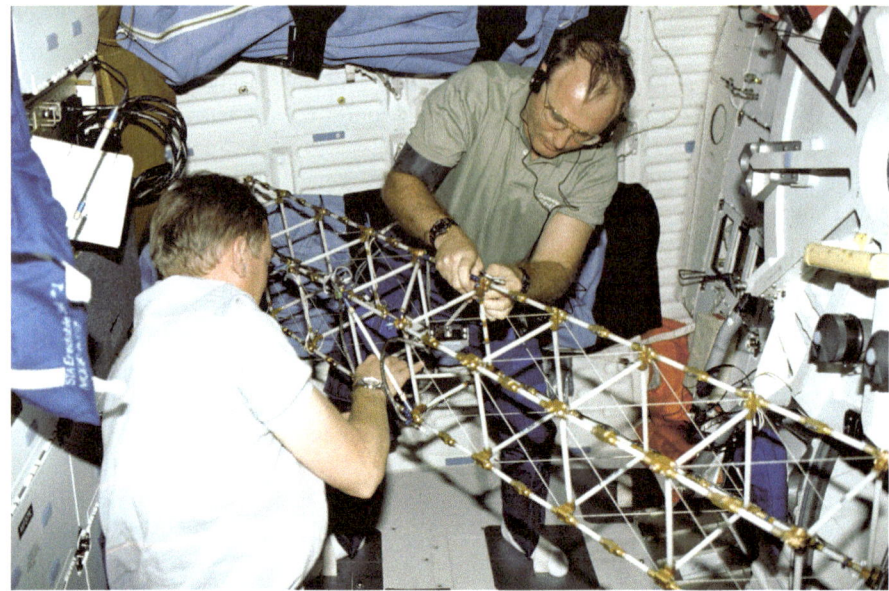

STS-48 astronauts Mark Brown (left) and James Buchli working with the structural test article which modeled a space station truss structure. Once assembled, the device was about 72 inches long with an 8-inch square cross section.

ON-ORBIT ASSEMBLY

The configuration of the International Space Station, as well most of the designs that preceded it (such as the so-called Power Tower and Dual Keel) has a long framework called the Integrated Truss Structure. Assembled from individual truss segments, this 230-foot backbone is the structural foundation of the station, supporting four solar power modules, two large heat dispersing radiators, and the pressurized modules that form the living and working facilities of the most complex engineering achievement in space.

Early in the 1970s, as concepts for space stations and other orbital outposts were being outlined, the advantages of truss style structures prompted detailed feasibility studies. Trusses provide excellent load bearing structures because they are light, stiff, and can be assembled to create large and complex facilities from a minimum number of components. One important question was how the individual segments were to be lifted into space. The predominant engineering opinion at that time was that carrying them fully pre-assembled would be inefficient and costly, even considering the large size of the Shuttle's payload bay. In fact, to fit the intended very large pre-assembled segments into the Orbiter's cargo bay they would require to be folded for launch and unfolded in space. It is easy to appreciate that this approach would have required the addition of mechanical devices, such as pulleys and gearboxes, to close and open the structure. Such complexity would not only have increased the cost of design but also the risk of deployment failure once on-orbit.

The fully assembled International Space Station. The four sets of solar arrays, the two large radiators, and the pressurized modules are all visible.

Manual on-orbit assembly was preferred. The Shuttle would ferry the individual rods and connecting nodes, and astronauts would manually assemble the structure by joining all the pieces in Meccano style. The cables, pipelines, and any other piece of equipment would be added to the open truss sections as construction progressed. By the time NASA got the presidential go-ahead to build a space station, a great many questions concerning on-orbit manual assembly remained unanswered. A number of techniques had been tried in the pools used for spacewalk training, but these still had to be tested in the real conditions of space.

This began with STS-61B in November 1985. Following the deployment of three communications satellites, mission specialists Jerry L. Ross and Sherwood C. Spring entered *Atlantis*'s payload bay to exercise two different methods for assembling large structures in space using small components. Leaving the protective confines of your spacecraft is undoubtedly one of the highest achievements in an astronaut's career. It is a particularly emotional experience if you are making your first spacewalk. Spring says, "We got suited up. I remember being impressed with [how much] you grow on-orbit…I grew about an inch and a half…But they size the suits to be basically an inch and a half 'too small' for you, for when you're on-orbit, so you've really got to work to get into those things on-orbit. But it feels good. It's nice to have your backbone, your spine, compressed back to where it's supposed to be. It really took some help, getting those pants up and cinched in. You've got to work to do that. I remember hanging on the wall of the airlock, waiting to go out. That was one of the two times, I guess, I had butterflies in my stomach; not before launch, but before I went outside. What's going through your mind is, 'Oh, I hope I don't screw up.' It's nothing more prosaic than that, for a couple of reasons. One, it's your big chance. Another one is, they've got all the video cameras in the world on you…and if you screw up, your friends will have [photos and videos] ready for you at the [post-flight] pin party."

Once outside, Spring and Ross propelled themselves towards a Mission Peculiar Equipment Support Structure (MPESS) payload carrier.[17] Spanning the full width of the payload bay, this held all the hardware which the two men would use during the space-walk. It included interface plates to allow easy mounting of four foot restraints and several hand restraints, to enable the astronauts to anchor themselves while doing the assembly work.

After confirming that all the equipment was in the proper configuration, the first experiment they set their gloves on was the Assembly Concept for Construction of Erectable Space Structures (ACCESS). Supplied by the Langley Research Center in Hampton, Virginia, this consisted of 93 tubular 1-inch-diameter aluminum struts of various lengths according to their intended positions, and 33 identical softball-sized nodal joints. The nodes and struts were carried in canisters mounted on the sides and top of the carrier. Prior to commencing the assembly, each astronaut placed his boots into a foot restraint, one mounted on the upper part of the MPESS and the other at its base. Next, they unfolded three vertical guide rails to serve as a jig to assist with the assembly of each bay (or cell). Then they opened the strut canisters and began taking them out.

Spring says, "We were allowed to take these things off and manipulate them…I remember sliding those struts out…They were in two canisters, and the nodes were in a rotating drum. The way to get a strut out was to just get it by the tip, just get the thing going, so it is flying on its own, and then grab it about midway. We learned that in the water and it worked fine, but you've got to concentrate on what you're doing. I remember Brewster [Shaw] expressing concern that we were getting a little cavalier in our handling of them."

Assembly of an individual bay was a straightforward task, as Ross recalls, "It was basically a matter of bringing a part out, putting it onto this assembly fixture, hooking the components together, rotating to the three faces, sliding the completed segment of truss up, and repeating the process for a total of ten bays, each bay being four and a half feet long. We knew that this would be a very satisfactory way of doing business because when a crew member's feet are anchored properly, that gives you both hands free to do work."

Once the ten identical bays were assembled, the structure resembled the frame of a 45-foot "high-rise" that towered impressively above the payload bay.

The beauty of the ACCESS technique was its simplicity. The only tools were the astronauts' hands. All they did was to snap together the prefabricated components at the end nodes and lock them in place. On average, an individual bay could be created in 2 minutes 15 seconds and disassembled in 1 minute 45 seconds.

---

[17] Refer to Chapter 6 for details of the MPESS.

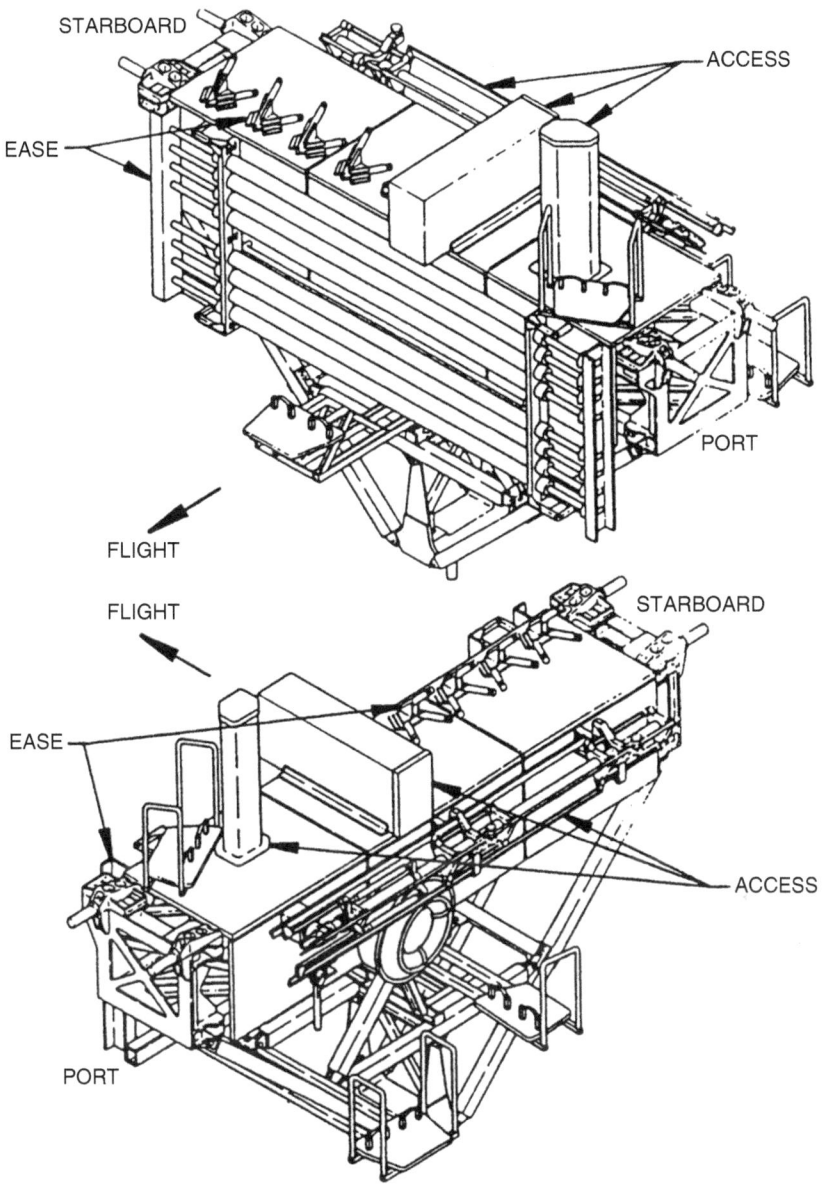

STARBOARD

ACCESS

EASE

PORT

FLIGHT

FLIGHT

STARBOARD

EASE

ACCESS

PORT

The MPESS in the STS-61B payload bay.

The second construction trial for Ross and Spring was Experimental Assembly of Structures in Extravehicular Activity (EASE) and it was developed by the Marshall Space Flight Center in Huntsville, Alabama. Whereas ACCESS used an "assembly line in space" technique with both astronauts standing at fixed work positions on the support structure,

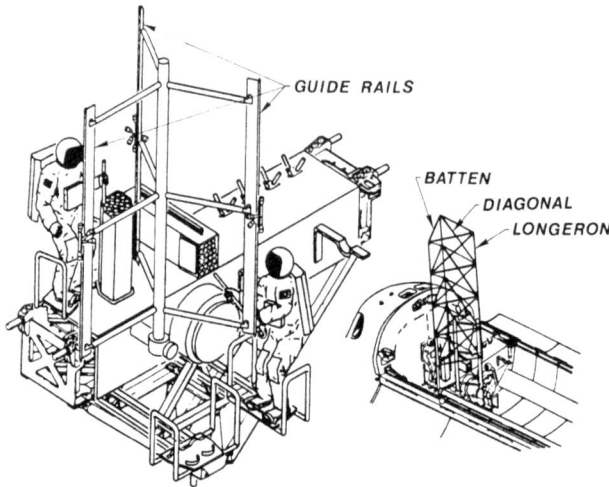

The main components of the assembled ACCESS experiment. To the left is the guide rail jig that was used to assemble the individual bays.

in the case of EASE they would be able to move about. As before, they would need only their hands to do the work. The hardware consisted of six 12-foot-long heavy aluminum beams (each with a mass of 64 pounds) and four identical joints.

As Ross says, "The EASE required one crew member to be floating at the top of the structure, just holding on with one hand and trying to maneuver these fairly heavy beams...kind of torqueing them into position, aligning them, and sliding the sleeve over to make the connections. We didn't think it would be a very desirable way of doing business, but we were certainly more than prepared to go investigate it and see what the deltas were between the water [tank] and on-orbit, and compare those with the ACCESS experiment."

The construction sequence consisted of connecting three vertical beams to a node attached to the top of the work platform and then connecting the free ends using three horizontal beams. On completion, the EASE structure resembled an inverted 12-foot-high pyramid. Owing to a fast working pace, eight assembly/disassembly cycles were achieved, which was two more than planned. As MIT investigator David Akin Jr, noted at a post-spacewalk press conference, "We feel there was clear evidence of a learning curve for both of the crewmen. There seemed to be a rapid adaptation to this free-floating assembly technique. They did an excellent job of controlling their body position. They did a superb job of aligning the massive beams without foot restraints or any other form and we saw evidence that this alignment task is faster in space than underwater."

Astronauts Ross and Spring during an assembly cycles of the EASE experiment. The inverted pyramid shape is clearly visible.

After a well-deserved day of rest, Ross and Spring donned their spacesuits again. Whereas the objective of their first excursion was to characterize human performance during construction work in weightlessness, this time they were to assess alternative assembly methods and work with the robotic arm to rehearse maintenance scenarios for the space station. As before, they started off with assembling the ACCESS tower. After quickly completing nine bays in the previous manner, Ross put the beams and nodes for the tenth bay into a component carrier on a workstation installed at the end of the robotic arm. In NASA parlance, this was called the manipulator foot restraint. As Spring explains, "What you've got when you step into the foot restraint, is you're being held in by little stirrups over your toes and there's a ridge on your heels that you just slide into. That's all you've got. There's

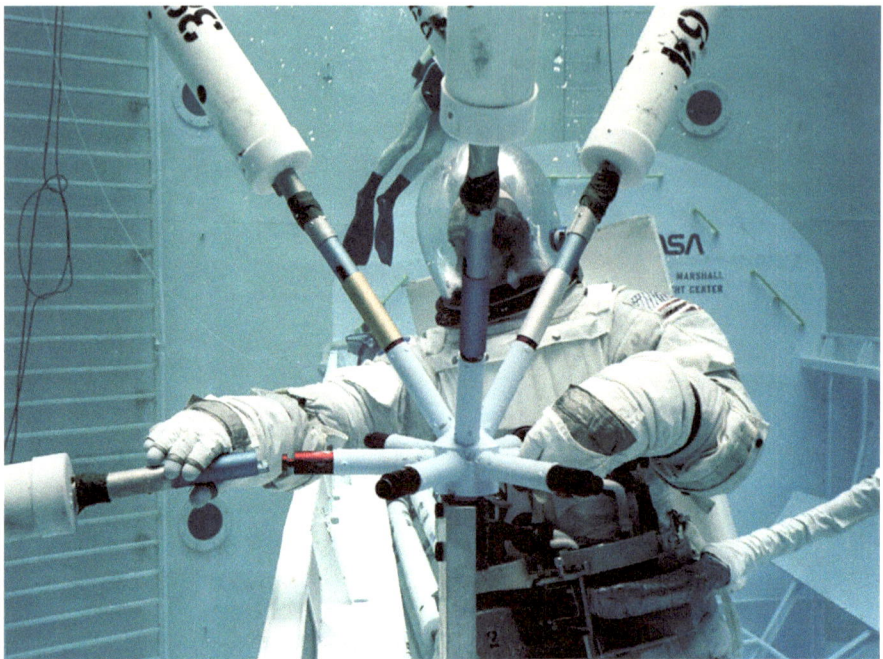

A demonstration of the EASE assembly technique during a training session in the Neutral Buoyancy Simulator at the NASA Marshall Space Flight Center.

also a little tether to your ankle, but that's about it when you're in that foot restraint." Ross climbed onto the platform and gave permission to arm operator Mary L. Cleave to move him to the top of the tower. Forty feet above *Atlantis*, Ross built the final bay and proceeded to clip a long piece of rope to the side of the framework to serve as a rudimentary simulation of how an electrical cable could be manually attached to such a structure. As the truss segments of the future space station were meant to support power and data cables, or pipelines of cooling fluid to and from the radiators, it was apt that the astronauts should try and replicate their installation.

The first spacewalk had proved that manual assembly of a large structure using small components was feasible. But what about building several of these structures and then joining them together? Some space station concepts envisioned creating a framework of considerable size to accommodate a variety of functions in addition to just performing scientific research in a few modules. For instance, there were plans to add hangars for repairing and servicing satellites or to refuel spacecraft bound for the Moon or Mars. Telescopes and other scientific instruments would also find a place in the farthest reaches of the facility, so that they could be serviced and upgraded. Free flying platforms that exploited the peculiar conditions of space to manufacture drugs and other products could dock at the station to deliver their output and replenish their raw materials before departing for their next run. Over time, the station would morph into a spaceport, and similar outposts would be created in orbit around the Moon and Mars.[18] Clearly, many frameworks would be required as the station expanded in size and scope.

Following the cable-laying demonstration, Spring, who had remained at the base of the support carrier monitoring his colleague, released the ACCESS tower from its support structure and handed it over to Ross, on the robotic arm. "Right now it feels very easy, but I don't have any rates on it. Very little force is required to start or stop motion," Ross reported as he rotated and translated the 190-pound frame into various positions. He was in full control of the structure's movement, and the task was even easier than during training in the pool.

Once the tower had been returned to its support structure, Ross and Spring traded places on the robotic arm's manipulator foot restraint. As he was moved toward the top of the tower, Spring experience a moment of anxiety, "The last time I remember having a little bit of anxiety was on the second EVA. Mary had worked Jerry first, so now she's feeling pretty confident on moving the arm at maximum speed, and we're allowed to work the arm pretty fast. So I said, 'Okay, Mary, take me up.' She did. She just did maximum elevation on the thing. She took me up at whatever the max speed was. It wasn't that much really, of course, but suddenly everything I knew was secure and safe and comfortable was going away from me very fast, and somewhere going through about 15 or 20 feet above the payload bay I had the sensation that I was way too high without a good handhold…I pause a moment and I say to myself, 'All right. This is cool. That's right. You're cool; you're cool; yeah, we're cool.' And then I was able to let go and work normally for the rest of the mission. But the first sensation was, 'Whoa. This is high.' The fact that you're 230 miles high and going 17,500 miles an hour matters not. Your focus, your security, your orientation is the Shuttle – a big 45 feet away."

Spring was not the only one to feel tense. Brewster H. Shaw, in command of the STS-61B mission, recalls "coming up behind Mary once, when she was operating the RMS and there was somebody on the end of the arm. I put my hand on her shoulder, and her whole body was quivering because she was so intent on doing this job right and not hurting anybody. She was so focused and so conscientious, not wanting to do anything wrong, because she had somebody out there on the end of this arm, and she was just quivering, and that just impressed the hell out of me, because I thought, you know, what a challenge, what a task for her to buy into doing when it was obviously stressing her so. You know, she was so nervous about it. But she did a great job, absolutely great."

At times, even the best training is unable to replicate the reality of space flight. As Cleave says, "We were facing the Earth and I was working the arm. This was when I had Woody and Jerry on it, doing the space construction experiment, and I had them up in the top window and we were payload bay to the Earth, like we usually work. So you had the Earth going at Mach 25 underneath you, and it was a moving target. So I was trying to move them against the truss, against the Earth, and ended up having to make a big mark on the window so that I had something to move them against. It was really distracting until we figured this out, putting them on the edge of the window, and that kind of stuff."

---

[18] NASA has never been short of wishful thinking, even more so in the 1970s and early 1980s. With the glory of the Moon landings still lighting the NASA engineering drawing offices, it was inevitable that grandiose plans for space exploration would be drawn up.

After dissembling bay number ten and returning the elements to the containers on the foot restraint, Spring was positioned at bay number eight. There he removed one beam and connecting nodes and restored them to assess the feasibility of a simple but effective repair by replacement. Then, as Ross had done earlier, Spring had his own go at manually maneuvering the tower. As he recalls, he fully enjoyed it, "I got this 45-foot grid structure moving kind of fast. I wanted to see how hard it was to get it moving. I'm a test pilot. I want to move things and see what it takes to do impulses. But I got it moving a little bit faster than Brewster was comfortable with, and it was just, 'Slow it down, Woody.'"

Anchored to the foot restraint on the robotic arm, Sherwood Spring is moving the ACCESS tower.

On returning ACCESS to the support structure, Spring joined Ross and together they dissembled the tower and erected EASE. This time they did it by taking turns at the foot restraint platform on the robotic arm while the other crewman remained on the MPESS support structure. Then the whole structure was manually manipulated. These exercises increased confidence in manipulating an entire framework as part of the construction of a larger structure. As NASA was considering using long external heat pipe loops to carry away the heat generated by the space station's equipment, the EASE beams were the perfect mockup to simulate such long and rigid apparatus. As Ross explains, "We had a little short coupling device and we coupled together two of these 14-foot-long beams that we made the EASE experiment out of. We put a little coupling in between to see how easy it would be to maneuver this longer beam from one end. It gave us quite a bit of understanding of what it would be like to assemble things in space."

Reflecting on what they accomplished, Spring offers the following observation, "Astronauts can do almost anything if you've got good foot restraints and a place to work from. Add the proper handholds, and you can manipulate huge objects, I mean multi-ton, quite precisely, as we've done with Hubble Space Telescope and a number of other missions."

However, when continuing funding cuts forced a dramatic curtailment of the size and functionality of the space station, the manual assembly approach was abandoned. On the other hand, despite having successfully completed all of the tasks assigned to the EVAs and in spite of the rationale that had prompted those spacewalks in the first place, it had become clear that on-orbit manual assembly of large structures was not a viable solution. In fact, as Ross explains, "When you think about having to integrate all the electrical and fluid lines and everything else into the structure once you have assembled this open network of trusses, it becomes hard to figure out how you'll do that and connect everything together and make sure it is tested and works properly."

Having discarded the manual assembly method, the International Space Station's backbone was assembled of pre-integrated segments, mechanically connected to each other on-orbit. The astronauts needed only to verify that those connections had been made correctly, and then join up the cables and pipelines between adjacent segments. It was a better and simpler approach that offered considerable cost savings because it reduced the number of missions and spacewalks devoted to the assembly of the space station.

Some 7 years later, further assembly testing was assigned to STS-49, which was *Endeavour*'s maiden voyage. This time the objectives were to assess mass transport and handling of large payloads such as those to be mated to the external structure of the space station. This involved the Assembly of Station by Extravehicular Activity Methods (ASEM) experiment jointly developed by the Goddard Space Flight Center, Langley Research Center, and Johnson Space Center. It comprised a set of struts and nodes to put together a truss pyramid structure spanning the full width of the payload bay. All components for the assembly were on an MPESS carrier, which in itself was part of the experiment. After building the ASEM pyramid, spacewalkers Kathryn C. Thornton and Thomas D. Akers were to manually unberth the MPESS and maneuver it into various positions with and without the assistance of the robotic arm. This they did. Later, they tried different methods to mate the MPESS to its fixture. The same exercise was duplicated using only the robotic arm, which provided an effective way to compare its effectiveness versus the manual handling. Throughout the EVA, crew comments, opinions, and photographic and television footage were recorded for later analysis. The experiment assisted in the development of payload handling procedures for use when assembling the International Space Station.

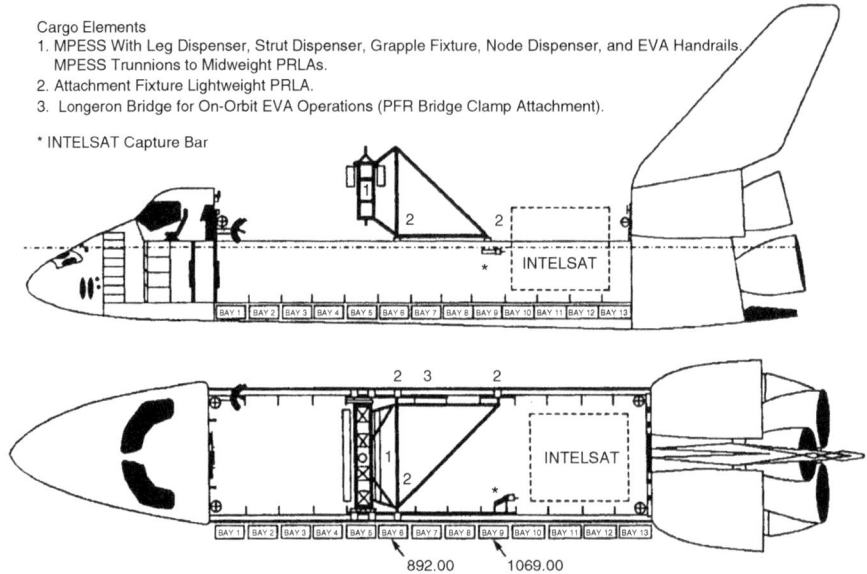

Cargo Elements
1. MPESS With Leg Dispenser, Strut Dispenser, Grapple Fixture, Node Dispenser, and EVA Handrails.
   MPESS Trunnions to Midweight PRLAs.
2. Attachment Fixture Lightweight PRLA.
3. Longeron Bridge for On-Orbit EVA Operations (PFR Bridge Clamp Attachment).

* INTELSAT Capture Bar

The on-orbit configuration of ASEM.

SPACEWALKING QUALITY TIME

Many people left NASA in the wake of the *Challenger* disaster. Among them were a good many trainers, flight controllers, technicians and engineers with knowledge and expertise in designing and planning spacewalks. To build a space station, NASA had to restore the lost basic skills. Astronauts Jay Apt and Jerry L. Ross were given this task. As Ross remembers, they convinced the Station program management that "we needed to start doing some EVA DTOs [Detailed Test Objectives], to conduct some actual EVAs to start building up that experience base again, both for our Office and the flight controllers and hardware designers." As the space station would be a very large structure, Ross proposed the concept of CETA as a small cart that an astronaut could use to readily move around the station during assembly and maintenance work. An acronym for Crew and Equipment Translation Aid (CETA), it received a positive welcome and within a month of its approval both Apt and Ross were assigned to the STS-37 mission that would test it. The release of the primary payload, the Compton Gamma Ray Observatory (CGRO), had required a contingency spacewalk to release a stuck antenna.[19] As ground control performed the

---

[19] Refer to Chapter 6 for an account of these events.

checkout of the satellite prior to the go-ahead for the robotic arm to let go of it, the two spacewalkers were allowed to remain outside in case any further problems with the satellite required their attention. Rather than spend the next 4 hours leisurely enjoying the view, the duo made a start on some of the tasks of the spacewalk that was officially planned for the next day.

In particular, they exercised the Crew Loads Instrumented Pallet (CLIP). As Ross notes, "This basically was a tape recorder and it had some batteries in it that powered the instrumentation on the pallet. The objective was to try to understand what kind of loads a crew member could react when he was in a foot restraint and when he wasn't in a foot restraint – the basic data that we needed to know in order to be able to start putting together design requirements for a space station." After affixing the hardware to the port wall of the payload bay, Ross went through a series of specific tasks, each representing a typical activity that would be required in assembling the space station, such as tightening a bolt or turning a knob. He did such a good job that he remembers "working up a sweat and the sweat getting down into my eyes at one point. It was a pretty exhausting period of time." Years later he would regret being so efficient in performing those tasks. "I ended up later – all of us did – living down that data. I put some pretty good forces into things, and they made those the design requirements for the station. It required them to put in load alleviators and all those other things. They said, 'Put the max amount in.' If they'd actually said, 'Put a reasonable amount in,' it would've been a totally different story."

Finally, with permission given for the robotic arm to release the CGRO, Apt and Ross headed back into the airlock.

The next day, they went out to resume the plan where they had left it. If it had not been for the contingency EVA to attend to the satellite, this would have been the first post-*Challenger* spacewalk. Ross and Apt started off by assembling a track along one side of the payload bay, on which the CETA cart would travel. The presence of the CGRO had hindered the installation on the ground of the full 46.8-foot-long track, as the satellite had taken up the full width of the payload bay. With the track ready, the duo began to evaluate three different configurations of cart to test the concept. Ross says, "Because the engineers in the Engineering Directorate wanted to build up their level of expertise…they agreed on a couple of different ways of propelling us up and down the track."

The first option was the so-called manual configuration. Once on a foot restraint, the astronaut would propel himself, hand over hand, along the length of the rail. The second configuration resembled the mechanism of a railroad car, where an astronaut would pump a T-handle to move. This motion was converted by a gear train into the continuous rotation of two wheel drives. Finally, in the third configuration, the cart would be powered by electric current generated manually using a crank handle.

Ross recalls, "Basically, you stood up on it and you had a hand crank which was driving a generator. The generator fed electricity down to a motor, which pushed you up and down the track." Both astronauts tested all configurations. Furthermore, Ross points out, "We also climbed onto the back of the other guy to add some mass, to see if that had much difference. A guy in a spacesuit weighs 350 pounds or so." During each testing session, the dynamics were measured, including the translation rates and crew loads induced in the carts and track. All configurations worked, but the manual one was found to be the most suitable, and years later it was added to the plan for the International Space Station.

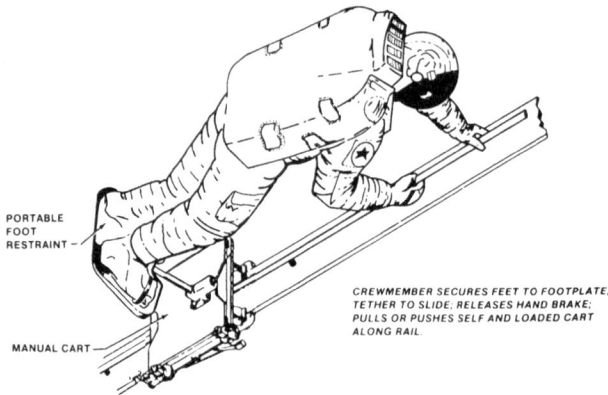

PORTABLE
FOOT
RESTRAINT

CREWMEMBER SECURES FEET TO FOOTPLATE,
TETHER TO SLIDE; RELEASES HAND BRAKE;
PULLS OR PUSHES SELF AND LOADED CART
ALONG RAIL.

MANUAL CART

A demonstration of the CETA cart manual configuration.

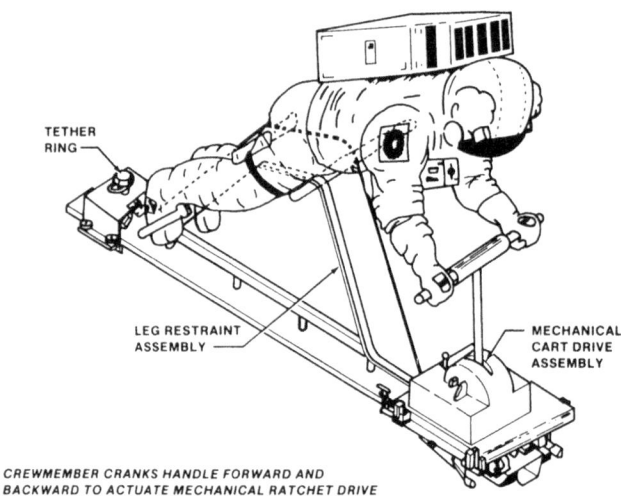

TETHER
RING

LEG RESTRAINT
ASSEMBLY

MECHANICAL
CART DRIVE
ASSEMBLY

CREWMEMBER CRANKS HANDLE FORWARD AND
BACKWARD TO ACTUATE MECHANICAL RATCHET DRIVE

A demonstration of the CETA cart mechanical configuration.

Ross expands, "We also had a tether shuttle, which was nothing more than a little knob mounted on a plate that had rollers that encompassed the sides of the tracks so that it could not come off. The concept was that you just tethered yourself to that and slid up and down the whole length of the track by yourself. That also worked nice." It too, was integrated into the design of the International Space Station.

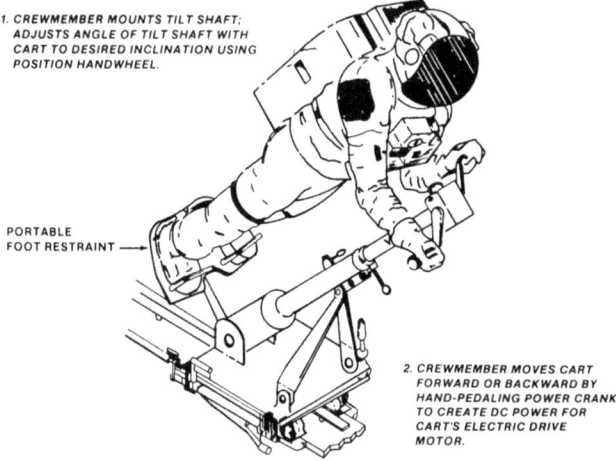

1. CREWMEMBER MOUNTS TILT SHAFT; ADJUSTS ANGLE OF TILT SHAFT WITH CART TO DESIRED INCLINATION USING POSITION HANDWHEEL.

PORTABLE FOOT RESTRAINT

2. CREWMEMBER MOVES CART FORWARD OR BACKWARD BY HAND-PEDALING POWER CRANK TO CREATE DC POWER FOR CART'S ELECTRIC DRIVE MOTOR.

A demonstration of the CETA cart electrical configuration with crank handle.

The final part of the 6-hour EVA was devoted to the EVA Translation Evaluation (ETE) aimed at getting to know better the Orbiter environment for spacewalks. For example, they strung a rope across the payload bay and used that as a way to move around. Ross also tested the feasibility of physically and manually maneuvering the robotic arm for moving payloads about. That is, while firmly standing on the payload bay, he grabbed the arm and tried to move it. Ross recalls, "I proved I could move it some in direct directions, but I couldn't do any rotations of the end effector, so that is probably not a really viable way of maneuvering the payload into position to attach it somewhere." To wrap up the spacewalk, "I rode on the end of the arm a little bit with Linda [Godwin] driving the arm, basically at maximum rates up and down and left and right. That was trying to understand what kind of speeds the crew member could tolerate, riding on the end of the arm."

As Ross and Apt had pointed out to Space Station managers, it was vital that the space agency log as much EVA time as possible, to gain the experience necessary for assembling the station. A number of scheduled flights were assigned tasks to refine training methods for future spacewalks, expand the experience of ground controllers, instructors and astronauts, and improve understanding of the differences between real weightlessness and the underwater training facility.

This development activity began with STS-54 in 1993, whose primary objective was the deployment of a TDRS satellite. On the fifth flight day, Gregory J. Harbaugh and Mario Runco Jr ventured into the empty payload bay to evaluate how well they could adapt to spacewalking, moving with and without carrying items, assessing the capacity to climb into a foot restraint without use of handholds, and aligning a large object in weightlessness. To simulate transporting a large object, one astronaut would carry the other. To explore the use of large tools, they worked with a device already aboard *Endeavour* that had been designed to manually adjust the tilt table used by the IUS/TDRS payload.[20]

---

[20] Refer to Chapter 2 for details on the Inertial Upper Stage hardware.

More station-related spacewalking time was logged on the fifth day of STS-57 by David Low and Jeff Wisoff. Their first task, however, did not concern station work but the safety of the Orbiter and payload. Two days earlier, the European EURECA free-flyer deployed by STS-46 the previous summer had been successfully retrieved. As *Endeavour* was approaching, the experiment-laden carrier was being deactivated. This process included folding two long radio antennas. While commands sent by the European ground control center were correctly received and executed, telemetry data showed that neither antenna had retracted into its stowed position. After unsuccessful troubleshooting, and with *Endeavour* in close proximity, it was decided to proceed with the grappling. But rather than stowing the spacecraft in the payload bay, it was to be held on the robotic arm. Pre-mission training had included such a scenario, and Low and Wisoff knew how to complete the retraction and locking of both antennas. But then, when the arm grappled the satellite a new issue arose. The platform was to receive electrical power from the Orbiter via the robotic arm in order to maintain the craft's thermal control system. Post-flight, it was ascertained that a connector on the end effector of the arm had been installed in the diametrically opposite orientation to where it was meant to be. With the batteries rapidly running out of charge,[21] Mission Control decided to stow EURECA in the bay immediately, to enable it to draw power using a remote umbilical connector that was directly linked to the Orbiter's electrical system.

Two days later, Low and Wisoff began their EVA by pushing the two antennas into place. If they had been left unlatched, the vibrations during re-entry could have caused the antennas to spring open and possibly damage nearby payloads and Orbiter structure. With this contingency action completed, the duo set about the prearranged tasks of the spacewalk. These included simulating the handling of large objects by having one crew member anchor himself to a foot restraint on *Endeavour*'s robotic arm and maneuver his colleague, serving as a piece of space hardware. In a similar fashion, they rehearsed the fine alignment of large masses. In this case, the man on the arm moved the floating spacewalker, attempting to position his boots close to a foot restraint. The next task was the insertion and removal of a bolt using a manual torque wrench, testing socket extensions of differing lengths, approach angles, and hand positions. Several improved EVA tools were also evaluated, including a short self-tending equipment tether designed to produce fewer snags, a safety tether that would stay closer to the spacewalker to enable him to perform tasks with minimum snagging and impact with surrounding structures, and a chest-mounted tool carrier and body restraint that had more than one hook.

The next spacewalk in this series was assigned to STS-51 mission specialists Jim Newman and Carl Walz, with an array of investigations of mobility in spacewalking. This time, the robotic arm was not used. In fact, on the second flight day *Discovery*'s crew had released the ORFEUS/SPAS free-flyer. This was to be retrieved later in the mission. To reduce the likelihood of a malfunction potentially stranding the platform on-orbit, the plan was not to use the arm at all between deployment and retrieval. An outing of 7 hours allowed the duo to test various types of rigid and semi-rigid tethers, as well as once again

---

[21] EURECA produced its own electricity by means of two large solar panels. However, at this point they had been folded for the retrieval of the satellite and therefore could no longer power the spacecraft.

moving about carrying each other. They also assessed how well a spacewalker could maneuver in weightlessness with a large object, and determined how well he must be restrained in weightlessness while applying a strong torque to tighten a bolt.

More spacewalk testing continued on STS-69, with Jim Voss and Mike Gernhardt evaluating thermal improvements made to their spacesuits and a variety of tools and techniques that could be employed in the assembly of the International Space Station. Each astronaut spent 45 minutes on the arm, located away from the warmth radiated by the payload bay. With the bay facing deep space, the astronauts were exposed to temperatures as low as minus 120°F. They continually provided subjective ratings on their comfort level, happily reporting that the bypass on the liquid cooling/ventilation garment and the heated gloves were performing as advertised. In fact, both astronauts confirmed that their thermal comfort was maintained throughout the excursion, both during the cold soak experiments and during a series of repetitive tool handling tasks.

Further training was achieved by two space station-related EVAs undertaken on STS-72. On the fifth flight day, Leroy Chiao and Dan Berry spent more than 6 hours in *Endeavour*'s payload bay assessing tools, techniques, and construction equipment. After a few minutes of acclimatizing themselves in the bay, first-time spacewalkers Chiao and Barry attached a portable work platform to the end of the robotic arm, in this case operated by pilot Brent Jett and mission specialist Koichi Wakata. Jett used the arm to grapple various bits of hardware that were designed to hold large modular components. This mimicked the way equipment boxes and avionics would be moved back and forth in assembling the space station. Chiao and Barry unfolded a cable tray diagonally across the forward portion of the payload bay. The cable tray, also known as a rigid umbilical, housed simulated electrical and fluid lines similar to those which would connect modules and nodes of the station.

As Chiao says, "Basically, this was a hinged cable tray that would be deployed to make connections from one piece of the station to another. You would deploy these rigid umbilicals with fluid and electrical umbilicals in them…We had to test this, to confirm it was a valid construction concept that we could easily make it work. One interesting thing we found was that in the electrical connectors, the plastic and the metal became very stiff in the cold in space. So one big lesson learned was we had to design enough extra slack in the cables to get enough bend in it. The connectors had to line up pretty perfectly for them to get mated together, and to be able to drive the bale and to latch them all up. So that was a pretty big deal."

Chiao and Winston Scott performed the second EVA. As Chiao recalls, "We had some toolboxes. We tested the body-restraint tether, which was really a great tool, something you wear on your waist…It was a stack of balls with a jaw on it and you could attach the jaw to a handrail and then tighten up that stack of balls with a cable in the middle, so that you were semi-rigidly attached. You could still grab a handrail and move yourself around, but you could use both hands to do whatever you needed to do because you were attached with that 'third hand.'" This successful test proved beneficial during the assembly of the ISS. As Chiao says, "That was a big time saver. Before the invention of that tool…they wanted to use foot restraints everywhere. In that case you'd have had to carry a big, heavy, bulky foot restraint around with you. You had to be careful not to bang it into anything and damage it, and you had to take the time to plug it into its socket and set it up. We ended up being able to eliminate that foot restraint for a lot of tasks. That saved a lot of time. But for the heavy duty tasks where you require both hands to really muscle on something, you still need that foot restraint. So we still have to cart that around with us sometimes. But this BRT as we called it, body-restraint tether, really helped streamline a lot of our operations."

On returning from *Columbia*'s 16-day STS-87 flight, mission commander Kevin Kregel could not have better summarized its achievements when he stated, "It was an action packed flight." The main objective was the fourth United States Microgravity Payload (USMP), a collection of nine experiments to gain additional understanding of the fundamental properties and behaviors of various materials and liquids in space. The second objective, the deployment and retrieval of the SPARTAN 201 free-flyer, gave the astronauts more interaction than expected. The fourth flight of this versatile platform was to investigate physical conditions and processes of the hot outer layers of the Sun's atmosphere, known as the solar corona. The autonomous spacecraft was released by the robotic arm on the third flight day, but it promptly failed to execute a pirouette maneuver as part of its startup routine. There was clearly a failure in its fine pointing attitude control system. Mission specialist and robotic arm operator Kalpana Chawla attempted to recapture the free-flyer, but did not receive any indication of a firm capture. She backed the arm away in order to attempt a second grappling, but to no avail. In fact, the free-flyer developed a 2° per second spin. At this point, Kregel fired *Columbia*'s thrusters to initiate a separation maneuver, because it was clear that a new plan would be required.

Mission specialists Winston Scott and Takao Doi were scheduled to perform an EVA on the sixth flight day to evaluate apparatus and procedures that would be used during assembly and maintenance of the International Space Station. This outing was to pick up all the tasks assigned to STS-80, where a fault with the airlock hatch had ruled out spacewalking.[22] Scott and Doi had been given training to manually capture the SPARTAN, and had been told to do so. Their entry into *Columbia*'s payload bay marked not only the first Japanese spacewalk, but also the first EVA for this Orbiter. Although the first spacewalk of the Shuttle era was assigned to *Columbia* on STS-5, faults in both spacesuits had obliged the spacewalk attempt to be canceled. Now, 83 mission's later, astronauts passed out through the airlock. Scott and Doi positioned themselves on opposite sides of the SPARTAN support truss, which spanned the full width of the bay, then *Columbia* maneuvered up to the free-flyer. When it was close enough, the two men took the satellite by hand, lowered it down onto the truss, and latched it into place. This done, they turned their attention to setting up and testing a crane device that would be used to move large Orbital Replacement Units (ORU) during the assembly and maintenance of the International Space Station. After the crane was installed in its holder on the port wall of the payload bay, Doi evaluated its operating characteristics. Meanwhile, Scott released a large battery unit and its carrier device from the starboard wall. The 500-pound battery/carrier unit was then hooked to the crane to assess its ability to move a large mass. Doi and Scott then stowed all the apparatus and re-entered the airlock.

Some consideration was given to making a second attempt to deploy SPARTAN, but eventually mission managers abandoned the idea because it would have disturbed the USMP-4 experiments and would have eaten into the Orbiter's propellant reserves against

---

[22] For more details refer to Chapter 3.

any possible contingency and landing scenarios. However, it was decided to send Scott and Doi out again in order to complete the tasks which remained from the first spacewalk. So, during an almost 5-hour spacewalk on flight day fifteen, the two men continued the crane trials, but instead of the large simulated station battery they used a different ORU simulator representative of small objects that would be moved around during station assembly. Scott also manually deployed the Autonomous EVA Robotic Camera/Sprint (AERCam/Sprint), which was a small, unobtrusive prototype of a free-flying television camera to assess the possibility of remotely inspecting the station. Using a joystick on the aft flight deck, pilot Steve Lindsey spent a little more than an hour maneuvering the small 14-inch-diameter sphere inside the payload bay on a pre-defined flight test scenario. At completion of the experiment, the sphere was manually grasped, secured, and returned to the airlock. All of the objectives of the experiment had been completed, and no system anomalies were experienced. It was a successful demonstration of a new capability for on-orbit operations and assembly of large structures.

When Scott and Doi re-entered *Columbia*'s airlock, it not only marked the end of the mission's final spacewalk but also of the conclusion of the training program that NASA had organized in order to learn how to assemble space structures. During the two previous decades, the agency had accumulated enough spacewalking experience to be confident that the International Space Station could be assembled using a series of elements ferried into orbit by the Shuttle.

One year later, STS-88 was to carry out the first assembly mission by mating the American Node named "Unity" to the Russian Control Module named "Zarya." The successful completion of the ISS in the summer of 2011 is a tribute to the ingenuity and hard work that NASA put into preparing for the biggest construction project yet undertaken in space.

5

Satellite Servicing

INTRODUCTION

Frank Cepollina began his career in the space industry designing satellites to explore the new frontier of space. After earning a degree in mechanical engineering from the University of Santa Clara, he spent time in the Army Security Agency in Warrenton, Virginia. In 1963 he joined the NASA Goddard Space Flight Center in Greenbelt, Maryland, and was assigned to the Advanced Orbiting Solar Observatory project for the development of space-based observatories to continuously monitor solar activity and its influence on the space environment in the neighborhood of the Earth. At that time the 5-year-old agency, and to a large extent the entire American space industry, was ascending a steep learning curve. In 1965 this project was split up into a number of less ambitious challenges.

Cepollina joined the new Orbiting Astronomical Observatory project in time to see the first spacecraft fail. As he wryly observes, "It lasted about 90 minutes in orbit and died because of a massive generic design problem." At a time when the failure rate for satellites was as high as 25% during their first 6 months of operations, this loss was par for the course. Cepollina kept seeing other breakdowns, "While I was on that program doing this, two or three other observatories failed. One of them was the most embarrassing – it was an observatory called Orbiting Geophysical Observatory. When they commanded it to go into operation, it spun backwards because they had hooked up the gyros backwards."

Clearly, this situation was unsustainable because each failure, despite providing a valuable opportunity to learn lessons, was a waste of taxpayers' money and point of criticism for the young agency. It was at this time that NASA Deputy Administrator George M. Low ordered Goddard to improve, and rapidly. "What he basically told us was, 'You guys at Goddard have got to find a way to build spacecraft cheaper and more reliable.'"

Cepollina assembled a team of engineers and rocket builders to tackle the issue, and they recommended a logical and elegant solution that was based on the principle of modularity. They said that reliability could be drastically improved if a spacecraft was designed with all of its subsystem components assembled as modules that could simply be plugged into the vehicle and connected to each other via a common set of standard interfaces.

© Springer International Publishing AG 2017
D. Sivolella, *The Space Shuttle Program*, Springer Praxis Books,
DOI 10.1007/978-3-319-54946-0_5

"The modules were rectangular and they went around the entire spacecraft system. Smaller modules were also used for electronics, sensors, and instruments. There is significantly less design iteration between [different] spacecraft and the science instruments if they are modular, since scientists don't have to become spacecraft experts and engineers don't have to become would-be scientists."

This concept was applicable to any type of spacecraft, ranging from space-based observatories to communications satellites. This Multi-mission Modular Spacecraft (MMS) concept spread quickly through the engineering community and was adopted by new projects, such as Solar Max and the Large Space Telescope (LST).[1] It even contributed to selling the emerging Space Shuttle program to Congress. At Goddard, they built a full-scale mockup of the LST and sent it to Downey in California, where North American Aviation[2] had assembled a plywood mock-up of the Space Shuttle Orbiter. It was used to demonstrate to the press how built-in modularity could allow a pair of spacewalking astronauts to service the large satellite within the payload bay of the Orbiter.

In the audience was NASA Administrator Rocco A. Petrone, who got hooked on what he saw. As Cepollina recalls, "He immediately got on the phone and called his associate administrators to come out and see [the mock-up]. They realized that we were part of their ability to sell the use of Shuttle." In fact, Cepollina and his team had developed the MMS with the Shuttle in mind. Also, when Low had told Goddard to improve on spacecraft reliability, Cepollina recalls that he added, "Oh, by the way, this thing called Apollo is coming to an end and we're going to have another vehicle coming along that could possibly provide astronauts to fix things for you." It all came together nicely. The modularity of the MMS was EVA-rated by developing power tools that did not require an astronaut to grab hold of a handrail in order to remain in place when unscrewing a box that needed replacing.[3] The MMS specifications also included an external grapple fixture to enable the robotic arm of the Orbiter to take hold of the satellite. All in all, rating an MMS for on-orbit man-tended servicing only negligibly increased complexity.

The first spacecraft to fully adopt the MMS ethos was Solar Max, manufactured by Fairchild Industries. It was the first solar observatory to be specifically devoted to studying solar flares, violent eruptions that regularly occur on the surface of the Sun. The satellite was 13 feet long and 4 feet wide, and weighed 5,105 pounds. It was split into two main sections connected by an adapter. The support module housed all the essential vehicle subsystems such as the attitude control system, and the instrument module held the payload of seven instruments to observe the Sun in the ultraviolet, X-ray and gamma ray regions of the spectrum. The instrument module also housed the sensors for pointing control. The adapter supported the two fixed solar panels that generated 3,000 watts of power, and a high-gain antenna for data transmission to the ground.

Solar Max was launched on February 14, 1980, to monitor solar flares during the coming "maximum" of the 11-year cycle,[4] when they are most frequent. Its findings were to be

---

[1] Several years later the LST would become known as the Hubble Space Telescope.

[2] NASA awarded North American Aviation the contract to build the Space Shuttle Orbiter.

[3] In the frictionless environment of space, Newton's third law of motion is visibly predominant in even the simplest of tasks. Fastening a screw results in the astronaut rotating in the opposite direction unless he has a way of retaining himself to counteract the rotation.

[4] This is why the mission was called Solar Max (maximum).

integrated with those from sounding rockets and ground observatories for what was expected to be a memorable International Solar Maximum Year. However, 10 months later the $77 million spacecraft was crippled by three blown fuses, all of which were hermetically sealed within the attitude control system. Unable to remain fixed facing its target, the satellite started to spin and wobble. It could operate only a coarse pointing regime by exploiting the interaction of torque bars with the Earth's magnetic field. This allowed three of the seven instruments to continue to take useful data. Thanks to the foresight of Cepollina and his team, NASA was able to call upon an ambitious plan to rescue the satellite.

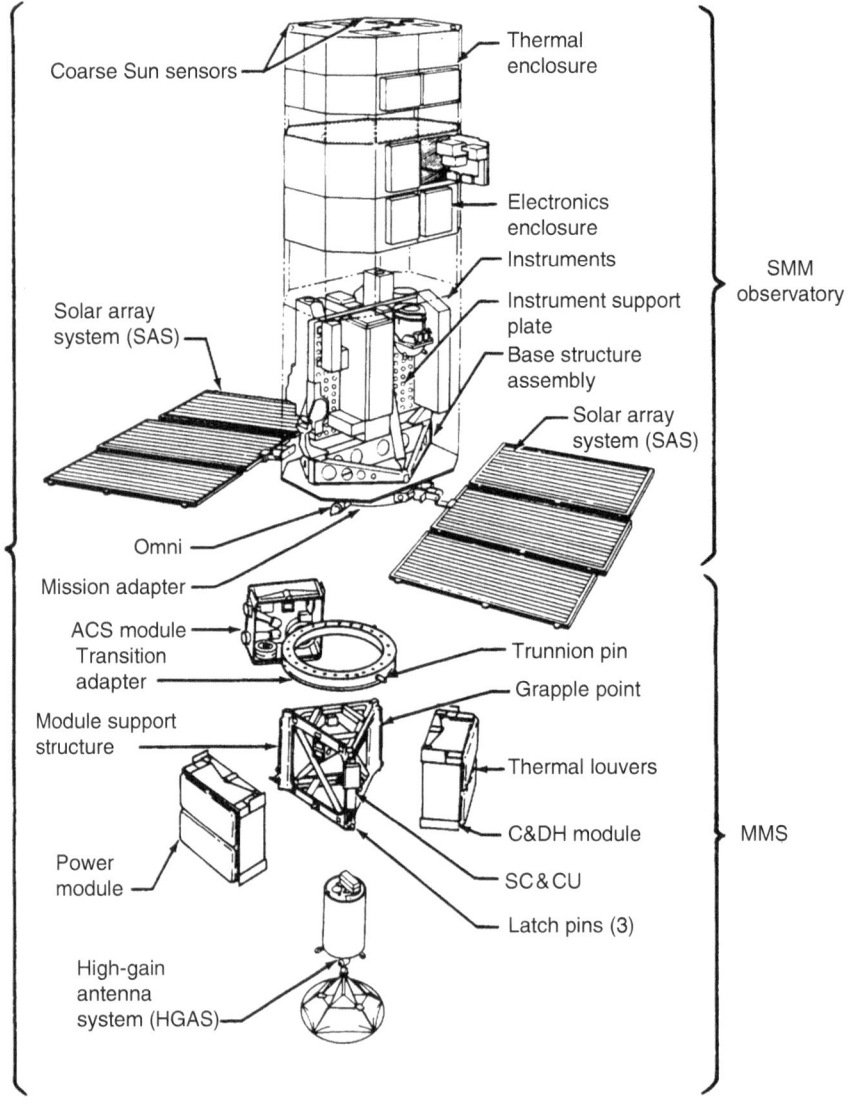

An exploded view of the Solar Max spacecraft.

THE FIRST ON-ORBIT SATELLITE SERVICING

When *Challenger* lifted off on April 6, 1986, STS-41C and its crew of five set about the space agency's most complex mission this far. Their first task was to deploy the Long Duration Exposed Facility (LDEF), which was a bulky 12-sided, open grid of aluminum rings and longerons that carried 57 scientific, applications and technology experiments. The plan was for another Shuttle to retrieve it several months later and return it to Earth to reveal how exposure to the space environment affected a variety of materials of interest to the designers of future spacecraft and in particular the space station. Since most satellites either remain indefinitely on-orbit or burn up during re-entry, with only a small number being deliberately recovered, the results of these experiments would help manufacturers to make better vehicles. LDEF also contained experiments in science, electronics, optics, power, and propulsion to minimize the risks of introducing new technologies in space. It was a simple passive structure with no systems for attitude control, power generation, data handling or communications. Individual experiments had to carry their own power source or tape recorder for data storage and be capable of operating without supervision from the ground.[5]

On the second flight day, Terry J. Hart deftly maneuvered *Challenger*'s robotic arm to engage a grapple fixture on LDEF, thereby activating an experiment-initiation system that turned on all those investigations that required power. He then skillfully hoisted the satellite out of its five anchoring points and straight up out of the payload bay. Being entirely passive, the only way that the satellite could maintain the stable attitude needed to provide an undisturbed microgravity environment was for Hart to release it in an attitude where its internal mass distribution would hold it stable in the gradient of the Earth's gravitational field. Once he had it in the desired orientation he withdrew the arm, taking care not to impart any rotational motion. Then the Orbiter backed off as the crew photo-documented the satellite.

Now the tricky part of the mission would come with the long-awaited attempt to repair Solar Max.

The first burn of the rendezvous sequence had been made 5 hours into the flight, to establish the correct phase angle with Solar Max. Other burns brought *Challenger* along-side Solar Max on flight day three. In addition to being the primary robotic arm operator, Hart helped mission commander Robert L. Crippen to conduct the complex activities of the rendezvous, "I was working with Crippen to use the radar and the optical star trackers to find the Solar Max, and everything was kind of just falling in right, as it should. And, of course, we had plenty of backup from Mission Control… We picked [the satellite] up

---

[5] Whilst it may seem a boring spacecraft, LDEF was in fact a good example of how NASA tried to lower the cost of running experiments in space. Depending upon its complexity, an experiment could cost from less than $10,000 to about $400,000, encouraging high risk/high return. It also made experiments particularly attractive to schools and research groups, because they would not require excessive amounts of investment or experience in space experimentation. In addition, LDEF was meant to be reusable, with flights lasting from several months to years. There could also be servicing missions to change-out some of the experiments. A 2-year mission was being planned for the end of 1986, with the main goal being to collect heavy nuclei such as those which occur in cosmic rays.

The LDEF being hoisted out of *Challenger*'s payload bay on STS-41C.

about 300 miles away with the star trackers, and then as we got closer with the radar. [We] processed all that, and the navigation system and everything was 'tickety-boo' as Pete Conrad used to say.[6] We came in, and Crippen took over manually as we got into a few hundred feet from the Solar Max."

Although the spin on Solar Max was slow enough for the robotic arm to grapple it, the first option was for an astronaut flying the MMU to attempt a manual capture. George D. "Pinky" Nelson left *Challenger* heading towards Solar Max, while fellow spacewalker James D. A. "Ox" van Hoften waited in the payload bay. Moving away from the Shuttle was a profound experience as Nelson recalls, "It was pretty exciting in retrospect, and the memories of the view from there are just amazing – the Shuttle against the Earth and jets

---

[6] Charles "Pete" Conrad flew Gemini, Apollo and Skylab missions.

firing and all that. What an extraordinary experience to be able to fly the MMU." The plan was to match the spinning motion of Solar Max and then use the TPAD[7] mounted on his chest to capture a trunnion pin protruding from the adapter section of the MMS satellite. At that point, Nelson was to use the MMU's thrusters to cancel the spin so that *Challenger* could move closer and Hart could grab the satellite.

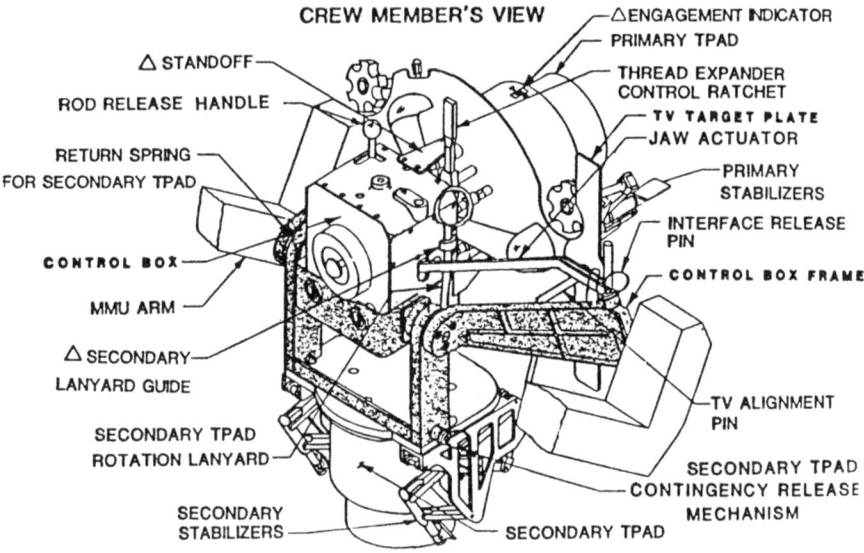

The TPAD in detail.

All went well up to when Nelson inched forward for capture. The TPAD engaged the pin, but an instant later he bounced back. He tried a second time, pushing a little harder, again without result. Nelson even sought to grasp one of the solar panels with his gloved hand, again in vain. As Hart observes, "Of course, we never even talked about this scenario over a cup of coffee. I mean, this was total freelance at this point. We didn't know what to do." Luckily for Nelson, the sharp panel did not cut open his glove, and happily for Solar Max his grasp didn't shatter the delicate panel. But each contact with the satellite perturbed what had been a stable spin, and the conservation of angular momentum imparted a tumbling motion which made any further efforts to grasp it more difficult. In fact, Hart took a shot at capturing it using the robotic arm, but failed – fortunately so, since the satellite's kinetic energy could have snapped the arm off the Orbiter.

---

[7] The acronym TPAD stands for Trunnion Pin Attachment Device. It was a T-shaped assembly measuring approximately 20 inches by 20 inches at its greatest height and depth.

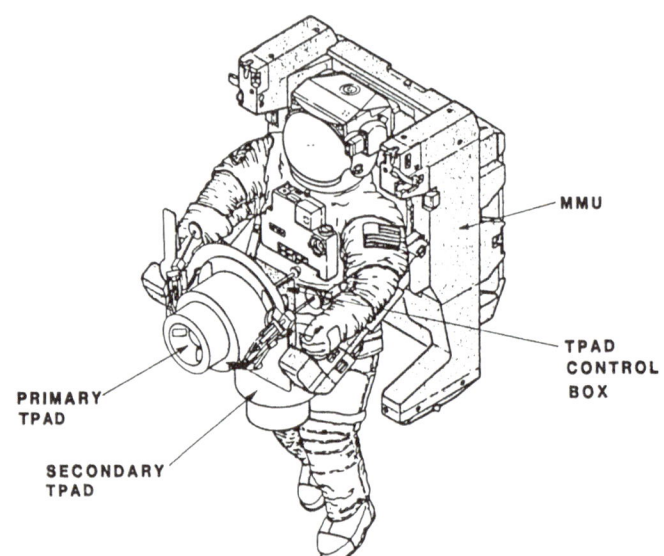

A space suited astronaut with TPAD and MMU.

George "Pinky" Nelson arrives at the ailing Solar Max.

Wisely, Mission Control canceled the retrieval and *Challenger* withdrew in order to station-keep at a safe distance while Goddard Space Flight Center worked out an alternative plan. As to what had gone wrong, Hart recalls that the problem was soon understood, "The pin they were trying to attach [the TPAD] to was a pin that held the satellite at launch, and around it there were some gold thermal blankets. The docking adapter on Pinky's chest was like a round shroud with a hole in it, and what no one had noticed was that one of the blankets had been put on with a little plastic standoff which the grommets on the blanket fit over. The drawings, the engineering drawings, didn't specify where those standoffs would be… But when they were designing the docking adapter, no one noticed that. They had [assembly] closeout photographs, but no one noticed that there was a pin there. So when he went to dock, the pin interfered with the docking adapter. They figured that out overnight and then told us the next morning, so we knew we couldn't use the docking adapter."

This was not good news, but at least the crew could now relax. In fact, Crippen recalls that he saw himself spending "the next six months in Washington explaining why we didn't grab that satellite," and likewise Nelson thought he was "going to get the blame for this, [get] the credit for not having it work." The astronauts were not at fault.

On the fifth flight day, *Challenger* again moved closer to Solar Max for a second retrieval attempt. Since the Orbiter's propellant supply was already approaching red limits, this would be the last chance to capture and repair the satellite. In fact, during the first attempt, Crippen had had to maneuver quite a lot in order not to lose sight of Nelson and avoid being struck by the tumbling satellite when the robotic arm tried to capture it. Now with less than 22% of fuel remaining in the forward RCS pod, only a tiny margin for error was left.

With an EVA retrieval out of the question, the only way to grab Solar Max was to use the robotic arm. Thanks to the excellent work of Goddard's flight controllers, the satellite was once again in a stable spin. As Hart recalls, "We came up and got right in front of the satellite, and we just watched as it spun around. And as it came around to the right place, I reached over and grabbed it. Everything was just fine. But it was a dramatic moment for Mission Control."

Hart berthed the satellite on the Flight Support System (FSS) that occupied the aft end of the payload bay. The large U-shaped structure was comprised of three cradles, designated A, B, and A-prime. Cradle A was 16 inches deep and had a latch beam that spanned the width of the cradle to serve as an anchor point for a satellite during launch or re-entry, as well as on-orbit servicing if access requirements demanded the satellite be placed in a horizontal orientation. Cradle A-prime was dimensionally the same, but had an extender and adapter section to hold and operate the Berthing and Positioning System (PBS) ring. Sandwiched between the two cradles was Cradle B, anchored to the Orbiter structure by four longeron trunnions and one keel trunnion. The main support structure had 12 motor-driven mechanisms to provide mechanical and electrical linkage between the Orbiter and the attached payload. The PBS ring was stored vertically inside the cradles for launch, then pivoted to horizontal on-orbit. A rotator on the ring could turn the satellite, and a pivoting mechanism could tilt it to any position from 90° (perpendicular to the bay) to 0° (horizontal in the bay).[8] The ring also contained three jaw-like berthing latches which would clamp

---

[8] The horizontal position would be used if it transpired that the satellite could not be repaired and needed to be returned to Earth.

onto pins near the base of the satellite's support module. Two umbilicals would be plugged into the satellite so that it could draw power and thermal conditioning from *Challenger*. The systems on the FSS provided for the electrical operation of the PBS and the electrical support services to the FSS and attached spacecraft while on the ground and on-orbit. These services included operating power, externally applied heater power, and relay of serial commands and telemetry via the Orbiter to the Payload Operations Control Center (POCC). The FSS was intended to be a reusable primary interface for MMS-based space-craft, and its design could be adapted to different missions. With all three cradles together, a satellite could be carried into space and deployed, or returned for a full service,[9] whereas only Cradle A-prime would be required for on-orbit servicing.

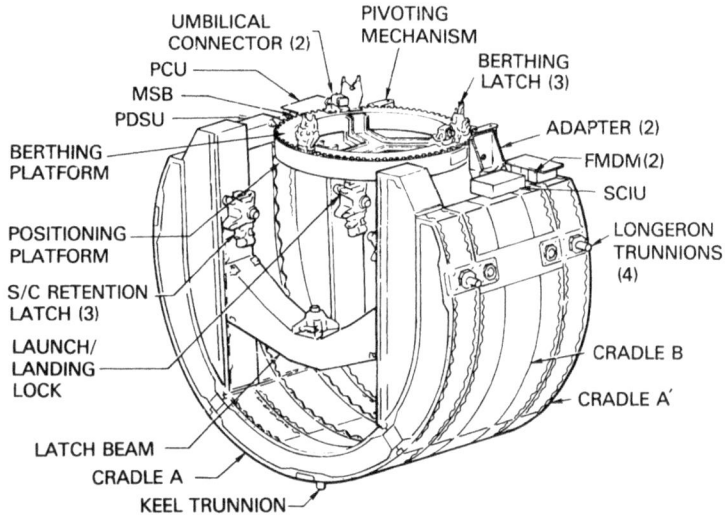

A detailed view of the Flight Support System for Solar Max.

It was now time for the repair work. The spacewalking duo began by closing the ring berthing latches to engage the pins on the satellite, firmly securing Solar Max on the sup-port structure. Hart was then free to withdraw the robotic arm in order to drive the astro-nauts about. In the time span of just 4 hours, they successfully completed the first ever on-orbit satellite repair. They began by replacing a faulty attitude control box. The substi-tution was carried out by van Hoften, who first covered the module's sensitive star track-ers. Using the so-called module service tool that was developed specifically for exchanging

---

[9] For STS-41C, the FSS was configured with all three cradles, in case Solar Max had to be returned to Earth.

the modules of the MMS,[10] he disengaged the module's two retention bolts[11] and then carefully eased the 500-pound box off the satellite. The replacement attitude control module was retrieved from the FSS and secured to the spacecraft using the same tool.

Next, they installed a baffle on the exhaust vent of the X-ray Polychromator. This instrument measured X-ray emissions from solar flares, and the baffle was required to prevent hot plasma leaking into the instrument and distorting the collected data. The repair did not require extra tools, because the baffle was designed simply to snap onto the vent with spring clips.

Finally, they moved to the instrument section to replace the main electronics box of the Coronagraph/Polarimeter that studied the outer atmosphere of the Sun. It had been malfunctioning ever since launch. This repair was perhaps even more pivotal in the history of on-orbit satellite servicing because, unlike the support section which was to be common to all MMS, the unique instrument section had not been designed for on-orbit repair. So the astronauts were being called upon to prove that equipment not specifically designed for orbital maintenance could be refurbished. The operation was carried out by van Hoften. He first cut an aperture into a panel in the satellite's shell, then through an insulation foil to access the box. He next removed a number of screws that had held a protective thermal blanket over the box and taped the thermal blanket and insulation out of the way, so that he could work. After that, he installed a hinge and removed the remaining screws in order to open the panel that covered the main electronics box. Now he could unplug the cables from the box, remove the box, and pass it to Nelson, who handed him the replacement box. This was then installed, all of the connectors remated, the door closed and secured, and the insulation placed back into place.

Later that day, Solar Max was released back into orbit and for the next 5 years it provided valuable observations of our parent star, then it succumbed to atmospheric drag and re-entered on December 2, 1989.

Compared to the intricate and elaborate servicing of the Hubble Space Telescope or assembly of the International Space Station, the repairs to Solar Max might appear trivial, and perhaps even of little value. But the significance of this truly pioneering mission must not be underestimated. First of all, it undeniably demonstrated that on-orbit repair is indeed feasible and can improve a spacecraft's operations by installing new state-of-the-art components, subsystems and experiments to make it "better than when it was new." It also showed that the MMS concept could reduce design costs and shorten production times, because manufacturers would no longer need to rely on designing expensive backup components into their systems to prevent failure. The accessibility to space offered by the Shuttle, it was argued, would guarantee recovery of the spacecraft and avoid an expensive write off in the case of failure. Perhaps even more far-reaching, it proved that on-orbit servicing made sense in both economic and scientific terms.

---

[10] The tool was controlled by two handles and two switches. Two latches on the tool fit into oblong holes on the module and held it in place as the socket wrench on the end of the tool loosened the retention bolt. The rather cumbersome tool was necessary because if a regular wrench had been used, the fasteners would have remained in place and the astronauts would have moved.

[11] Each box-like module was attached to the MMS by two retention bolts.

Solar Max in *Challenger*'s payload bay being repaired. Clearly visible on the lower end of the FSS are two large boxes to carry repair parts and tools.

For instance, the mission press kit estimated the cost of the repair at $48 million, as compared to the reported cost of $235 million for replacing and launching a brand new Solar Max. It was an almost five-fold difference in money to earmark for just a single mission. And it eliminated the uncertainty of having a replacement spacecraft approved. Time and time again, insufficiency of funding and lack of orbital servicing capabilities have forced the abandonment of an otherwise successful spacecraft. With on-orbit servicing an entirely new approach was possible. And, of course, as NASA happily pointed out, the servicing of Solar Max had been conducted during a Shuttle mission that had already accomplished its primary objective of deploying the LDEF satellite and so, in a sense, was "free."

An invaluable engineering lesson was also brought home, as astronaut Jerry. L Ross says, "The satellite was never expected to be handled by humans in space, even though some of it was designed for on-orbit servicing. One of the lessons which we learned from that mission, was you need a very carefully documented configuration of the hardware that is in space, so that you can properly design the hardware that is going to interface with it." The prime example of this was the failure of the TPAD to mate with the satellite.

In addition, Solar Max proved that servicing an expensive scientific tool in space pushed the science objectives and results beyond those originally envisaged. In fact, NASA was able to extend the observatory's mission, giving scientists an unexpected and very welcome set of observations beyond the period of solar maximum and on into the period of minimum activity of the 11-year cycle. Of course, the five highly successful servicing missions to the Hubble Space Telescope are the best illustrations of what on-orbit maintenance can deliver to the scientific community.

STS-41C projected the Shuttle program into the realm of new grand possibilities. As we shall see, other dramatic recovery and servicing missions would follow, but it will also be shown that, despite its benefits, on-orbit servicing was not exploited as it should have.

"SOMETHING THAT NOBODY'S EVER DONE BEFORE"

The flight plan for STS-51A in November 1984 proved be one of the most exciting of the Shuttle program, thanks to STS-41B, which, earlier that year, had deployed the Palapa B2 and Westar 6 communications satellites. Both of these failed to reach their assigned orbits because of a systematic fault in their PAM perigee kick stages.[12] As a result, hardware worth millions of dollars and in perfect working conditions had been stranded in useless orbits. But something good arose from this debâcle, namely the successful introduction of the MMU.[13]

As STS-51A astronaut Joe Allen recalls, even before *Challenger* returned home from STS-41B, the space agency, "very quickly evolved a plan to use the MMU and spacewalk-ing crewman to move out to the satellites...grapple them...and bring them home." Hence STS-51A would attempt the first space salvage operation in history. Despite some initial doubts, particularly concerning the fact that neither satellite was designed to be retrieved, NASA agreed to the plan when the insurers of the satellites requested retrievals. As Allen points out, "This plan was encouraged from the outside by business entities that now owned the satellites...insurance companies, one being Lloyd's of London and the other being International Technology Underwriters."

STS-51A astronaut Dale Gardner, a Former Navy test engineer, soon realized that the problem was similar to that of in-flight refueling. The way the US Navy refuels an aircraft in midair requires the receiving aircraft to insert a probe into a drogue at the end of a long flexible hose that is reeled out from the aircraft tanker. He thought a similar concept could

---

[12] An investigation revealed a manufacturing defect in the walls of the nozzles of these two PAM stages, with the result that after only a few seconds of firing there occurred a burn through which shut down the motor.

[13] For further details of STS-41B refer to Chapter 3. For details of the PAM upper stage see Chapter 2.

be applied. If the nozzle of the satellite's apogee kick motor[14] could be treated like a drogue, and a probe installed on the front of the MMU could be inserted into the nozzle, the astronaut could fly the satellite into the payload bay of the Orbiter. This led to the development of what became known as the "stinger." As Allen further explains, "It resembled a folded umbrella which one could put up inside the rocket [nozzle] and then open, such that the tines of the umbrella would now stick against the side, and essentially the stinger would now be locked in the engine of the satellite."[15]

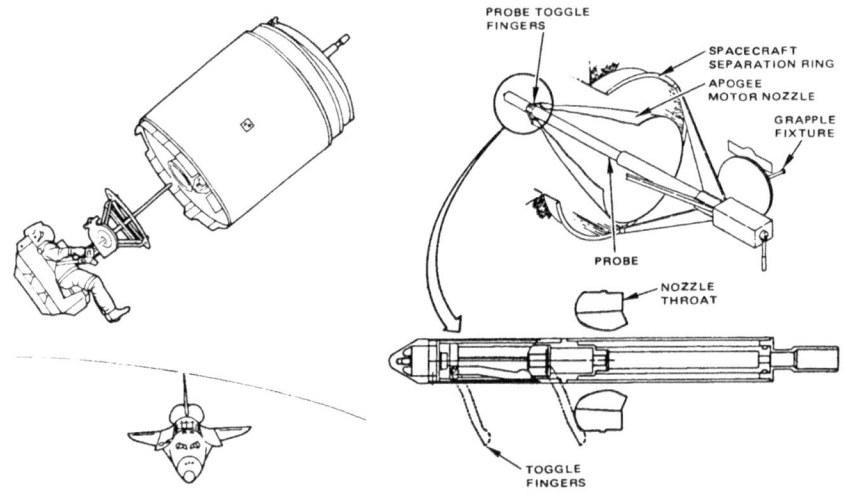

The "stinger" concept.

On flight day five, Joe Allen entered *Discovery*'s payload bay, donned the MMU on the left side of the payload bay, and propelled himself towards the slowly spinning Palapa B2, approaching along its cylindrical rotational axis towards the apogee kick motor.[16] He

---

[14] By this time, the PAM stages on the satellites had been jettisoned, leaving the apogee kick motors exposed. The PAM stage was meant to serve as a perigee kick motor to insert the satellite into geosynchronous transfer orbit. The apogee kick motor would circularize the trajectory to enter geostationary orbit.

[15] To appreciate how the "stinger" would work, imagine opening an umbrella inside a chimney.

[16] By this point Hughes Aerospace, which built the satellites, had commanded them both to perform a series of about 130 maneuvers by their apogee motors and their attitude control thrusters to stabilize their orbits at an altitude reachable by *Discovery* and positioned just several hundred kilometers apart. In addition, they were spun down from 50 rpm to 2 rpm (the slowest rate that would still provide gyroscopic stability) and the remaining propellant was jettisoned to eliminate a safety hazard during the retrieval operation and to make them lighter for the trip home. Such a large and complex sequence of engine and thruster firings prompted the implementation of ground control automation to provide the required firing and attitude commands at the right time.

readily inserted the stinger directly into the nozzle and deployed the tines in order to achieve a so-called soft docking. He then operated a hand crank to extend a jackscrew through the stinger and tighten a padded ring at the stinger's base against the satellite's separation ring[17] to accomplish a hard docking. There was now a rigid connection of the satellite, stinger, and MMU. When Allen engaged the attitude hold mode of the MMU this canceled the residual spin on the satellite.

Anna Fisher maneuvered the robotic arm to grapple a fixture that projected from the side of the stinger and moved Allen and the satellite down into the payload bay, forward end first.[18] Dale Gardner was waiting alongside a cradle built by McDonnell Douglas, ready for the second act of the play. In this configuration the arm could not put the satellite on the cradle, as that operation would have required a lot of twisting, complicated by the risk of violating the angular limits of the joints. Instead, Gardner was to manually install a second grappling tool called an antenna bridge structure, or more simply an A-frame. This consisted of a horizontal bar with a grappling fixture, shaped to be placed above the dish antenna and clamped down onto two diametrical opposite places on the top side. Gardner would have then kept the satellite in position by manually holding the A-frame while Fisher transferred the arm from the stinger to the grapple fixture on the A-frame. Next, both spacewalkers would move to the aft end to attach an adapter that would engage with a retention mechanism on the cradle. At least that was the plan.

After removing some thermal insulation that obstructed the installation of the A-frame, Gardner received a nasty surprise. There was a rigid box at the position where one of the A-frame extremities was to be attached. A small piece of metal, never seen before in the satellite drawings and therefore not taken into account in designing the A-frame, had ruined months of planning and training. Without the A-frame, the crew would not be able to install the satellite in the cradle. Or so it seemed…

The situation required rapid improvisation. Allen disengaged the stinger from the MMU, leaving the stinger in the nozzle at the bottom end of the satellite. He flew the MMU back to its station on the left-side wall of the payload bay and doffed it. Then he stood on a foot restraint placed on the right-side longeron. In the meantime, Fisher turned Palapa B2 to offer its top end to Allen, who took ahold of the antenna.[19] He was now manually holding in position above the payload bay some 2,000 pounds of space hardware, an impossible feat on Earth but entirely possible in the perpetual free fall of orbital flight where all you have to concern yourself about is the inertia of the object. Indeed, because the mass had a natural tendency to remain in position, inertia worked in his favor and at times he was able to hold the satellite with one hand while he relieved his other hand and arm of muscle fatigue. Nevertheless, in the long run it became uncomfortable, as Allen says, "A spacesuited crewman is naturally cooled all over his body, except for the hands – the hands are not encased in the liquid-cooled garment. My hands were getting quite warm

---

[17] This was the connecting interface between the satellite and the PAM stage.

[18] In this description, the forward end of the satellite is the housing for the dish antenna (still folded) and the aft end accommodates the apogee kick motor.

[19] It is helpful recall that at this point the satellite was held upside-down by the arm connected to the grapple fixture on the stinger. Therefore, the top end was towards the payload bay and the bottom towards deep space.

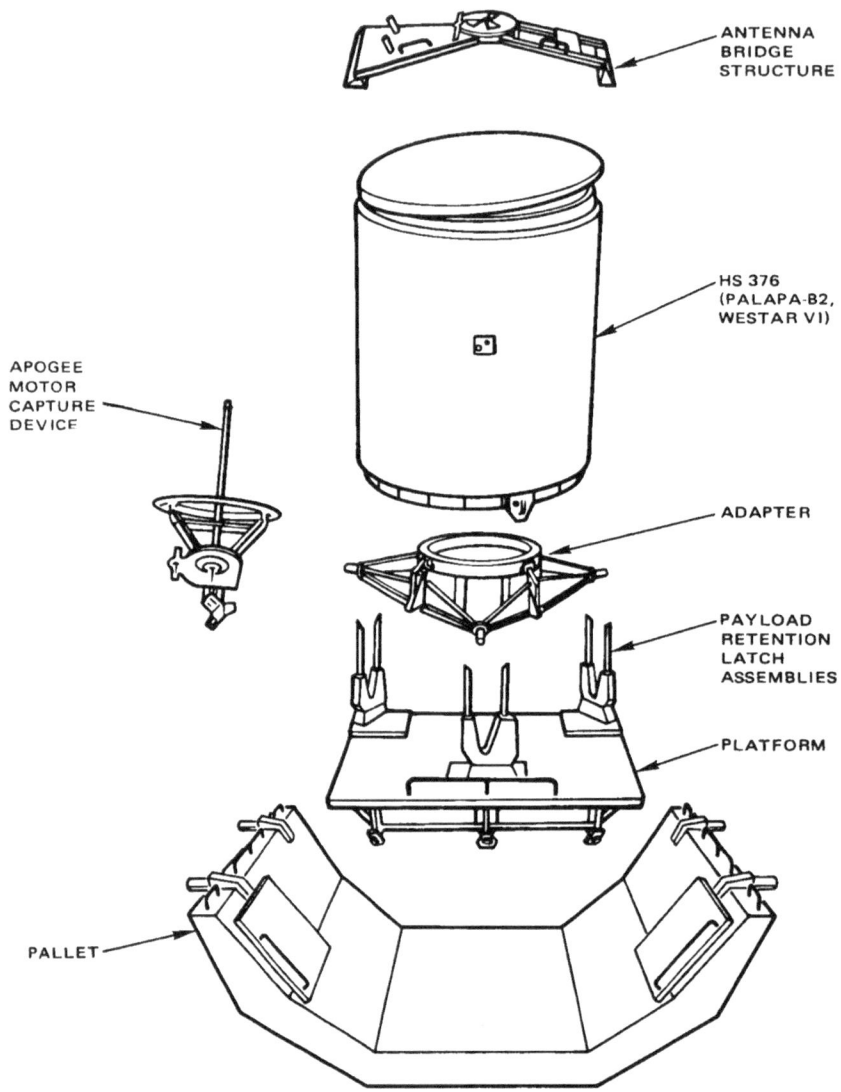

ANTENNA
BRIDGE
STRUCTURE

HS 376
(PALAPA-B2,
WESTAR VI)

APOGEE
MOTOR
CAPTURE
DEVICE

ADAPTER

PAYLOAD
RETENTION
LATCH
ASSEMBLIES

PLATFORM

PALLET

STS-51A's recovery mission hardware.

in places and very cold in other places. It was uncomfortable, but I didn't think it was life threatening."

Fisher ungrappled the arm from the stinger, leaving room for Gardner to remove it and stow it in the payload bay. Next, Gardner returned to the satellite to affix the large adapter to the satellite's separation ring. There were nine guide rails on the top of the adapter. Three of the guide rails held spring loaded latches (soft dock clamps), to provide loose capture (soft docking) of the satellite, thus freeing the astronaut to tackle other tasks. If the adapter was misaligned with the satellite during this initial mating, the astronaut would release the latches and try again. Alternating with the guide rails were nine clamp shoes, each controllable from the bottom of the adapter. The clamp shoes were to achieve a firm grip on the satellite. In fact, the separation ring was quite flexible, and so a proper combination of the number of contact points (the clamp shoes) and contact point preload was needed in order to hold the satellite fixed during worst-case re-entry loads. Nine shoes were found to be the best solution, with six being sufficient if two good ones flanked each failed shoe. The shoes were tightened by a torque wrench in a carefully defined sequence. A passive indicator on each drive train gave verification of a proper preload. The shoes of the clamps were designed to allow for manufacturing tolerances in the separation ring and also for dimensional variations due to temperature. The adapter had three legs, each of which had a trunnion at the end. It was required to provide attachment points beyond the perimeter of the satellite to allow visual verification of the connection to the retention platform. This was a 6.89×10.71-foot platform installed onto a Spacelab pallet by a mounting surface such that the solar panels on the cylindrical body of the satellite would not contact the sloping sidewalls of the pallet. On the platform were mounted three payload retention latch assemblies. These held the adapter ring during launch and the adapter/spacecraft during re-entry. The pallet/platform also contained fittings for the stinger, the antenna bridge structure, and various other tools.

If the A-frame had been able to be installed, Fisher would have used the robotic arm to maneuver Palapa B2 to its resting place in the payload bay. Without it, the spacewalkers used their elbow grease. Carefully, they turned the satellite vertical and lowered two of the protruding clamp trunnions into the V-guides that projected above the cradle base, hoping the third trunnion would line up with its guide. It did, and the cradle clamps engaged with the satellite. This seemingly simple task ought not to be underestimated. Mating a satellite with an adapter on Earth, with excellent visibility and easy access, is a laborious task. In the weightlessness environment, within the confines of the payload bay where access and visibility were greatly restricted for an astronaut wearing a spacesuit, the task proved rather difficult and even tedious.

With Palapa B2 safely installed in *Discovery*'s cargo bay, the now mentally and physically exhausted spacewalkers set about stowing all the apparatus they had used in achieving their remarkable feat.

Mindful of the problem encountered with the A-frame, the astronauts, in concert with Mission Control, decided to capture Westar 6 using the improvised procedure. While the first spacewalk had lasted 8.5 hours from start to finish, this time the task took less than 6 hours.

Whereas a Shuttle carrying a satellite on a PAM stage into orbit had a clamshell shield to protect the hardware from the thermal environment, the retrieved satellites would be exposed to the repeated temperature variations when passing in and out of the Earth's shadow during each orbit, so there was concern that the repeated freezing and thawing cycles would cause fuel lines to rupture and leak what remained of the corrosive propellants. An extensive analysis had been performed to mitigate this risk, and a flight profile drawn up that would ensure that the propellant temperatures could not fall below freezing.

Despite its success on STS-51A, the MMU was never used again. This mission marked the last time that an astronaut performed an untethered spacewalk until 1994, when the SAFER package was tested in space for the first time.

Palapa B2 and Westar 6 were safely returned to their manufacturer, refurbished, and relaunched years later on expendable launchers to finally supply the services for which they had been designed, albeit now owned by different commercial operators.

Joe Allen (top) with the "stinger" device and the Palapa B2 communications satellite. Dale Gardner waits in the payload bay of *Discovery* to assist in the berthing of the retrieved satellite. The end effector of the robotic arm, controlled by Anna Fisher, holds a special grapple point to Allen's right.

Allen is the sole anchor for the upper portion of the retrieved Palapa B2 satellite. Note the difference between the two stinger devices stowed on *Discovery*'s wall (right). The one nearer the vertical stabilizer is spent, having been inserted by Allen earlier in the day to stabilize the satellite. The one nearer the camera would be used in the later capture of the Westar 6 satellite.

"SOMETHING WE WEREN'T PLANNING ON DOING"

On the first flight day of STS-51D in April 1985, *Discovery* deployed Anik C-1 (also called Telesat-I) on a PAM-D stage. It was the first in a series of Canadian satellites to create a pay-television network. The next day it was the turn of Syncom IV-3 (also known as Leasat-3 because its services were to be leased to the US Navy). Once the large drum-shaped satellite had rolled out of the right-hand side of the payload bay in the Frisbee deployment mode, the crew waited for a series of events to be enacted. As Jeff Hoffman recalls, "After 2 minutes there was an antenna that was supposed to pop up. Of course, it always did in the simulators." Other satellites in this series had been successfully deployed by STS-41D and STS-51A in September and November of the previous year, respectively.

Riding the Manned Maneuvering Unit (MMU) Dale Gardner prepares to dock with the spinning Westar 6 satellite.

This was the so-called omnidirectional antenna, an appendage folded on top of the satellite that was meant to pop up once the drum was outside the envelope of the payload bay. The satellite would then spin up to increase its gyroscopic stability, in preparation for firing its internal engine about 45 minutes later. However, neither the antenna deployment nor the spin-up took place. Mission Control ordered the crew to separate from the satellite, in the case the engine were to ignite at its pre-set time and damage the Orbiter.

A fault tree analysis of the failure quickly identified an external switch on the side of the satellite that had not moved to the ON position upon leaving the payload bay. Houston explained the findings to the crew, and since the motor had not ignited they added, as pilot Donald E. Williams recalls, "We're thinking about a rendezvous with the

Dale Gardner (left) and Joe Allen work together to bring Westar 6 into the payload bay of *Discovery*. Allen is on the foot restraint attached to the robotic arm, which is controlled by Anna Fisher. Gardner is working to remove a stinger device from the now stabilized satellite.

satellite to do an inspection and see if we can figure out what we can do about it." He and his commander, Karol J. "Bo" Bobko, couldn't have been happier. "Bo and I are thinking to ourselves, 'Great. We're going to get to do something we weren't planning on doing.'" Houston continued, "You might be able to do something about it. We're planning now to figure out how we might do that. It may involve an EVA." Suddenly two more crew members got extremely excited. In fact, Hoffman and his colleague David S. Griggs had received contingency EVA training.[20] Now they were likely to attempt an EVA task for which they had never trained. It would be the first unplanned spacewalk of the Shuttle program. So we can readily appreciate Hoffman saying, "When we heard the word EVA, it was just electric."

The original idea was to have Hoffman take up position at the end of the robotic arm and manually flip the switch, but it was decided instead to attach a makeshift device to the end effector of the arm and use that to flip the switch. The rationale for this change was that the two contingency spacewalkers and the two pilots were out of practice with EVA and rendezvous procedures, respectively. A teleprinter uploaded instructions for the crew

---

[20] This minimal EVA training would have allowed two crew members to go outside and perform vital tasks, such as manually closing the payload bay door.

to use readily available materials to create a makeshift tool that resembled a combination of a fly swatter and a lacrosse stick. Two days later, the spacewalkers attached this improvised device to the end effector using speed tape. As Hoffman says, the task was relatively easy, "The problem was that since we hadn't planned an EVA, we didn't have any foot restraints, we didn't have any special tools, so each of us had to use one hand to restrain ourselves. We could only work with one hand…Our bodies were flapping all over the place. But we were pretty well trained."

Before the duo returned inside *Discovery*'s cabin, Houston requested verification that the arm could be stowed without the appendage interfering with the aft section of the payload bay. This was because the planners did not want to have to send the men out again later in the mission to remove the swatter from the arm; they were to return home with it in place. The astronauts confirmed that despite a slight interference with one of the cameras, the arm could indeed be safely stowed.

On the sixth day, the Syncom was again in sight, still spinning at a slow rate of 2 rpm. Bobko and Williams did an excellent job of closing in with the ailing satellite, despite their limited training in rendezvous and not having a rendezvous checklist on board![21]

As they approached closer, the astronauts reported that the switch that supposedly was the cause of the satellite's problem was in the ON position. As Hoffman recalls, "Rhea [Seddon] reached out with the arm and snagged it a couple of times." This was not an easy task. The end effector had to be placed exactly where the flyswatter and lacrosse stick could engage the protruding switch, as the satellite was spinning. Also, this contact caused the satellite to move away, obliging her to continue to move the arm to maintain contact. The switch was engaged at least three times in this way, but nothing happened. The omni-antenna did not deploy, and nor did the satellite start to spin up. The malfunction was clearly not as simple as a failed switch, but even before *Discovery* landed two smart astronauts on the ground were figuring out a solution.

"WE THINK WE COULD GO UP AND FIX THIS THING"

As the drama of STS-51D was unfolding, James van Hoften and J. Michael Lounge were at Ellington Field, near Houston, for a 2-day tour of duty on Air National Guard alert to scramble an F-4 Phantom jet to deal with any threat over Texas. Having time to kill, they set out to think through a possible rescue plan which, right from the start, took the shape of a manual capture by an astronaut. Van Hoften says, "I immediately said, 'Well, I'll go out and get on the end of a remote manipulator arm and grab this thing. I wonder how hard that would be.'" Some back-of-the-envelope calculations showed that a force in the range of 20–30 pounds would be enough to grab, stabilize, and then release the satellite manually.[22]

---

[21] This was sent up by teleprinter, so the pilots could familiarize themselves with the procedure in advance of attempting the maneuvers.

[22] In fact, later detailed analyses showed the force to be precisely 27.36 pounds.

Discovery's robotic arm is maneuvered close to the Syncom satellite to attempt to activate its deployment process. Clearly visible at the end of the arm is the makeshift flyswatter that was installed during a contingency EVA a few days earlier.

Van Hoften and Lounge were scheduled to fly on *Discovery* for STS-51I later in 1985. Joe Engle and Richard O. Covey, the commander and pilot of the mission, as well as NASA HQ and Hughes Aerospace, which supplied the satellite, got quickly involved in the plan. Van Hoften says, "They explained to us that they kind of knew what was wrong with this thing, and that they had a fix, and that…we were going to go repair it on-orbit and relaunch it." This was a sensible decision, if one thinks how risky it could have been to bring back a satellite full of propellant! Space missions, whether manned or not, always require a considerable amount of time to develop, but with STS-51I scheduled for launch in late August, NASA had just short of 4 months to firm up the rescue plan. In fact, as Covey explains, "We didn't have any tools to do the job that we needed to do, because basically Ox could get up there and maybe get ahold of this spacecraft, but we needed a way to get it on the end of the RMS and to get it stabilized so that we could do this other work on it and redeploy it. So what happened was a very simple approach…We came up with three tools that had to be developed in this short time period."

Starting with the assumption that the satellite would still be spinning, the first tool to be developed was a so-called "capture bar" that van Hoften would snap into two of the four trunnion pins located along the length of the cylindrical satellite and that had held it in place during its ride to orbit. Then he would grab the bar in order to halt the satellite's rotation and manually move it over to fellow spacewalker William Fisher, who would snap the so-called "handling bar" onto the bottom of the drum so that he could hold the satellite in position. Van Hoften would then move to the top to fit the third tool, the "grappling fixture" that would enable the robotic arm to take ahold of the satellite, freeing the two spacewalkers to do other work. If at first glance it seems a fairly complicated plan, it must be remembered that although the Syncom had been designed to be launched from the Shuttle, it had not been fitted with any features for a possible retrieval operation. Thus the need for this highly choreographic ballet.

The tricky part of the repair operation was to consist of a one-of-a-kind electrical surgery. As Covey says, "Basically, what they came up with, and it cost them about a million bucks to develop…was a box that had a timer on it, and it had a set of cables that would go around to a test port on one side of the spacecraft and another set that would go around the spacecraft over to another set."

After deploying three satellites, *Discovery* reached Syncom on flight day five. As Engle says, "I gave Dick the responsibility of doing the rendezvous of the Syncom, so he did that. He trained in the simulators and practiced in the simulators and did an absolutely superb and excellent job of that rendezvous." And Covey did such a good job that, as Engle continues, "When he had completed the rendezvous maneuver and had stabilized…I kind of expected to see it somewhere in the field of view in the window, but he had flown that rendezvous and perfectly nailed it so that the satellite was right behind the pippers. I had to look a couple of times, in fact, to see where it was, because it was right in the center of the pipper." Engle then took command for the proximity operations which would maneuver the Orbiter within a few feet of the satellite.

Despite *Discovery* being already in good attitude for flying this final part of the rendezvous, Engle still had time to be taken by surprise. To that moment the satellite had been in the field of view of the overhead windows of the aft station on the flight deck. But now, for flying in close proximity, the commander had to look at it through the windows that faced into the payload bay. On the aft station, there were two hand controllers to command translations and rotations. However, such inputs would cause translations and rotations about different axes and in different directions, depending upon which window the target was being sighted upon as a result of changes in the reference system used by the flight software. This is further explained by Engle, "As you go from the overhead window, and it comes down into the aft window view, you change the orientation of the vehicle, and you also change perspective, where lifting and pushing and pulling take on different axes as you transfer from one window to another." This could be disorienting, even for a former X-15 pilot. "In that transition, my first [command] input was one using the old system, so it was the wrong input and the satellite started drifting toward the window instead of down into the payload bay." Engle rapidly reacted to this scary moment. "I reversed controls, to get it back out of the way so that we wouldn't impact it. Then I had to kind of restart and bring it down into the payload bay." Looking out the windows, the astronauts immediately understood that the rescue mission would be more complicated than anticipated. The plan, tools, and training all presumed that the satellite would still be slowly spinning about its longitudinal axis. In fact, nobody either at NASA or Hughes Aerospace had predicted that the rotational energy drained off while Rhea Seddon was attempting to trigger the micro-switch using the flyswatter had been further reduced by the metal of the satellite interacting with the Earth's magnetic field. As a result the satellite was now slowly tumbling.

Riding on the robotic arm, van Hoften was maneuvered over to the Syncom. He waited for the rotational motion to place the two trunnion bolts in front of him, then installed the capture bar and stabilized the satellite. Watching the video footage of the mission, it is evident just how dramatic this moment was. As van Hoften fought the inertia of 15,000 pounds slowly rotating, it is possible to see the flexing and bending of his legs, arms, and torso, as well as the robotic arm oscillating. "I was holding this thing, I wasn't going to let go for anything. I didn't care what happened, but I was not going to let go of that thing, and it was starting to translate downward toward the Shuttle. It would have hit it if I hadn't stopped it. I was holding on as hard as I could."

The fact that in those days, with only one TDRS satellite available, there was not full ground coverage might have contributed to the success of the effort. As Covey says, "Finally we went LOS [loss of signal] and Ox said, 'Fly me up to it.' And he went up and he just grabbed it. If the ground had been watching, we wouldn't have done that, I'm sure, like that. Then with AOS [acquisition of signal] we announced, 'Well, we got it, Houston.' They didn't ask why. They didn't ask how."

Once back on Earth, van Hoften, mindful of his experience, approached Hamilton Standard, the spacesuit manufacturer, to ask whether they had ever done any tests on the strength of the suit boots. "They never had. They did after that, because I said, 'You know, I think I came close to ripping those things off, and that would've killed me.'"

The operation was made even more arduous by the inability to operate the robotic arm in automatic mode, owing to a short circuit in the elbow joint caused by a piece of loose scrap wire. Having to maneuver it in single joint mode, Mike Lounge was really restricted in the help that he could offer his colleagues.

The remainder of the capture operation went to plan, with the installation of the handling bar and the grapple fixture. Due to the longer-than-planned time required to catch Syncom and the degraded state of the robotic arm, it was decided to complete the repairs over two spacewalks.

To redeploy the now fixed satellite, Fisher took position on the right payload bay longeron and held the drum in place while van Hoften was maneuvered to the other side to remove the grapple fixture and replace it with a spin-up bar. After that, Fisher let go of the satellite and *Discovery* took up the correct attitude for deployment. As Engle says, "It was a matter of positioning Ox again on the arm and letting him give some pushes on the bar. The satellite started to spin up, and every time the bar came around he would push it. The satellite, of course, not only took on rotational, but also translational energy and it started to move away. It was a matter of flying Ox, to keep him close in so that when the bar came by again, he could give it another push and spin it a little faster." After three or four shoves the satellite had a rotation of 2.7 rpm, as if it had just been deployed Frisbee style from the payload bay.

Discovery moved away to allow the repaired Syncom to belatedly start its journey to geosynchronous orbit.

"Ox" van Hoften has just given a shove to the Syncom satellite.

"I'M GOING DOWNSTAIRS TO GET READY"

Spacewalking is one of the most exciting activities an astronaut can perform during a mission. All the more so when it is done in response to an unplanned contingency condition which requires physical intervention. This was the situation faced by Jerry L. Ross and Jay Apt on the third day of the STS-37 mission.

Launched on April 5, 1991, the five-member crew of *Atlantis* had as their primary objective the deployment of the Compton Gamma Ray Observatory (CGRO).[23] As its name indicates, this spacecraft, the second of the Great Observatory series, had a variety of instruments to observe cosmic gamma ray emissions, a powerful source of information on some of the most exotic objects in the universe, such as supernovae, neutron stars, black holes, and quasars. Although these objects are very prominent in the highest energies of the electromagnetic spectrum, they cannot be observed by the human eye. In fact, their

---

[23] The spacecraft was named after Arthur Holly Compton of Washington University in St. Louis, who won the Nobel Prize in 1927 for his research in high-energy physics.

radiation cannot penetrate the Earth's atmosphere. Cosmic rays were discovered early in the 20th century by sending instruments to high altitude on balloons. The first measurements of gamma rays in space were by the Explorer 11 satellite in 1961. Later measurements were made by satellites of the Orbiting Solar Observatory series. In the late 1960s and early 1970s the Vela satellites launched by the US to detect the flashes of gamma rays issued by fission reactions during nuclear tests by the Soviet Union, also detected, by chance, intense but brief bursts of gamma rays from objects in deep space that are among the most powerful events occurring in the universe.

Although it was a telescope, CGRO did not look at all like one. Gamma rays can be detected only indirectly, by monitoring scintillations – flashes of visible light that occur when gamma rays strike a detector made of liquid or crystal materials. Because gamma rays are revealed when they interact with matter, the number of gamma ray events detectable is directly related to the mass of the detector. CGRO was equipped with four instruments, each specialized to detect gamma rays within a specific range of energies. At 33,923 pounds, this was the heaviest science satellite deployed by the Shuttle to that time. Its imposing mass was matched by its size. As Ross remembers, "My first thought about the observatory when I saw it, was that this thing looks like a diesel locomotive. I mean, the thing was huge. Everything on it was real bulky, real thick, real heavy, and the stoutness of the satellite was very impressive. Most times, you go up to a satellite and you're almost afraid to breathe on it because it may fall apart on you. This thing had huge beams that were the center part of the structure of it, and several of the experiments were great big devices, pretty heavy devices, and that's why they needed that much structure to it."

The deployment started well. After the robotic arm had hoisted the satellite out of the payload bay, it was commanded to unfold the two long solar arrays. As mission commander Steven R. Nagel recalls, they "unfolded one at a time, very slowly. With those unfolded, the wingspan [of the satellite] was 70 feet. From tip to tip, the Orbiter is only 78 feet. So it was a pretty big thing hanging out there on the end of the arm." Next was the release of the 10-foot-long high-gain antenna that was required for the fast transmission of the large volume of data that the observatory was expected to produce. The command was sent and received, but nothing happened. After several more tries, it was evident the antenna had somehow got stuck. As the antenna was on the rear of the satellite, the astronaut were unable to view it to assess the glitch. It was a critical moment because, with the solar arrays deployed, the satellite could not be placed back in the payload bay. The satellite would still be able to return data without the high-gain antenna, but the pitifully slow transmission rate would severely restrict the science.

Procedures had been drawn up and rehearsed in training to address a wide variety of contingency situations. For instance, a plausible technique for freeing the jammed antenna was to shake the satellite by sharply moving the robotic arm or by firing the Orbiter's reaction control system thrusters. Both methods were tried, unsuccessfully. In fact, it is rather easy to understand why they would not work. As the robotic arm is a flexible appendage, even substantial vibrations are rapidly damped and the attached payload doesn't feel much motion at all. In addition, the arm was not designed to be moved at the rate required to vigorously shake a payload.

As Jerry Ross says, "After about the second or third iteration of the things that we were trying in a sequence, I looked over at Steve, I took off my wedding band, and I said, 'Steve,

I'm going downstairs to get ready'… And it wasn't 45 seconds later the ground called us and said, 'Why don't you send Jerry and Jay downstairs.'"

Ross and Apt duly prepared for the second contingency spacewalk of the Shuttle program and, as it happened, the first outing since the loss of *Challenger*. Thanks to their having spent 2 days checking their suits in preparation for a spacewalk assigned to the following day, and having completed the pre-breathe protocols and reduced the cabin pressure to 10.2 psi in readiness for a contingency, they were soon heading out through the airlock.

In the payload bay, Ross worked his way around the satellite in order to reach the troubled antenna and find a good support for himself to start pushing. "It took some pretty good pushes on the boom. The first two had felt pretty solid, but I could tell it was starting to loosen up a little bit. I was probably putting in, I don't know, 45 to 50 pounds of force, and I could tell it was starting to walk out. Finally, it came free and swung out about 30 or 40 degrees from the stowed position." It was concluded that a portion of an adjacent thermal blanket had dislodged during launch and snagged the head of a bolt, and the deployment motor was too weak to overcome that resistance. Having freed the antenna, Ross locked it into place using a specific tool. In less than 20 minutes the job was done, and space-based gamma ray science was back on track. As Ross delightedly says, "That was really a good feeling, demonstrating where the man in the loop can help a robotic system and let it go off and do some really great science, which it did for years before it finally got de-orbited."

Because ground control had to perform checks on the satellite prior to giving the go-ahead for its release, the two spacewalkers were allowed to remain outside in case an unexpected situation required manual intervention. However, they did not simply spend the next 4 hours enjoying the view. They made a start on some of the tasks that were assigned to the spacewalk that was scheduled for the following day.[24] As they were re-entering the airlock, CGRO was released and the Orbiter moved away from the satellite.

THREESOME SPACEWALK

Following the *Challenger* disaster, it was recognized that using the Shuttle to deploy commercial satellites was not worth the risk of losing a crew. In future, such satellites would be reassigned to expendable launch vehicles. The lives of the astronauts would only be risked for missions dedicated to science and technology. However, STS-49, the first mission for *Endeavour*, was an exception.[25]

On March 14, 1990, the second launch of the new Commercial Titan III rocket[26] failed to deliver the Intelsat 603 communications satellite into geosynchronous orbit. Once again, a healthy and expensive satellite was stranded in low orbit. When NASA was invited

---

[24] Refer to Chapter 4 for more details of this spacewalk.

[25] *Endeavour* was the Orbiter built to replace *Challenger*.

[26] This was a version of the well-known Titan 34D launcher offered to soak up the launch opportunities that were opened up by the withdrawal of the Shuttle, post-*Challenger*, as launcher for communications satellites.

by the owner, it was decided that a rescue mission would be attempted with the objective of attaching a new perigee kick motor to start the satellite towards its intended orbit.[27]

Endeavour rendezvoused with Intelsat 603 in May 1992, and as had occurred on all previous rescue and servicing missions, things did not go as intended. Because the satellite had not been designed for retrieval by spacewalkers or by robotic arm, it was necessary to develop a handling bar which an astronaut would attach to the ring that had mated with the now discarded upper stage.[28] There were four major components: the central beam, a right and a left beam extension, and a steering wheel. The bar had features to facilitate multiple latch attachment attempts. The rescue plan called for a spacewalker to ride at the end of the robotic arm in close proximity to Intelsat and, at the right moment, snap the capture bar at the right position on the clamp ring to avoid unwanted contact with obstructing structures. Short V-guides surrounding the active latches at both ends of the bar would assist with manual alignment and installation. Triggers located below the V-guides would then activate the capture latches to grip the clamp ring with a force of 50 pounds on both sides of the ring. The astronaut would then use the large steering wheel to despin the rotation of the satellite by brief, light impulses. Once the satellite was stationary, he would drive a power tool to hard dock the latches. In the event of a latch misfire, the astronaut could use two levers to rearm the latches.

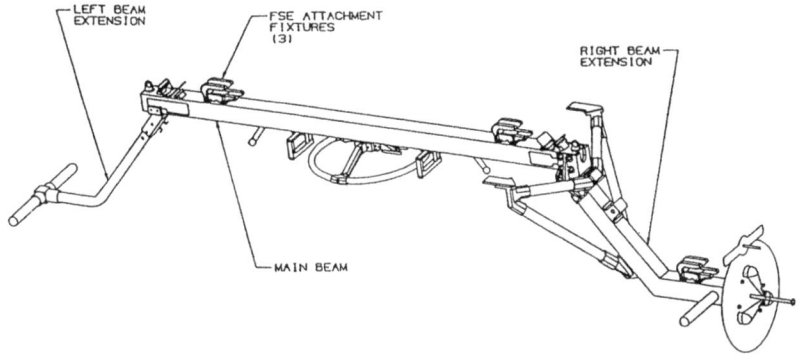

The capture bar assembly.

With the Intelsat captured and stabilized, the spacewalker would attach a grapple fixture to one end of the capture bar so that the arm could grip the satellite and, with the astronaut still riding the arm, position it directly above the new upper stage in the payload bay. The astronaut would then dismount from the arm and join his colleague, who had been preparing things in the payload bay. To ensure the correct alignment, the left and right capture bar extensions were to slide through tall V-guides on the support cradle that held the motor. Once the satellite was attached to the motor, the capture bar extensions would be removed and the center bar connected to the stage's adapter in order to minimize the impact on the satellite mass properties and reboost propellant.

---

[27] It is interesting that this satellite was originally scheduled to be deployed by the Shuttle.

[28] This was the interface between the structure of the satellite and the perigee kick motor.

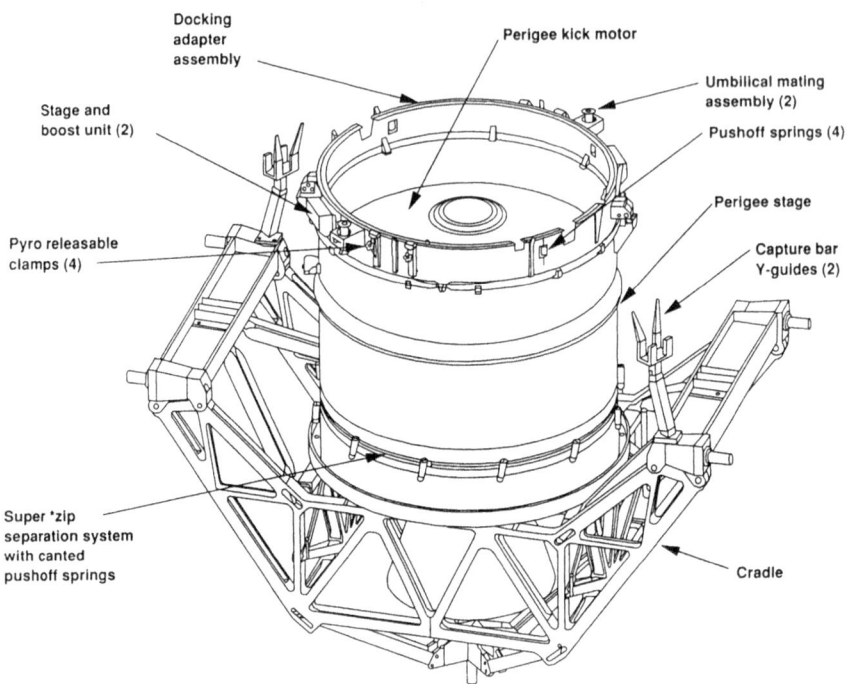

The Intelsat upper stage and docking adapter assembly combination.

The first attempt at capture was an utter fiasco. The extensive training had failed to adequately represent the inertia of the satellite, and particularly the sloshing of the propellant inside. The attempt was called off. *Endeavour* returned the following day for a second try, which was frustrated by the failure of the capture bar latches to fire. *Endeavour* retreated, then returned two days later for a third attempt. This would be the last opportunity, since the Orbiter did not have propellant for a fourth rendezvous. The availability of a spare spacesuit, a standard procedure to avoid having to cancel a spacewalk because of a failed suit, prompted the most audacious spacewalk exercise ever attempted.

With approval from Mission Control, three astronauts passed through the airlock into *Endeavour*'s payload bay and, upon placing themselves around the upper stage, they manually grasped Intelsat and halted its spin. The capture bar was then installed and the robotic arm mated the satellite to its new upper stage. It was then deployed as if new and successfully achieved geostationary orbit.

Thus far, the three-man spacewalk of STS-49 remains the only such EVA ever to be performed by any space mission.

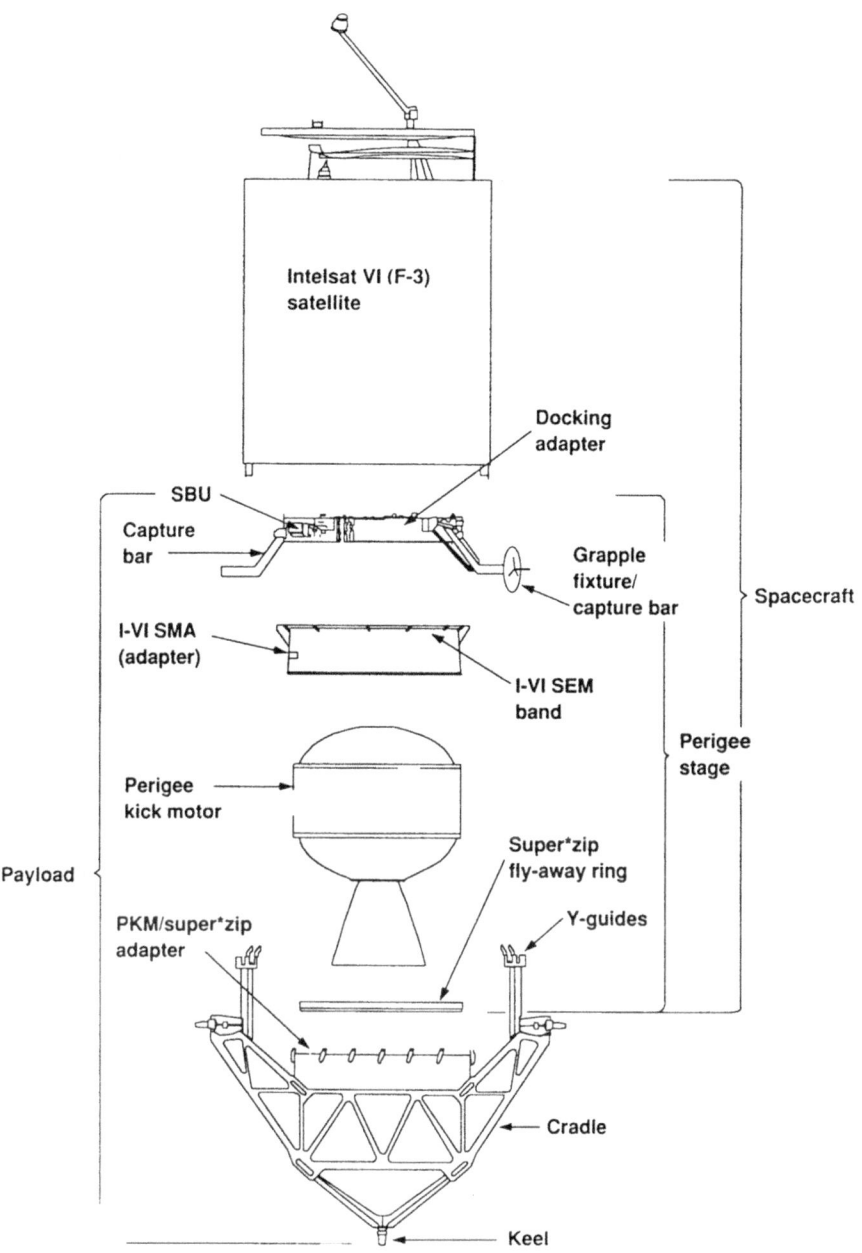

The Intelsat reboost payload layout.

Intelsat 603 being captured by three pairs of human hands. Visible beneath the satellite is the replacement perigee kick motor (painted black) and the support cradle (coated with golden thermal blankets).

SERVICING THE HUBBLE SPACE TELESCOPE

The Hubble Space Telescope and the Space Shuttle will always be linked together by a sort of invisible umbilical. The constant stream of astronomical and astrophysical discoveries that Hubble is still yielding after a quarter of a century would not have been possible if not for the hard work of the five Shuttle crews who, between 1993 and 2009, maintained and serviced this superb eye on the sky, several times saving it from an otherwise inevitable premature demise.

The story really began in 1946 when Lyman Spitzer, a theoretical physicist and astronomer, published an intriguing article entitled "Astronomical Advantages of an Extra-Terrestrial Observatory."[29] He started out by explaining how the atmosphere hinders astronomical observations: X-rays, ultraviolet light, infrared radiation, and everything else that lies outside the visible portion of the electromagnetic spectrum is almost, if not completely, absorbed by the atmosphere, thus preventing a broad array of astronomical and astrophysical phenomena from being detected. Also, the passage of light down through the atmosphere is far from being a smooth ride, as the optical properties of the air along with the continuous turbulent condition of the troposphere seriously degrade the quality of what astronomers call "seeing," in particular of very distant and faint sources.[30] Spitzer offered a revolutionary solution to the problem by proposing that a telescope be placed into orbit around the Earth. By virtue of being above the atmosphere, such a telescope could observe a wider range of wavelengths, opening entirely new windows on the universe. If one recalls that Sputnik, the first artificial satellite, was still a good decade away, is not inappropriate to regard Spitzer as a true visionary.

Fast forwarding 30 years to 1977, the US Congress approved a NASA request for funding to develop an orbiting astronomical observatory. In 1983 the blandly named "Large Space Telescope" was renamed the Hubble Space Telescope in honor of the astronomer Edwin P. Hubble.[31] With a far-sightedness typical of NASA engineers, and keeping in mind the substantial costs involved in developing and building such a telescope, it was soon appreciated that to best exploit its potential, the design should encompass regular on-orbit maintenance and servicing. In this way, as technologies advanced, it would be possible to upgrade the observatory with the newest and best instruments in addition to maintaining and improving its onboard systems. As NASA was also deeply engaged with the Shuttle program, it was inevitable the design of the telescope would be fully compatible with this new spaceship, both for launch and for servicing. As events would show, it would be one of the best marriages in the history of space exploration.

In all likelihood the last humans to see Hubble up close were the crew of the final servicing mission in 2009. It had once been hoped that the telescope would be able to be returned

---

[29] De facto, this article is the father of any space-borne astronomical telescope.

[30] This is why ground-based astronomical observatories are built atop mountains, thousands of feet above sea level. In this way, they are placed above most of the weather turbulence in the lower atmosphere and starlight is less impaired by optical distortions. Infrared observations are not possible, because water vapor is a formidable sponge that prevents any useful amount of this radiation from reaching the ground. The only way to do infrared astronomy without going into space, is to fly a telescope at altitudes even higher that those regularly flown by a commercial airliner. In this way, the telescope is well above the densest regions of atmospheric water vapor and good quality observation is practicable. As of writing this book, NASA has for several years been operating the Stratospheric Observatory For Infrared Astronomy (SOFIA), which is a modified Boeing B747-SP with a 98-in-diameter telescope in the aft fuselage and a fully-fledged control center in the forward fuselage. It flies regular night sorties each month.

[31] Edwin P. Hubble is best known for discovering that the universe is expanding, with vast agglomerates of stars called galaxies moving away from each other.

to Earth at the end of its mission and placed on display in a museum, but the retirement of the Shuttle made this impossible. Instead, the final servicing crew installed a docking system to enable an unspecified robotic vehicle to link up and de-orbit the telescope.

If you visit the National Air and Space Museum in Washington DC, you can see a replica of the Hubble Space Telescope. It is in the same hall that displays the backup Skylab space station. Although it is a test article that was used for static and dynamic testing of the telescope's structure, it has been configured to accurately represent the telescope in space and readily shows how its design made it maintenance friendly for astronauts. At the base of a long cylinder in which the primary and secondary mirrors are housed, there is a larger cylinder with four large doors to allow internal access to the several instruments and onboard systems, with sufficient room for a spacewalker to operate. Handrails and anchor points were strategically placed to assist in reaching any spot on the telescope while wearing a bulky spacesuit. Other provisions included the use of captive fasteners, the use of wind tab or rack/panel electrical connectors, and sufficient accessibility to allow astronauts wearing suits with limited visibility to see what they were doing, and to operate tools. In addition, attention had been paid to EVA safety requirements.

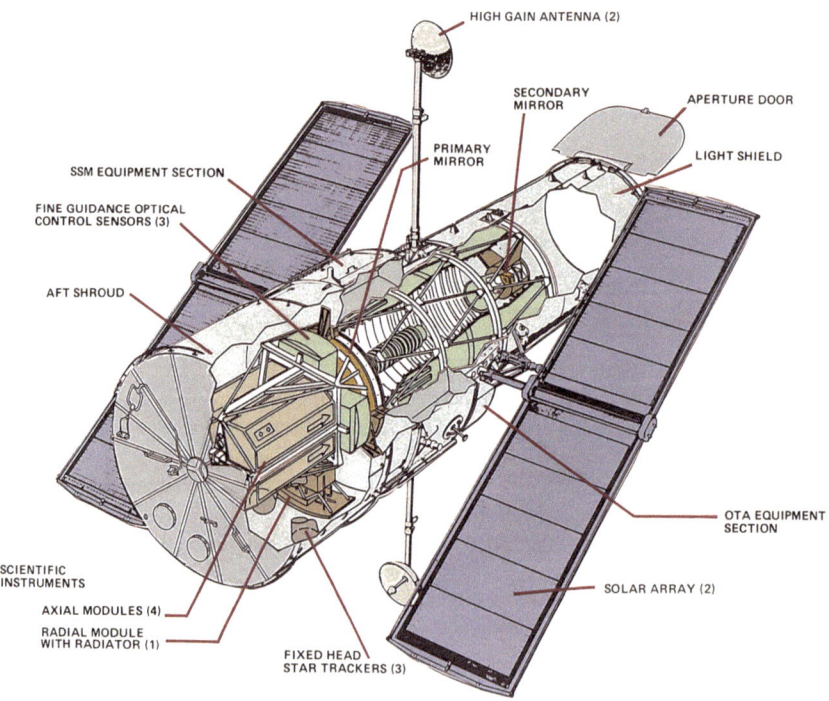

A cross section of the Hubble Space Telescope showing its main components.

Initially scheduled for launch in the autumn of 1986, Hubble was grounded by the loss of *Challenger* at the start of that year. When the Shuttle resumed operations in September 1988, Hubble remained in the hangar while higher priority payloads were launched. It was finally carried into space by STS-31 on April 24, 1990.

Releasing satellites into orbit was a tried and tested business for the Shuttle, but Hubble was not an ordinary satellite; so much so that it required NASA once again to push the envelope of the Shuttle's performance.

A trait common to space-based telescopes is that in order not to contaminate the optics and instruments the reaction control system cannot employ thrusters, but rather gyroscopes or magnetic torque bars. For the same reason, they cannot employ a large rocket engine to adjust their altitude. This limitation requires that a space telescope be inserted into a high orbit where the atmospheric drag is negligible. Despite orbital space being a vacuum by human standards, for satellites traveling at speeds of tens of thousands of miles per hour, each sporadic atom and molecule in its path generates a feeble but incessant braking drag. If not adequately counteracted, this drag can cause a satellite to fall into the atmosphere and burn up.[32] The greater the spacecraft's cross section in the direction of travel, the greater the drag and the greater the effect on its attitude and pointing stability. Gyroscopes and magnetic torque bars are an effective reaction control system, but at a low orbital altitude where drag is greater, they might not be strong enough to guarantee the pointing stability for a telescope. Therefore, it is wise to place a telescope at an altitude where atmospheric drag will not cause any significant impairment.

To deploy Hubble, *Discovery* had to reach a much higher altitude than was usual for the Shuttle. As STS-31's Kathryn D. Sullivan says, "The deployment altitude for Hubble was quite high for a Space Shuttle…We were pretty high, 340 nautical miles. The standard design orbit is 160, so we're over twice as high." It is not difficult to appreciate that the higher the orbit, the greater propellant that is required. In fact, as Sullivan explains, "When you put all the numbers together and run it against Orbiter performance and consumables, when you arrive on-orbit about half of your onboard propellant will already have been consumed."[33] Hence, on a flight scheduled to last 5 days, half of the propellant would be gone in just a few hours! With the remaining propellant, *Discovery* had to station-keeping in proximity to Hubble after its release, until Mission Control confirmed the satellite was

---

[32] It is worth recalling that the mathematical expression for aerodynamic drag depends on the density of the fluid that the vehicle is traveling through, on the cross-sectional area in the direction of travel, and on the square of the velocity. It is easy to appreciate that if the density of space is extremely low, then a spacecraft traveling at thousands of kilometers per hour is indeed subjected to a braking aerodynamic drag that is far from negligible. In determining a suitable orbital altitude for a space telescope that lacks a propulsion system, it is also necessary to consider the changes that the Earth's atmosphere goes through due to variability in solar activity and the position of the planet along its orbit. For instance, when closer to the Sun or during a particularly intense period of solar activity, the planet's atmospheric envelope becomes physically larger and threatens spacecraft at higher altitudes.

[33] Here Sullivan is referring to the propellant that each Orbiter allocated to the Orbital Maneuvering System (OMS) and Reaction Control System (RCS) – the former for orbital changes including the de-orbit burn, and the latter for attitude control. For details of the workings of these subsystems see Chapter 6 of my previous book *To Orbit And Back Again: How the Space Shuttle Flew in Space*.

operating properly. This way, they would be on hand to capture the satellite and try to fix the malfunction. After finally withdrawing, there had to be sufficient propellant remaining to carry out the de-orbit burn to return to Earth. An error in propellant management or a leak would mean the crew couldn't come back! Sullivan says, "We put a lot of energy into how to respond to propellant leak alarms. If you've got fatter fuel margins and you get a leak alert, a first prudent step is to find indications that confirm it's not a false alarm. If it's not a false alarm, you act on it. For STS-31, just the risk that it might be a genuine leak had to trigger action, because the risk that it could deplete the fuel we needed for de-orbit was too great. So for any indication of a leak, we needed to launch parallel paths of action. We had to start with the assumption that it is a leak, and that we're going to button the heck up really fast. I'd rather burn the propellant lowering my orbit than spew it out the side. So I want to quickly get ready to do a de-orbit burn. It may be a complete de-orbit burn, it might only lower the altitude, but I need right away to be ready to do the burn and change the orbit. I might be going home right away, so I'd be starting to button up the vehicle. Someone had better be checking whether it really is a leak, what part of the tank system is leaking, and how much we have to burn off. All of that has to get acted on in parallel." Never had a Shuttle crew trained to cope with such nominal low margins,[34] but as Sullivan says, "We got awful darn good at that."

Had *Discovery* needed to return home before Hubble could be released, the flight plan called for quickly getting rid of the telescope. As Sullivan explained to scientists who wanted the telescope to be released as per nominal procedure, regardless of any emergency, "If a leak alarm goes off, all bets are off…We are punching you off ASAP…We're getting out of there. This was just incomprehensible. 'No, no, no, no, you can't, we need this and this.' They would start back through the rehearsal of all the deployment constraints. Just in conversation, you'd go back and forth. 'Yes, we understand the deployment constraints. We intend to do that, but if this circumstance arises, forget it.' 'No, no, you don't understand. You have to – .' It was this endless circular argument."

Eventually, a long simulation involving all interested parties was scripted and run through. It had the astronauts coping with a sudden leak just about when Hubble was lifted out of the payload bay by the robotic arm. As per contingency procedures, the astronauts got into the "button up and get down" mode. For the crew, the simulation was over when they made it home safely. In the meantime, the teams at the various control centers figured out how to make the telescope operational, despite its hurried deployment. The simulation produced valuable lessons and raised confidence in the procedures.

Sullivan wryly notes, "That exercise forced us all to think and come up with new insights, and dig deeper into what do we understand and how can we help each other; all those kinds of things that you would imagine. That process revealed some greater smarts, some better approaches to normal ops for the telescope. So the program got two good

---

[34] Other flights had seen their crew cope with low fuel quantities. During Gemini VIII a malfunctioning valve in the reaction control system forced commander Neil Armstrong and pilot David Scott to expend a lot of their propellant to maintain and stabilize their spacecraft. The planned 5-day flight was curtailed to just 8 hours. In this and similar cases, low fuel margins were the result of an emergency condition. But for STS-31, low fuel margins were planned from the beginning as part of the nominal mission itself.

things out of it. It got a telescope flight operations control team that had learned some important lessons that would let the telescope mission go forward and succeed even if we had to throw it overboard and 'Get out of Dodge,' and they even learned some improvements to the normal course of business. They deployed with better operating skills than they would've otherwise, and they were ready to cover that contingency if it happened. But it was pretty funny at the time."

Fortunately, neither STS-31 nor any of the subsequent five servicing missions had to cope with a propellant leak when in the telescope's very high orbit.

Once hoisted out of the payload bay, Hubble was commanded to deploy its two large solar array wings. The first opened up as planned, but the second did not. This argued for a contingency EVA on which spacewalkers would try to open the jammed panel. It was not the first time a Shuttle crew faced an unscheduled EVA, but this one posed a significant time constraint. Once Hubble was disconnected from *Discovery*'s services, which occurred when it was hoisted out of the payload bay, it had only its internal batteries to keep itself alive until it could draw power from the solar panels. It can take astronauts up to 4 hours to prepare for an EVA. The batteries on Hubble would last 3 hours. By normal procedure, Hubble would be lifeless even before the astronauts exited the airlock. In a new procedure specifically written for STS-31, as soon as *Discovery* arrived in space, the crew reduced the cabin pressure from 14.7 psi to 10.5 psi. This allowed the two nominated spacewalkers to adapt their bodies to a lower atmospheric pressure, thus shortening the time necessary for EVA preparation. In addition, the spacesuits were checked out on the first day, as if a spacewalk had to be made. These precautions would enable Sullivan and astronaut Bruce McCandless to exit the airlock in only 2 hours. In the end, Mission Control was able to work out a solution that opened the jammed solar panel and the contingency EVA was canceled.

After releasing Hubble, *Discovery* remained nearby in case of a malfunction that would require the crew to intervene. But the satellite operated nominally, and 5 days later the mission was concluded at Edwards Air Force Base in California.

The rest, as the saying goes, is history. A few weeks later, the world learned that the most expensive and sophisticated telescope was shortsighted.[35] The ambitions of thousands of astronomers of seeing the universe in unprecedented detail were lost in blurry images.

But people in NASA do not give up easily. Hubble was designed to be serviced by astronauts, so there was hope. The original plan had envisaged performing orbital maintenance every 2 years and returning the telescope to Earth for a major overhaul every 5 years. But concerns about the fatigue life of the structure of the telescope if repeatedly subjected to the loads of launch and re-entry, obliged NASA to adopt only on-orbit servicing. A solution to Hubble's myopia was soon available in the form of an elaborate optical system to adjust the focus of the light that came from the primary mirror so that "corrected" light reached the instruments.

---

[35] An investigation to determine the cause of this embarrassing flaw, found that the equipment used to check the shape the primary mirror was wrongly calibrated. The mirror was perfectly made, it was simply to the wrong profile.

STS-61 was assigned one of the most important and challenging missions of the Shuttle program. Once again the performance envelope was to be pushed to the limit. For instance, two teams of spacewalkers would train for an unprecedented five back-to-back spacewalks to tackle a multiplicity of complex maintenance tasks.[36]

STS-61 spacewalker and astronomer Jeffrey A. Hoffman says, "It turns out to be rather complicated and time-consuming because you don't just get your spacesuits out, use them, and then put them away. After each EVA, you have to clean them, you have to change-out the lithium hydroxide carbon dioxide scrubbers, and you have to make sure that the batteries are charged, which means that some batteries are being charged and other batteries have to get put in the suits. You have to keep a record of everything." To exploit every single minute of EVA time, the crew came up with an innovative proposition to mission planners, as Hoffman explains, "The original plan for doing the repair was that all our tools would be in a toolbox out at the aft end of the payload bay, so every day we'd have to go out and load up the tools. That'd take 15 or 20 minutes…We wanted to eliminate any EVA time that was devoted to things that were not necessary, including preparing our tools. We suggested that if we could bring those tools inside we could save a half hour by getting them all ready before we went out. We were several months from launch and they had done all the weight and balance [work]. But this was Hubble and the tools were [relocated] inside. Again that saved us a lot of time every day."

In planning the best attitude for *Endeavour* during the spacewalks, one issue was how to prevent direct sunlight from entering the telescope when its access doors were opened, lest the strong light evaporate any organic contaminants inside it and pollute the ultraviolet optics. The first proposal was for the Orbiter to keep its belly towards the Sun as a shield. In this inertial orientation, the telescope would be facing Earth during orbital daylight, with the heat radiated by the planet providing warmth for the spacewalkers, and at night it would be facing deep space and they would not receive any radiant heat. As Hoffman says, "We knew we were going to get very cold, so we spoke to the mission planners about this. They got the thermal people to sharpen their pencils a little bit and figured that yes, some of the metallic equipment was going to get down to about 150 below zero." This was not good news. Firstly, it was known that the spacesuit gloves were prone to coldness, causing discomfort to the astronaut. Secondly, there was the issue of whether the repair tools would work at all in such condition. Since spacewalks had never been performed in such freezing conditions, NASA arranged for a series of simulated spacewalks to be made in a thermal vacuum chamber.[37]

On the first simulation run, Hoffman soon found that he could not use most of the tools because they were frozen in the toolbox. On a real spacewalk, this would have been a disaster. As he remembers, "The engineers went to work. They took out all the residual grease. And I guess they filed away to increase some of the clearances." His colleague Story Musgrave made the second trial. He was able to use all the tools, but reported that

---

[36] An initial idea to split the flight into two separate missions was quickly rejected in favor of a single flight, as NASA and its astronauts gained confidence in their capabilities.

[37] These are large chambers where the thermal and vacuum environment of space are reproduced in order to test hardware and procedures for spacewalks, or to test the design of a spacecraft prior to launching it into space.

his hands were very cold. After an inquiry from a doctor, Musgrave reported that he could no longer feel pain in his hands, presumably because they had warmed up. When he removed his spacesuit, he found that his fingers were purple and black owing to severe frostbite. He was urgently airlifted to Alaska for specialist attention that managed to save his fingers (and thus his place on the mission). It was clear that if NASA did not want to injure its astronauts any further, it would have to develop a new Orbiter attitude for servicing Hubble.

An alternative was to keep the telescope in the shade during orbital daylight and facing towards Earth for warmth in darkness. This would require *Endeavour* to make two attitude change maneuvers per orbit. One at sunrise to position its belly towards the Sun and the second at sunset to face the belly to deep space and Hubble towards Earth. Fuel margins were a concern, as Hoffman explains, "We didn't have enough propellant to do that, so they had to come up with a new way of working the reaction control system jets so they could do the maneuvers with less propellant."

More than usual care was spent in planning the astronauts' training. There were to be seven astronauts, four of whom would perform the five spacewalks.

On any other mission, the flight deck seated four astronauts for the ascent and re-entry phases. The mission specialist who sat behind the commander and pilot trained to assist with the procedures in case of an emergency. For STS-61, this would mean that one spacewalker would have to undergo training as a member of the flight deck team at the expense of spacewalk training. As Hoffman explains, "We decided that because of the criticality of the EVA and the amount of time it was going to take us to train for it, that the flight deck crew were just going to be the three – Dick Covey, Ken Bowersox, and Claude Nicollier.[38] They'd prepare for the ascent and re-entry. We didn't go through that training. That gave us more time for our EVA training."

To maximize the training opportunities, NASA for the first time selected an all-veteran Shuttle crew, to delete the basic training of a rookie in order to focus on the mission-specific training. As Hoffman says, "We spend a lot of time rehearsing flight day one, for instance, just because that's the first day you're in space. A lot of people are disoriented. Some people are not feeling good. It's such a new environment that it just is a really good thing to have gone through all the activities you are planning to do, over and over again. We usually have three or four run-throughs of that first day, such as removing your suits, reconfiguring the Orbiter, and getting things ready for orbital operations. That way, if you are not feeling good, if you are disoriented when you get into space, you can fall back on your training and just look at the procedures and, 'Oh yes, I know what I'm doing.' You don't really have to think about it much. That can make a big difference. We didn't have to do that because we all knew what we were doing. The fact that we were all veterans allowed us to really concentrate on Hubble."[39]

---

[38] They were respectively the commander, pilot, and robotic arm operator.

[39] This was the first and only time that NASA required that a Shuttle mission be flown by veterans alone. As a rule, the only person on a mission who must have flown in space previously was the commander. Interestingly, the two STS-2 crews were both rookies but this was allowed because mission commander Richard D. Truly had earned USAF astronaut wings flying to the edge of space on the X-15 experimental rocket plane.

Even more training time was made available for spacewalking by eliminating all objectives unrelated to Hubble. It was routine for a mission to include a number of secondary objectives.[40] Despite the secondary importance of such tasks, every crew is eager to achieve all of their assigned work, and this involves setting aside time for training. Deleting the secondary objectives from STS-61 enabled that crew to focus exclusively on Hubble.

STS-61 lifted off on December 2, 1993 and triumphantly returned home 11 days later having accomplished one of the most complex missions ever performed by a Shuttle. Four more servicing missions followed, each of which serviced and updated the telescope.

The final servicing mission by *Atlantis* on STS-125 pushed the Shuttle program to new boundaries by requiring plans to be set in motion for a possible rescue mission. Named STS-400 this was the only mission to be prepared in the hope of never being flown.

After the *Columbia* accident in 2003, NASA imposed a rule that Shuttle missions would only be flown in order to finish assembly of the International Space Station. A damaged Orbiter would be able to offload its crew at the station. This would not be possible for a Hubble mission because its orbit was incompatible, so there was some reluctance to add a fifth servicing mission to the manifest. STS-400 was designed as a standby rescue mission and assigned to *Endeavour*.[41] Since it typically took three months to prepare a Shuttle for flight, the processing of *Atlantis* and *Endeavour* was done in parallel.

When *Atlantis* lifted off from pad 39A on May 11, 2009, *Endeavour* was on pad 39B, ready to be launched if it should prove necessary to rescue the stranded crew of *Atlantis*. This marked the final time that two Space Shuttles were simultaneously on the pad. In the nominal plan, the rescue Orbiter would perform all of the maneuvers to approach *Atlantis* from below. Just before the newcomer made the last rendezvous burn, *Atlantis* would maneuver to face its payload bay to Earth and its nose towards orbital south. Capture would be performed by the robotic arm of the rescuing Orbiter grappling the berthed Orbiter Boom Sensor System[42] installed on *Atlantis*. Next the arm of the rescuing Orbiter would align *Atlantis* so that the two spacecraft were nose-to-nose for optimum attitude control in this mated configuration. Finally, *Endeavour* would maneuver the stack into a gravity gradient attitude for stability. Contingency plans were also developed in case the rescuing Orbiter proved unable to execute the nominal rendezvous profile owing to either

---

[40] Usually these are experiments to be performed inside or outside of the pressurized cabin or for testing new procedures.

[41] After several revisions to the flight manifest due to technical delays, *Endeavour* was eventually chosen as the rescue vehicle and it would have been flown by the four flight deck crew members of the STS-123 mission.

[42] The Orbiter Boom Sensor System (OBSS) was a 50-foot extension of the robotic arm which contained a suite of sensors for inspecting the Orbiter's thermal protection system. The boom was installed on the right-hand payload bay sill. The arm would retrieve the boom using a grapple fixture on the forward side of the boom, and maneuver it to enable the battery of imagining sensors on the other end of the boom to inspect areas of interest. For additional details refer to Chapter 8 of my previous book *To Orbit and Back Again: How the Space Shuttle Flew in Space*.

a velocity shortfall at orbit insertion or a subsequent propellant failure.[43] Whereas it was preferable to fly the mission with the rescue Orbiter approaching from below and behind *Atlantis*, provisions were made to approach from above and ahead. In this case, *Atlantis* might have been called upon to perform altitude and phasing adjustments to enable the rescuing Orbiter to fulfill the rendezvous. The propellant margins on both spacecraft would have been managed to ensure that the rescue Orbiter had sufficient remaining propellant after the rescue to perform a safe de-orbit.

The Hubble servicing missions were among the most complicated and expensive assignments that NASA and the Shuttle program were called upon to perform. Since risk rapidly increases with complexity, it is legitimate to ask whether it was worth the effort. Detractors of on-orbit servicing often quote the higher development cost of a satellite that is essentially a reusable vehicle, to which must be added the associated costs of on-orbit servicing by astronauts. Their logic is that it is economically more sensible to build a new satellite to replace one in space whose useful life has come to an end. But this is a false perception. Would NASA have built five Hubble telescopes instead of flying servicing missions? The answer is: No it would not have done that. Flagships such as Hubble are built once, and when a critical malfunction occurs, as it did multiple times, they become among the most expensive pieces of space junk ever paid for by the taxpayers. And of course, it is significant that each servicing left the telescope substantially improved, with state-of-the-art instruments and systems. As a result, the Hubble Space Telescope that is on-orbit today is a very different machine from that which was designed and built in the 1970s.

CONCLUSIONS

When *Atlantis*'s wheels stopped on Runway 22 at Edwards Air Force Base on May 24, 2009, it not only concluded STS-125, it also ended the era of on-orbit satellite servicing. In the years that followed, all Shuttle missions were to complete assembly of the International Space Station and prepare the outpost for the post-Shuttle era.

As of today, despite talk of on-orbit servicing as a future profitable activity for the expanding private and commercial space industry, there are no definite plans to do it and satellites are still being built as disposable assets. Surely, had the Shuttle fulfilled the hopes and promises of its designers and proponents, we would have seen more satellite rescue and servicing missions. One reason why this did not happen lies with the rather rarefied launch manifest that saw a Shuttle lifting off only several times a year,[44] especially with

---

[43] This was a condition in which the Orbiter would attain orbit at a lower speed than planned. In other words, it implied that the Orbiter had reached a lower orbit running on its main engines and therefore had to burn more propellant in the Orbital Maneuvering System (OMS) to reach the desired orbit. A propellant failure implied OMS propellant leakage, which clearly would have required even more careful propellant management in order to arrive at *Atlantis* and still have enough fuel for the de-orbit burn.

[44] This is a far cry from the one-flight-a-week manifest that had been touted by NASA to sell the Shuttle to Congress and avoid a catastrophic cancellation of the program in the mid-1970s.

increased safety concerns following the *Challenger* accident. With the Shuttle launching sporadically, at a high cost, after years of preparation and repeated delays, it was soon apparent to satellite operators that planning for on-orbit servicing would not be economically justifiable. If a communications satellite were to fail, the operator would have wanted a repair mission to be carried out as soon as possible. But in reality it would have had to wait years, perhaps suffering slippage in the schedule due to technical problems within the Shuttle fleet or a reshuffling of the manifest. It would make more sense to write off the malfunctioning asset and launch a replacement.

Yet it must be recognized that the absence of an on-orbit servicing infrastructure has been, and still is, endorsed by the very industry that such an infrastructure sought to serve. In fact, the principles of modularity established by Cepollina and his team at Goddard were applied to only a handful of spacecraft.[45]

Satellites and space telescopes are mostly still one-of-a-kind, each designed from scratch without leveraging on existing hardware or configurations.[46] Awful as this sounds, it is actually the best way for spacecraft manufacturers to raise their financial gain because they can bill their customers higher development and construction costs and keep their assembly lines operating at high capacity. Servicing and refurbishing satellites clearly removed the need to build a replacement, eliminating an opportunity for profit. This state of affairs is difficult to challenge, since it requires a U-turn in the way the global space industry creates its orbital assets.

Perhaps concerns about the rapidly growing population of space debris, strides in space robotics, and the inevitability of reusable launchers that will slash launch costs and improve reliability, might stimulate a move towards on-orbit satellite servicing, and build upon the legacy left by the Shuttle program.

---

[45] In addition to Solar Max, the MMS concept was applied to Landsats 4 and 5 and the Upper Atmosphere Research Satellite.

[46] Commercial communications satellites often share common onboard systems, especially if they are made by the same manufacturer. This helps to lower the costs, but still none of them is designed for on-orbit servicing and refurbishment.

6

Science Laboratory

FROM SORTIE CAN TO SORTIE LABORATORY

Space is not the exclusive domain in which NASA operates its assets and undertakes research. In fact, over the years the agency has enjoyed several research opportunities well within the Earth's atmosphere, carried out by a number of aircraft equipped as airborne laboratories. In the 1960s and 1970s, the Ames Research Center carried out astronomical and Earth observations from a Convair 990 and a Lear Jet, respectively an airliner and a business jet. Such platforms permitted short lead times, a rapid turnaround of experiments, multi-discipline research with direct involvement by the scientists, a rapid high-quality data return and, last but not least, a comfy shirt-sleeve environment. Ames managers and scientists argued that these benefits could easily be extended to a man-tended space-borne platform that offered the additional benefits of weightlessness, a very high quality vacuum, and higher altitudes for comprehensive coverage of the atmosphere and surface. They argued that the generous payload bay envisaged for the Shuttle would permit the installation of one-of-a-kind laboratories on sortie missions.[1]

This logic was so convincing that from September 1971 to January 1972, a small team at the Preliminary Design Office of the Marshall Space Flight Center explored in depth how such an idea might work in practice. Within 5 months, they had come up with a "sortie can laboratory" that would be carried in the Shuttle's payload bay. The study recommended the facility consist of two sections: a cylindrical pressurized module 15 feet in diameter and 25 feet in length, and an open-truss of variable length on which to mount experiments which required to be directly exposed to the vacuum of space. Additional features for the pressurized module would include internal rack mountings for experiments, a work bench, observation and optical windows, a small airlock and boom for temporarily placing experiments outside, and a pointing system for astronomical instruments.

---

[1] These were flights in which the primary objectives were related to perform research and experimentation rather than engaging in activities of satellite deployment/retrieval and construction of orbital infrastructures.

© Springer International Publishing AG 2017
D. Sivolella, *The Space Shuttle Program*, Springer Praxis Books,
DOI 10.1007/978-3-319-54946-0_6

The adjective "sortie" was intended to signify that the laboratory would be flown for relatively short periods (up to 30 days) and to emphasize the low-cost nature of its operation. The latter was meant to be a high selling point for the sortie can laboratory concept. The excessive costs associated with the development of space hardware are mostly due to the expensive expendable launchers[2] and the lack of infrastructure for on-orbit servicing and refurbishment of assets when they suffer a fault and operations are either prematurely terminated or considerably reduced in scope.[3]

One means of increasing mission success and delaying the onset of a catastrophic malfunction is to raise the reliability of a spacecraft and its payload by rigorous but costly ground testing. This subjects the hardware to the most severe load conditions (e.g., vibrations, magnetic fields, extreme temperature variations, and so on) they are likely to encounter during their intended orbital life, to "guarantee" that nothing will fail. Also, backups for critical components can be included to provide alternatives in case failures do occur. As one would expect, space hardware is manufactured using high-specification materials whose development and certification adds an additional cost. There is nothing intrinsically wrong with using the finest possible materials and testing machinery to prove that it will reliably last for its scheduled operating life. In fact, this is good engineering practice. However, costs escalate when the design does not incorporate the possibility of maintenance and repair. This approach is called fail-safe design, and for critical components is often the only approach allowed by safety concerns.

For instance, the timing belt of a car engine is designed with such a long life that a failure will not occur before the car is sent to a scrapyard. It would be possible to make a car entirely fail-safe, so that it will never need an oil filter change or a brake replacement, but the development costs and the consequent retail price would quickly close the assembly lines and bankrupt the manufacturer. Yet this approach seems to work for space hardware, or at least it is widely tolerated.[4] It makes spacecraft and payloads with the highest probability of success, but combines exorbitant costs with intrinsic inflexibility.

It is standard practice for laboratory equipment and experiments to be upgraded as the results of an investigation prompt the investigators to expand the objectives of their research. This approach enables the new instruments to use the most up-to-date technology to maximize the scientific yield. For an Earth-based laboratory, this is a relatively undemanding process because an experiment or device can be redesigned, rebuilt, and immediately put to work. However, for a space-based laboratory, such as a telescope or an Earth-sensing satellite, the intermediate step of thorough testing and fail-safe design imposes a long waiting time before a new or upgraded apparatus can be flown. And without an on-orbit serviceability for the existing spacecraft, a new or enhanced investigation will also require a new spacecraft and an expensive launcher. Worse, by the time the experiment is ready for launch, the instrument could well be already technologically

---

[2]Think of using your car for just one trip, throwing it away and buying a new one for the return trip and you will readily appreciate the ludicrous manner in which space exploration is still being carried out.

[3]Again think of writing off your car every time it suffers a mechanical malfunction because there are no car mechanics or breakdown recovery trucks.

[4]This paradigm is currently being challenged by two private companies, SpaceX and Blue Origin, who are developing partially or fully reusable launchers. Other companies have ambitions to do likewise.

outdated. This is not the kind of environment in which the scientific community likes to work, as research ought to be a continuous process that exploits state-of-the-art technology, not a spasmodic exercise that incurs long breaks and relies on apparatus that is approaching obsolescence.

Reusability and the high annual flight rate projected for the Shuttle were poised to elevate space-based science to the standards of its terrestrial counterpart by reducing costs and allowing flexibility. For instance, a high flight rate would give investigators the opportunity to fly experiments after only a short period of development and then as often as they desired. In the case of mechanical failures or malfunctions the crew could perform a degree of in-flight maintenance; otherwise the experiment would be returned home, repaired, and assigned to the next available Shuttle. This meant that experiment hardware would no longer need to undergo such extensive certification or to comply with stringent fail-safe design requirements. In fact, cheaper off-the-shelf commercially available equipment could be approved for space applications, as long as it met outgassing and flammability requirements. Thus a sortie can mission would offer an experiment and the supporting science community an unprecedentedly short lead time and rapid turnaround. Likely fields of analysis and technology applications included physics, Earth observations, communications, navigation, and the materials and life sciences. And prior to the assembly of a space station, sortie missions would transform the Shuttle from a simple truck for on-orbit deliveries into a mini-station. Since the design of the space station project was mired in controversy, the research opportunities offered by the sortie can missions would enable NASA to exploit the Shuttle to strengthen its ties with the scientific community.

Based on the positive response to the Ames study of the sortie can missions, Dale Myers, the Associate Administrator for Manned Space Flight at NASA headquarters, authorized the Marshall Space Flight Center to continue in-house studies and plan for the definition phase of a Sortie Can project. By March 1972 a development project timeline was written that envisaged a 1-year period for the definition of requirements, then 5 years of design and development that would culminate with a maiden flight no later than 1978. Total funding was estimated at between $90 and $115 million, on the basis that in-house workers and resources would be used whenever possible.

Even as the Sortie Can concept was taking shape, an alternative was under study by the Convair Division of General Dynamics, whose intention was to create a series of laboratories suitable for carriage in the Shuttle payload bay and attachment to the space station. The rationale for their Research and Application Module (RAM) was the assumption that the space station would have a core section and a complement of research facilities. The core section would take care of the proper functioning of the outpost by generating power, maintaining attitude, providing crew accommodations, and so on. The research facilities would provide a laboratory-like environment whose fixtures would accommodate a range of research requirements. Each RAM would be delivered to the space station by Shuttle and subsequently retrieved and brought back to Earth for refurbishment.

Although at first sight these two concepts may seem identical, they could not be any more different in terms of mission capabilities. While the Sortie Can proposed by Ames would operate only in the Shuttle payload bay, the RAM of General Dynamics was a laboratory that could be configured for a range of space infrastructures such as the space station, free-flying platforms, or man-tended telescopes. For instance, since astronomical observations are impractical on a space station due to stringent pointing accuracy requirements, which the vibrations generated by a crew would compromise, concepts were under study for large free-flying telescopes that would be periodically refurbished by Shuttle crews. Some concepts envisaged having a RAM at one end of the telescope to provide a pressurized shirt-sleeve environment for a servicing crew, their tools, and a stock of spare parts. Alternatively, a RAM could be berthed to the telescope upon the Shuttle's arrival and retrieved upon completion of the servicing. RAM modules could also be part of free-flying platforms for research that required an environment devoid of vibrations and other interference typical of a space station. If necessary, free-flying RAMs could be picked up by the Shuttle for refurbishment or be returned to the space station. And of course, a RAM could be used in the Sortie Can mode for missions where the Shuttle would operate as a mini-station on its own.

The RAM would be flexible and accommodate a large variety of missions. To be cost-effective and yield the best scientific return, it was suggested that such missions should last at least 30 days. NASA's interest in this concept was demonstrated by the award of a Phase-B study to Convair to run from May 1971 to August 1972.

Sortie Can and RAM were both intended to satisfy the need to initiate research into science and technology applications prior to the introduction of a space station. As a matter of fact, by the time the Shuttle program was authorized by President Richard M. Nixon, the ambitious plans for a space station and other man-tended outposts had been mothballed. Sortie Can and RAM would offer a practical and sensible means of kicking off space-based research. While both were aimed at the same goal, the Sortie Can won the contest, largely because, in the absence of any man-tended outposts, the operational flexibility of the RAM concept was no longer required.

The space agency's proposition was put to the scientific community at the Space Shuttle Sortie Workshop hosted by the Goddard Space Flight Center between July 31 and August 4, 1972. The intent was to involve NASA scientists in defining how the Sortie Laboratory[5] would operate, and to help the scientific community within the agency to think through innovative ways in which experiments could be run in space. Chaired by the Associate Administrator for Space Sciences, Dr. John E. Naugle, the workshop produced a 2-inch-thick report which detailed potential uses for the Sortie Laboratory, suggestions for research and new application technologies that could be investigated, and how the presence of astronauts would be beneficial. Subsequently, the working groups established by the Goddard meeting were extended to encompass scientists outside the agency, and by March 1973 several other recommendations and suggestions for experiments and hardware were published. On the whole, however, the scientific community was rather lukewarm to the Sortie Laboratory idea. In fact, because experiments had been flown on sounding rockets and unmanned satellites, it was felt that the introduction of a "man in the loop" would complicate developments and increase their cost due to the need to man-rate the experiment hardware. Despite a rationale for man-tended experimentation in space, the scientific community wasn't eager to forgo their way of conducting space science, particularly if they would have little involvement while the investigation was in progress.

---

[5] By this point the word "can" had been dropped in favor of the more professional "laboratory."

Once again, the capabilities offered by the Shuttle were challenging an old school of thought with exciting new possibilities.

EUROPEAN COLLABORATION

On January 5, 1972, NASA was authorized to create a Space Transportation System. By the end of that year, it was clear that this would involve three main elements: the Shuttle, the space tug, and the Sortie Laboratory.

On September 24, 1973, NASA and ESRO[6] signed an agreement by which the European aerospace industry would take the lead in the development and fabrication of the Sortie Laboratory. In fact, the chief points of this agreement had been penned months earlier during a series of meetings held in America and Europe that explored various ways in which Europe could participate in the Shuttle program.

By this agreement, ESRO would develop the Sortie Laboratory and deliver a fully functional engineering model and a flight unit. Their ownership would be assigned to NASA. The American agency would have full control over the flight unit, including the right to specify how it should be used and any modifications. However, a clause was added to say that Europe would provide components and services. It was further decided that ESRO would deliver the flight unit by 1978, then support the first two flights.[7] If NASA required additional flight units, as then seemed highly likely, they would purchase them. Another important provision concerned technology transfer, a rather delicate point which in the end meant the Europeans were able to work only on the Sortie Laboratory.

Even before Apollo 11 returned from its historic Moon landing, President Nixon had expressed the notion that NASA ought to seek international cooperation for the post-Apollo era. A report called "The Post Apollo Space Program: Directions for the Future" issued in September 1969 by the Space Task Group provided further impetus for the agency's Administrator, Dr. Thomas O. Paine, to seek collaboration with the countries of the Free World.

Canada answered by proposing a marvel of engineering which became the iconic robotic manipulator system. Traveling in Europe, Dr. Paine was pleased to find that the European space industries were up to the challenge of manufacturing some of the components for the overall program, possibly the space tug. If Europe could obtain a slice of the program, not only would it gain expertise and skills in a major manned program, it would also become a fully-fledged service provider to NASA. However, although extensive European collaboration could reduce the development cost to the US taxpayers, some leading members of Congress did not share the enthusiasm with which Paine was seeking such partnerships.

The Department of Defense was likewise dissatisfied. The Shuttle was being sold as the only means of launching into space for decades to come, including for the all-important

---

[6] The European Space Research Organization (ESRO) was one of two progenitors of the European Space Agency (ESA); the other was the European Launcher Development Organization (ELDO).

[7] At that time, the Shuttle was scheduled to start operations in 1979 and begin flying sortie missions shortly thereafter.

DoD classified projects. Placing critical parts of the Shuttle program into Europeans hands would create a dangerous reliance on foreign countries for reliable access to space. For instance, Europeans might be late in supplying spare parts, thus delaying national security missions. Another worrying issue concerned the transfer of technical know-how. The Shuttle was to be extensively used to deliver commercial satellites as a source of revenue for the agency. If the Europeans were involved in the development of the Shuttle they might discover how to build a competitive launcher that would steal a large slice of that market from NASA. Indeed, ELDO, another of the progenitors to ESA, was planning an all-European expendable launcher named Europa. This was one of the main reasons for not only excluding European industry from the Shuttle but also from any work on the space tug that was meant to serve as an upper stage for the delivery of defense payloads into higher orbit. Once again, the DoD was determined to eliminate any threat to its independent access to space.

At a meeting between the International and Scientific Affairs of the Department of State and a delegation from Europe held in Washington DC on June 12–14, 1972, it was made clear that Europe would be asked to participate only in the development of the Sortie Laboratory. Despite understandable dismay, the Europeans recognized that this development, although less glamorous, could still provide a good foothold in the Shuttle program because it was an element that NASA needed and would use. Also, it would be less challenging to develop, cheaper, could be sold in volume, and could provide a concrete opportunity for Europe to fly their experiments and possibly their astronauts. The European delegation was given August 15, 1973 as the deadline to accept the work package. After a series of meetings within the European aerospace community, with NASA, and with the American government, the historic agreement was ratified on September 24, 1973.

Despite the issue of technology transfer, it was eventually agreed that both sides would reserve the right to move hardware rather than know-how, although specific know-how would be made available on a case-by-case basis to help Europe execute its responsibilities in the program.

Also, after a series of discussions over a lost-in-translation issue, the project was also given an official name. While the English word "sortie" was meant to reflect the short duration of such missions, in the French language it meant "exit." As it did not have anything to do with a research module project, the name Spacelab was proposed instead. Although NASA thought the name sounded too similar to Skylab, the issue was eventually settled and the official agreement specified that the project would be known as Spacelab.

SPACELAB DEVELOPMENT

It was now time for European industry to transform their Spacelab studies into a real piece of space-worthy hardware. On June 5, 1974, the German manufacturer ERNO was chosen as the winner of a competition set by ESRO to develop and construct the system. The German manufacturer was awarded a 6-year contract for the equivalent of US$226 million. April 1979 was set as the deadline for delivery of the first flight unit to NASA. In addition, ERNO was to supply an engineering model for ESRO and NASA, three sets of ground support equipment, and appropriate spares. ESRO would remain ultimately responsible for the design, development, and construction of the module and would hand it over to NASA. By now, though, the scheduled inaugural launch of the Shuttle had slipped into 1980.

In order to facilitate the widest possible range of scientific investigations, ERNO elected to build considerable flexibility into the Spacelab concept. As the astronauts were to conduct experiments, it needed a pressurized section that was connected to the cabin of the Orbiter. When performing the battery of investigations selected for the mission, the astronauts would spend most of their time in that module. It was also decided to develop an unpressurized platform called a "pallet" to accommodate those experiments that required direct exposure to the vacuum of space or were capable of fully autonomous operation.

The overall designs of the two segments gave Spacelab additional flexibility. For example, the pressurized module was conceived as a pair of cylinders, one called the core segment and the other called the experiments segment. They were both 13.3 feet wide and 8.8 feet long.[8] The core segment would serve as the control center for all of Spacelab's subsystems and it could accommodate a minimal stock of experiments. The experiments segment, as its name suggests, would carry only experiments. It was possible to use the core segment alone, but not the experiments segment alone. With the two connected, the forward and aft ends were closed by truncated cones bolted to the outer shell. To run utility cables and lines to and from the Orbiter, both cones had three cut-outs 1 foot in diameter, two located at the bottom and one at the top. In this way, mission planners could select a configuration to match the type and number of experiments. For instance, only the core segment could be used along with pallets, as could the core and experiments segments when they were joined. And it was possible to fly pallets without a pressurized module.

Aeritalia, an Italian aerospace manufacturer based in Turin, was subcontracted to build the load-bearing structure of the pressurized segments. Although it was a rather simple structural configuration, Aeritalia faced a number of challenges. For instance, special ground support equipment had to be developed to safely handle the unstable unsupported cylindrical shells to preserve their circular profile during manufacturing and assembly, because otherwise the shell would have required thicker walls, which would have imposed a weight penalty. The design was further complicated by the presence of two large apertures at the top of both the core and experiment segments. These necessitated buttressing of the opening supporting flanges. In fact, to allow for optical experiments, photography, or general observations of space and the Earth, the ceiling of the core segment was cut to accommodate a window adapter assembly that consisted of a circular aluminum reinforced frame with a 1.35 x 1.80-foot rectangular aperture to accept a high-quality window of an almost distortion-free, low-reflective glass.[9] Next to it was a circular cutting 1 foot in diameter to house a window made of conventional glass that the crew could use for observations and photographic tasks that did not require such high optical fidelity. Window adapter plates on the ceiling served as the mounting base for experiments or photographic apparatus to undertake such observational tasks. Cover plates mounted on the outside were

---

[8] This diameter was selected in order to take full advantage of the width of the Shuttle payload bay.

[9] An interesting piece of trivia is that this high-quality window was a spare from the Skylab program.

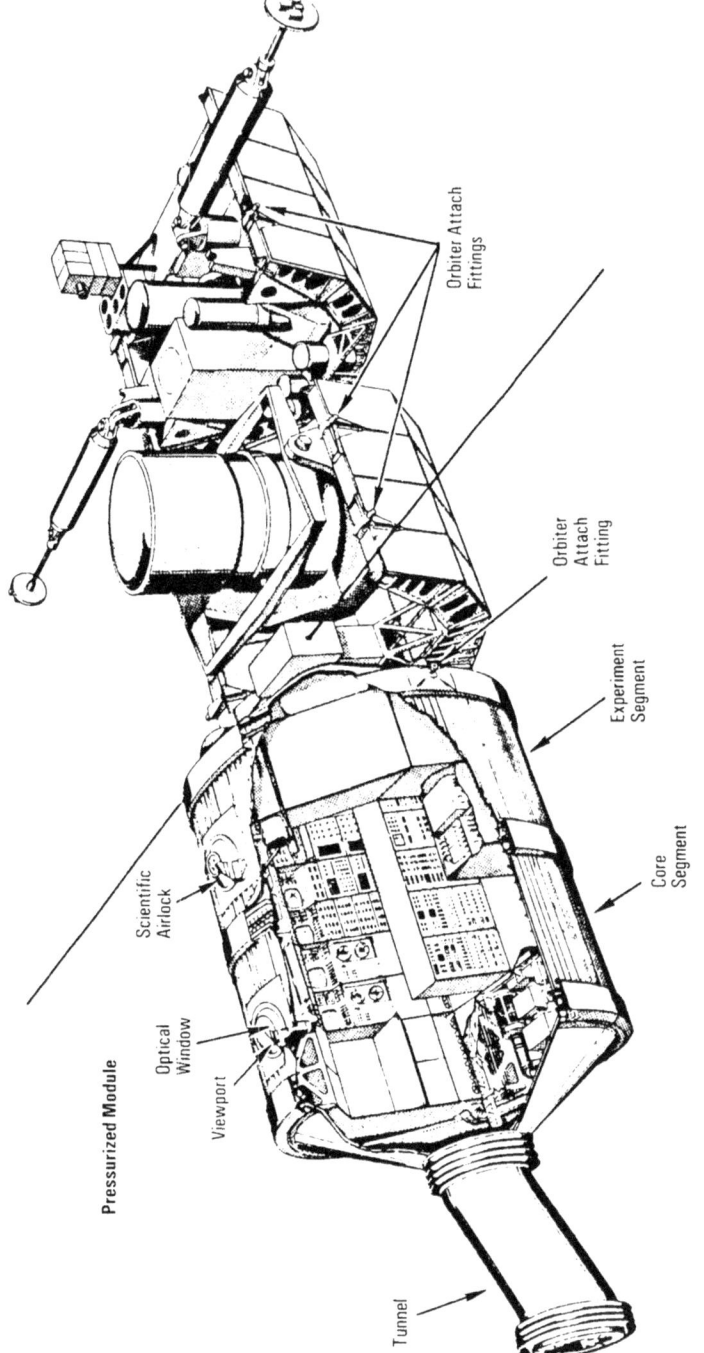

Orbiter Attach
Fittings

Orbiter
Attach
Fitting

Experiment
Segment

Core
Segment

Scientific
Airlock

Pressurized Module

Optical
Window

Viewport

Tunnel

The main components of Spacelab.

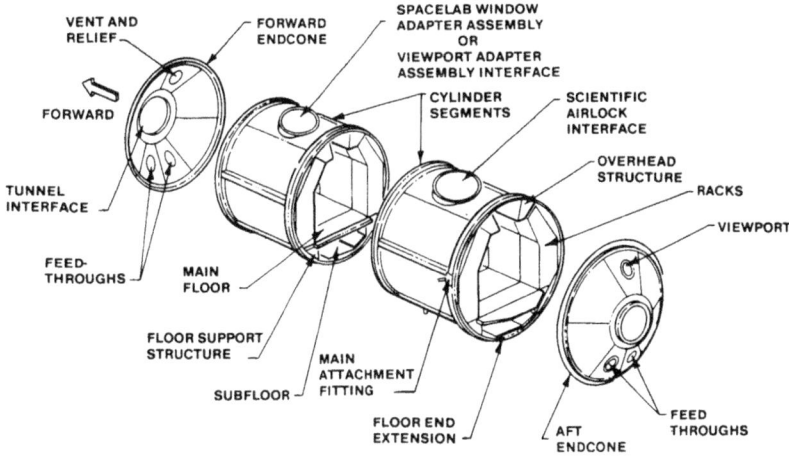

The main structure of the Spacelab pressurized module.

to be kept closed for additional shielding from micrometeoroid impacts, and opened when required by the flight plan.

For experiments that required exposure to the vacuum of space, the ceiling of the experiments segment had a circular aperture to accept a small scientific airlock. This was designed as a self-contained, manually operated unit, capable of accommodating a 3 x 3-foot experiment with a maximum mass of 218 pounds. This volume could also be used to store up to 132 pounds for launch and 209 pounds for re-entry. The airlock structure was a cylindrical shell fabricated from two large aluminum roll ring forgings which were machined into shape and welded together. Each end of the airlock was closed by a honeycomb sandwich panel hatch. The inner one was fully removable, and included a viewport with a diameter of 7 inches. A table could be manually extended 2.2 feet inside the pressurized segment to simplify mounting and dismounting of equipment. Airlock depressurization would take place by venting to space for some 16 minutes, then the outer hatch could be opened and the table extended outward. On completing the investigation, the table would be retracted into the airlock, the outer hatch closed, and the compartment pressurized with gaseous nitrogen from the local environment control system. Although repressurization would take only 10 minutes or so, several hours would have to pass before the inner hatch could be opened. This was to allow full thermal reconditioning of the airlock and the apparatus that had been exposed to space.[10]

---

[10] After having spent days exposed to the cold of space, the airlock and experiment required to warm up to a comfortable temperature to allow manual handling by the crew.

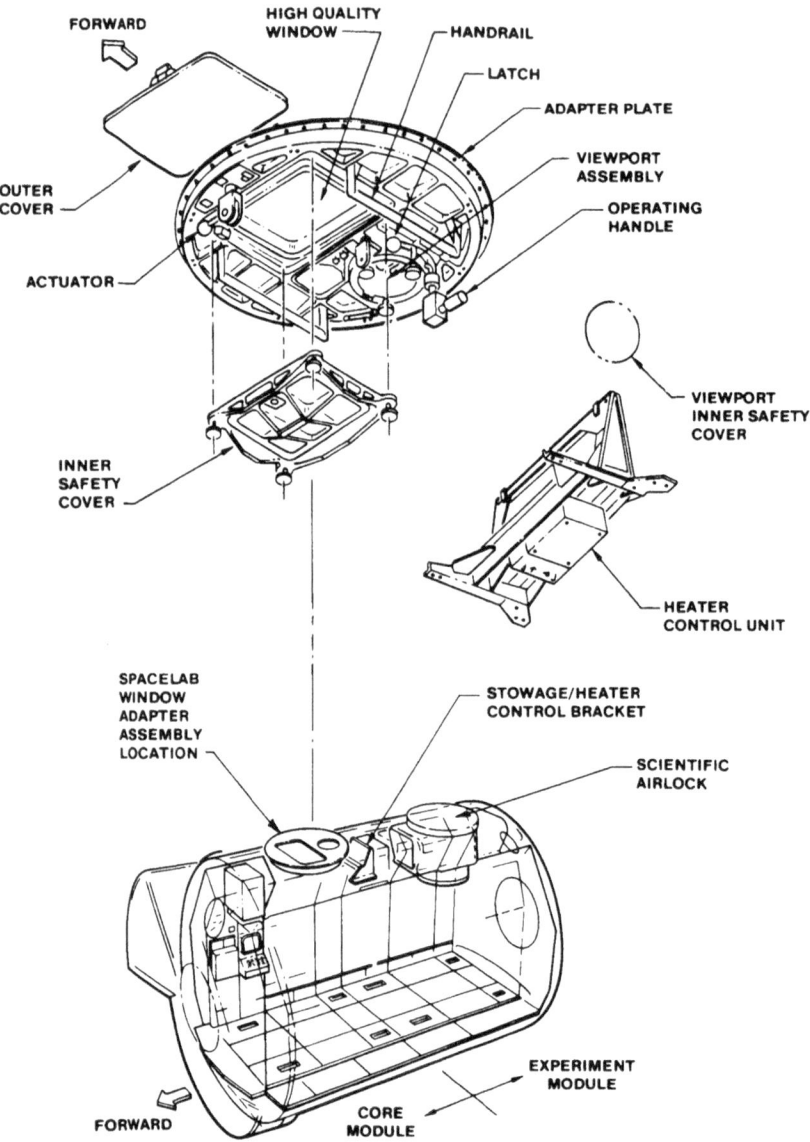

The Spacelab window adapter assembly.

Aeritalia had also to work hard on the design of the laboratory's external fittings, which played the dual role of providing support during ground handling and then as attachments in the payload bay. They had to withstand the high-g and vibrations of launch as well as the extreme thermal stresses of space. Also, their surface finish had to provide a very low

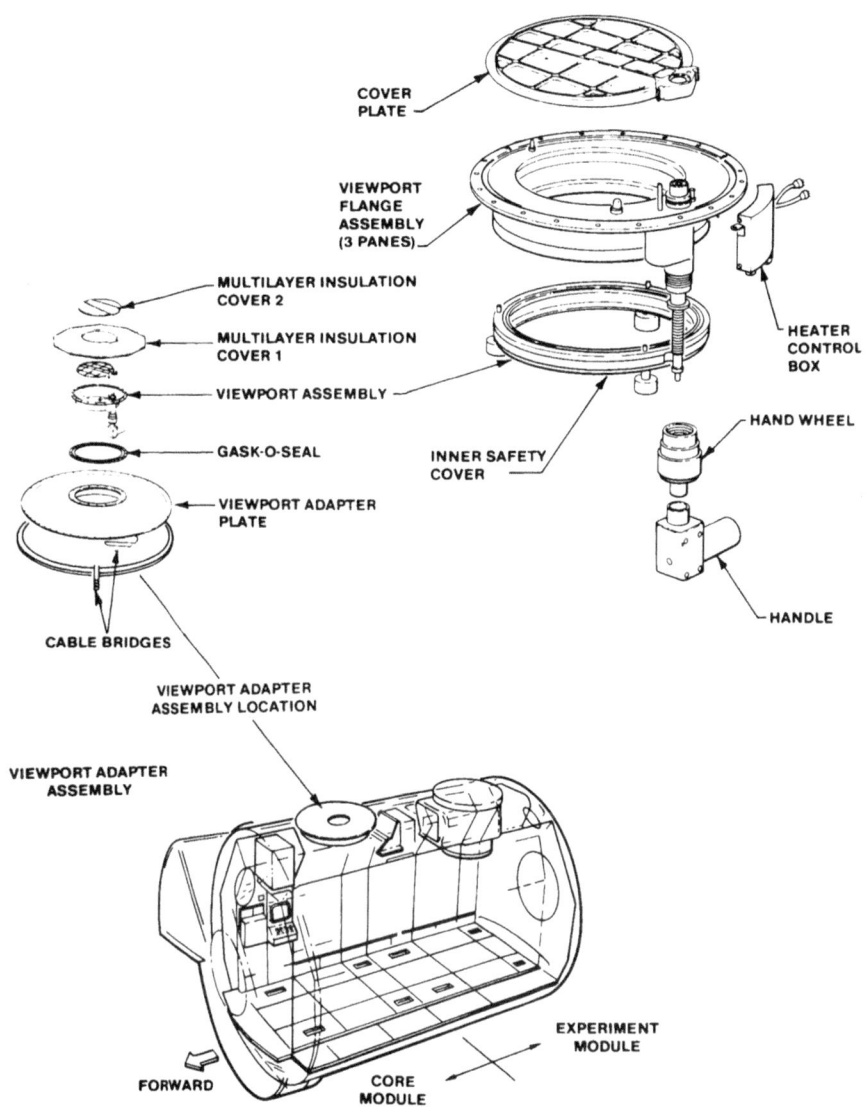

COVER
PLATE

VIEWPORT
FLANGE
ASSEMBLY
(3 PANES)

MULTILAYER INSULATION
COVER 2

MULTILAYER INSULATION
COVER 1

VIEWPORT ASSEMBLY

GASK-O-SEAL

VIEWPORT ADAPTER
PLATE

CABLE BRIDGES

VIEWPORT ADAPTER
ASSEMBLY LOCATION

VIEWPORT ADAPTER
ASSEMBLY

INNER SAFETY
COVER

HEATER
CONTROL
BOX

HAND WHEEL

HANDLE

FORWARD

CORE
MODULE

EXPERIMENT
MODULE

The Spacelab viewport adapter assembly.

friction coefficient to minimize the pressurization and thermal loads imposed by Spacelab on the Orbiter structure. And there was the susceptibility to fatigue and stress cracking of the 2219-T851 aluminum alloy that was selected as the material for the shells and end cones. With a design life of some fifty missions, the structure had to be able to withstand

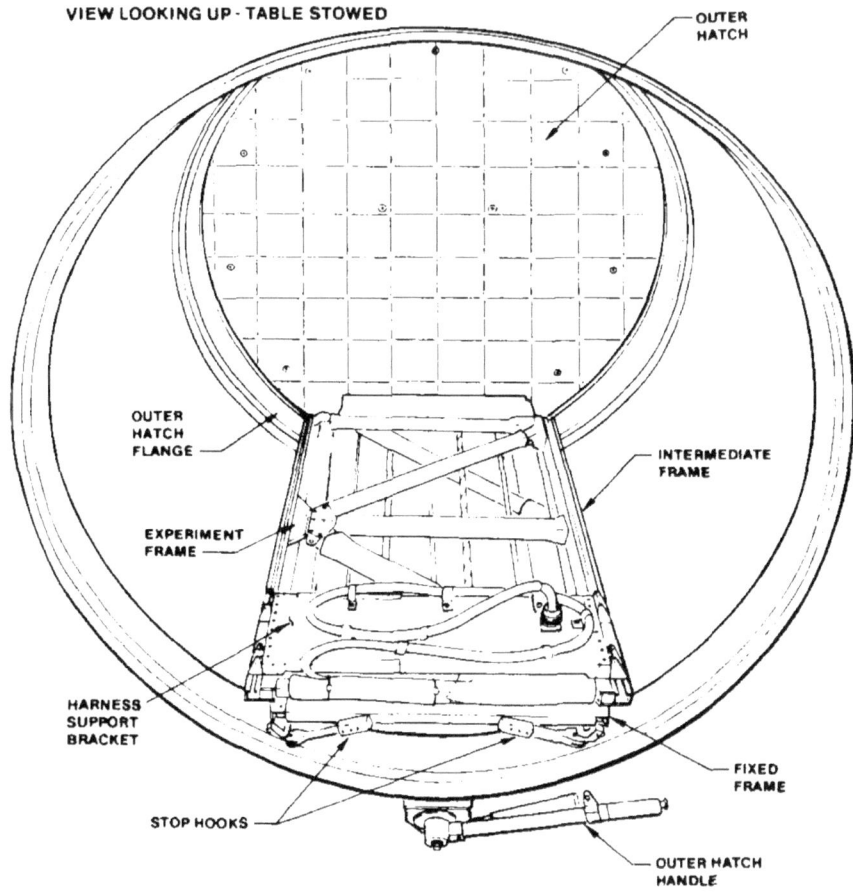

The Spacelab airlock experiment table assembly.

the repeated loads of launch, re-entry, orbital flight, and ground processing which included assembly and disassembly.

Because Aeritalia lacked the expertise and computational power to carry out such sophisticated stress analysis, technical experts from the Marshall Space Flight Center and US industrial consultants gave the Italian manufacturer considerable assistance. One aspect of this problem concerned the design of the seals to fit between the shells and end cones. They would be the largest ever built for space applications and would have to endure repeated assembly and disassembly without any degradation of their airtight and pressure-bearing capabilities. The design consisted of a double butyl seal molded into an aluminum ring base. Clamping of the ring flanges would deform the elastomer that filled the seal cavity and thus ensure airtightness. Upon completion of the shells and cones,

A view of the Spacelab scientific airlock.

static and dynamic testing confirmed the structural design and qualified the hardware as space-worthy.[11]

Within the protective shells of the pressurized Spacelab, subsystems components and experiments alike would be lodged into universal support structures called racks, each a 7075-T73 aluminum alloy truss structure with a framework of panels on the front and sides. There were two sizes, referred to as single and double. A single rack was 1.58 feet in width, had an available volume of 31.8 cubic feet and could carry up to 951 pounds of

---

[11] Thanks to the expertise that Aeritalia gained in working for the Spacelab program, years later they would gain a primary role in the development and constructions of most of the modules that would form the habitable section of the International Space Station.

apparatus. A double rack had twice the width of a single one. Racks for experiments would be open at the front, but closed at the sides with removable back panels and front panel attachment provisions. They also had standard patterns of holes for attaching payloads at each corner and on the sidewalls. The core segment would be mostly given over to double racks modified to accommodate subsystem controls. These were referred to as control center racks. They held components of the command and data management subsystem, the environmental control subsystem, the caution and warning/fire suppression subsystem and experiment power switches.

In the main, the core segment contained two control center racks and could also carry two single and two double experiment racks. The experiments segment could accept a full complement of four double and two single racks. For both segments, the racks would always be installed in the "double-double-single" order, extending from the forward end along each side. Each rack would be bolted to the main floor and attached by fittings to the overhead support structure. The main floor was designed as a framework of aluminum beams covered by aluminum honeycomb core panels with anodized aluminum face sheets. The panels of the center aisle were fixed, but those in front of the racks were hinged in order to provide access to the subfloor area. Side floor panels covered the openings left when experiment racks were not installed. The main floor spanned the full width and length of each segment to serve as a platform for experiments which were not on racks, as a flat surface for on-orbit activities, and as a walking surface during ground processing.

A subfloor made of aluminum sandwich panels located underneath the main floor provided a cavity for components of the environmental control and electrical systems in the core segment and additional research hardware or apparatus in the experiments segment. Together with attachments on the top of each rack, the overhead support structure provided additional storage space, attachments for the environment control subsystem air ducts, handrails, and lights. For this reason, the support structure was conceived as two sets of beams, one on either side of the centerline and composed of a quadrilateral and triangular beam made of aluminum sheet and plates which were riveted to create a continuous member.

Spacelab would be open only during orbital flight to cater for the requirements of the science program. Daily crew activities such as sleeping and eating would be done in the Orbiter's cabin.

In 1977, NASA hired McDonnell Douglas to develop and fabricate an 18.86-foot-long, 3.34-foot-diameter cylindrical Spacelab Transfer Tunnel to be connected at one end to the outer hatch of the middeck airlock and at the other end to an opening at the center of the forward end cone of the pressurized module. As these two opening were not in line, in 1974 NASA had established that the 3.51-foot vertical offset would be accommodated by introducing a vertical joggle section in the tunnel.

The need for a transfer tunnel stemmed from early design choices concerning the Orbiter. As Max Faget, who served in a multitude of key roles in NASA starting with the Mercury program, says, "When the solid rockets stage[12] at 4,000 feet a second, the Orbiter

---

[12] Here Faget is referring to when the solid rocket boosters separate from the external tank, some 2 minutes into the ascent phase of a Shuttle mission, an event known as "staging."

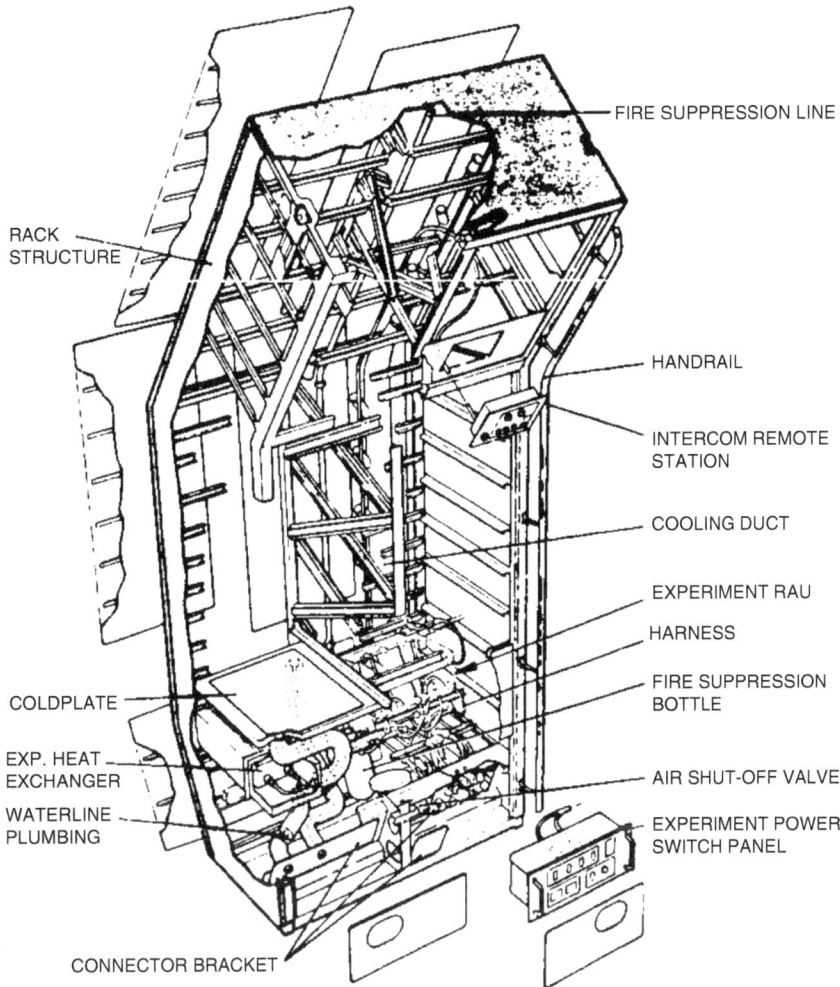

RACK
STRUCTURE

FIRE SUPPRESSION LINE

HANDRAIL

INTERCOM REMOTE
STATION

COOLING DUCT

EXPERIMENT RAU

HARNESS

FIRE SUPPRESSION
BOTTLE

COLDPLATE

AIR SHUT-OFF VALVE

EXP. HEAT
EXCHANGER

WATERLINE
PLUMBING

EXPERIMENT POWER
SWITCH PANEL

CONNECTOR BRACKET

The framework of an experiment double rack.

had to have an awful lot of propellant left for [orbit] insertion,[13] which meant that you had to have enough thrust to carry the weight of that propellant, and that increased the size of the engines in the Orbiter, which moved the center of mass of the orbiter farther back than we would have liked…As a consequence of that, the center of mass of the Orbiter was well

---

[13] After jettisoning the external tank, the Orbiter had still to perform one or two OMS firings to achieve the desired orbit. For further details, see Chapter 11 of my previous book *To Orbit and Back Again: How the Space Shuttle Flew in Space.*

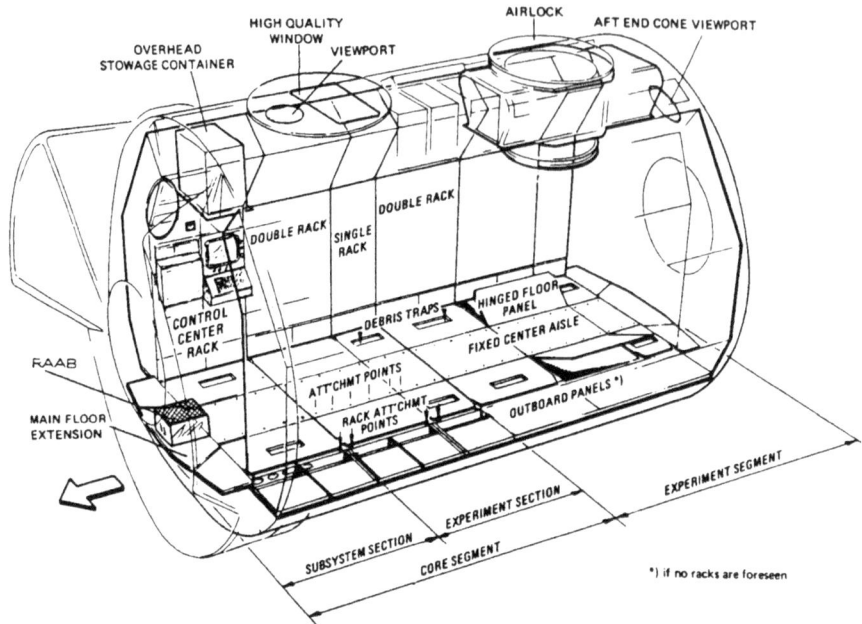

A typical rack internal accommodation layout.

behind the center of volume of the payload bay…As we didn't have a tail on the damned thing, we had very little capability to move the center of mass and maintain stable, manageable flight. As a consequence, a good part of the front end of the payload bay is virtually useless." In other words, it was necessary to position Spacelab towards the aft end of the payload bay in order to bring the payload's center of mass closer to that of the Orbiter for better aerodynamic stability during re-entry.[14]

In addition, NASA had specified that the tunnel must run along the floor of the payload bay in order to satisfy two essential requirements. Firstly, it would permit the installation of an external airlock. With the middeck airlock attached to the tunnel, it was essential that astronauts be able to exit to make a contingency spacewalk in case of a problem in the payload bay. Secondly, it would be possible to install lightweight payloads on structures that spanned the volume above the tunnel.

Although an external airlock would enable astronauts to attend to experiments in the payload bay, either on unpressurized Spacelab pallets or carried independently, it was soon

---

[14] An internet search will easily show that most payloads were installed towards the rear of the payload bay, leaving the front end either empty or with just a lightweight payload.

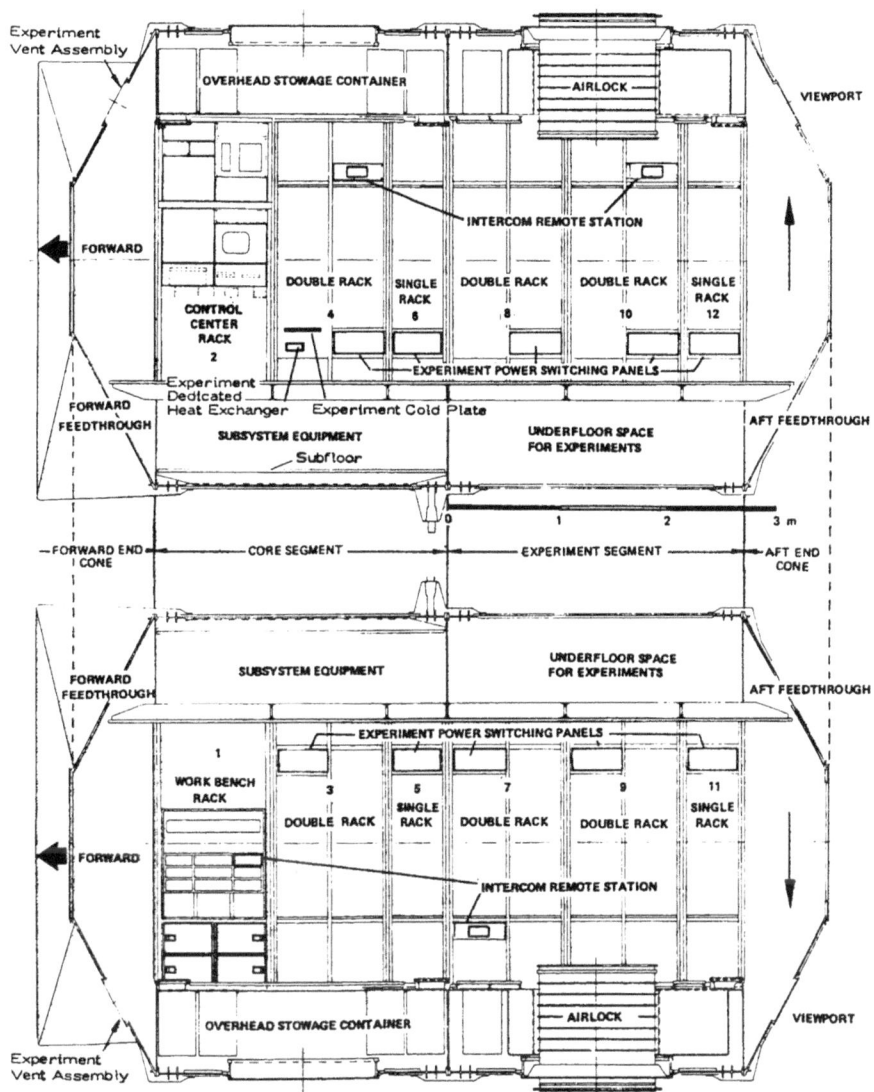

A sectional view of a Spacelab pressurized module.

realized that undertaking spacewalks in parallel with activities in Spacelab would impose too great a demand on consumables. For this reason, the concept of an external airlock was discarded. As a matter of fact, none of the pressurized Spacelab flights ever needed EVA as part of the manifested scientific work. Nevertheless, the requirement for contingency EVA capability remained. To achieve this, the so-called removal tunnel adapter, also

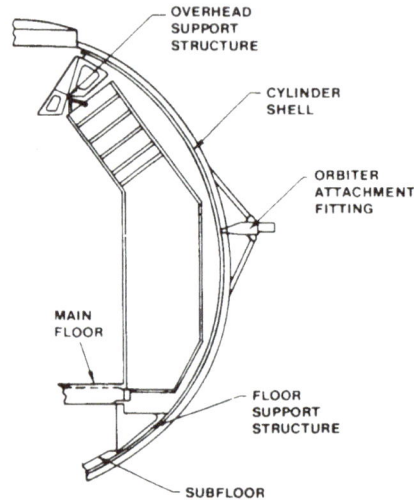

The rack support structure.

A group of racks is being slid inside a pressurized module during ground processing. The tight fit inside the module is clearly evident.

known as the forward extension, was added close to the forward bulkhead of the payload bay. This had a hatch at each end and a third at the top. If an EVA became necessary, the spacewalkers would move from the middeck airlock[15] into the adapter and exit from the top hatch. As part of the preparations for such a spacewalk, research in Spacelab would be halted and the module vacated and sealed off.

The simplicity of the transfer tunnel configuration proved to be deceptive. When it was pressurized, it had a tendency to rotate and straighten out like a lawn sprinkler, imposing severe stress not only on its cylindrical wall but also on the joggle section. The addition of elliptical reinforcement rings provided the extra needed strength. To reduce the load transfer between the Orbiter and the pressurized module, two flexible toroidal sections were added at each end of the tunnel. However, their development proved arduous, with McDonnell Douglas struggling to find a structural solution that would withstand longitudinal deflections, torsion, shear, and misalignments. In the end, NASA hired Goodyear to develop a single convolute bellow using a composite material of two plys of Nomex cord embedded in Viton rubber.

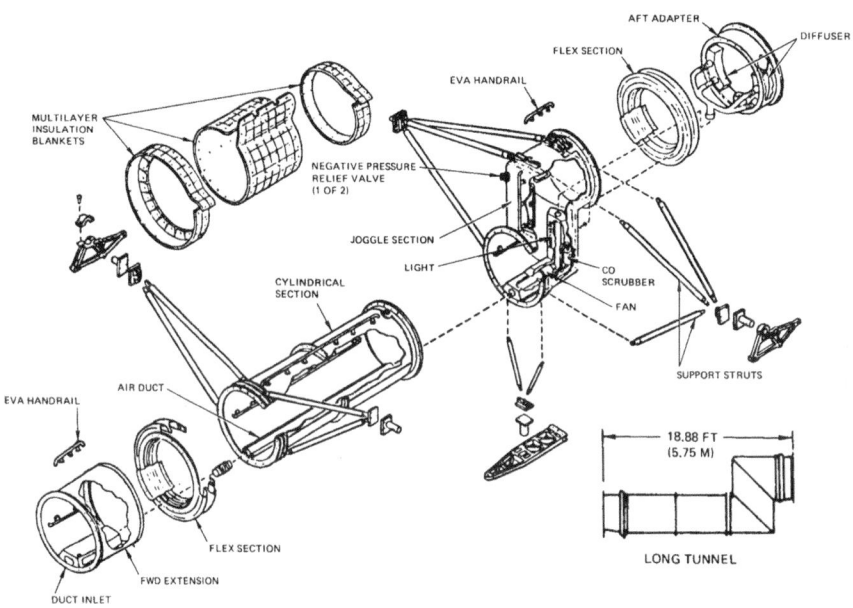

The main components of the Spacelab Transfer Tunnel.

---

[15] This means that the Orbiter's airlock was left in the crew cabin. The tunnel adapter was therefore a simple extension of the airlock and the top door simply replaced the airlock outer hatch.

A Spacelab pressurized module and tunnel being installed in the payload bay. The vertical joggle and tunnel adapter section are clearly visible at the aft and forward ends of the tunnel respectively.

To provide a shirt-sleeve working environment in the pressurized Spacelab, it was necessary to create an Environment Control and Life Support System (ECLSS). This comprised an Atmosphere Storage Control Subsystem (ASCS), an Air Revitalization System (ARS), and an Active Thermal Control System (ATCS). The ASCS was to draw gaseous oxygen from the Orbiter's reserve[16] and mix it at the correct pressure and flow rate with gaseous nitrogen from the laboratory's external tank supply. The ARS ensured that the mixture was evenly distributed within the habitable volume and provided the driving force to circulate the air for removing heat, humidity, odors and carbon dioxide before returning the air to the module. Carbon monoxide was instead scrubbed in the transfer tunnel, and avionics-generated heat was carried away by one of two dual-redundant fans which blew air over the electronic equipment. The water loop of the ATCS was activated in full mode

---

[16] Early studies envisaged Spacelab carrying its own oxygen bottles, as it was thought that the Orbiter's air revitalization system would not be adequate for both pressurized environments. However, as the Orbiter design evolved it became apparent that it would indeed have a surplus of oxygen that could be used to serve Spacelab as well. This expedient saved some precious weight from the pressurized module design.

on-orbit and in a reduced performance mode for ascent and re-entry.[17] Once on-orbit, and depending on the mission, heat from specific rack-mounted experiments was removed by a dedicated fluid loop that would dump its thermal load into the laboratory's water loop. The water loop would then release its load to the Freon coolant loops of the Orbiter. Heat generated by any experiment on an external unpressurized pallet was instead removed by a Freon loop which transferred its load to the water loop. In the case of an all-pallet configuration, this circuit would directly interface with the Freon coolant loop of the Orbiter.[18]

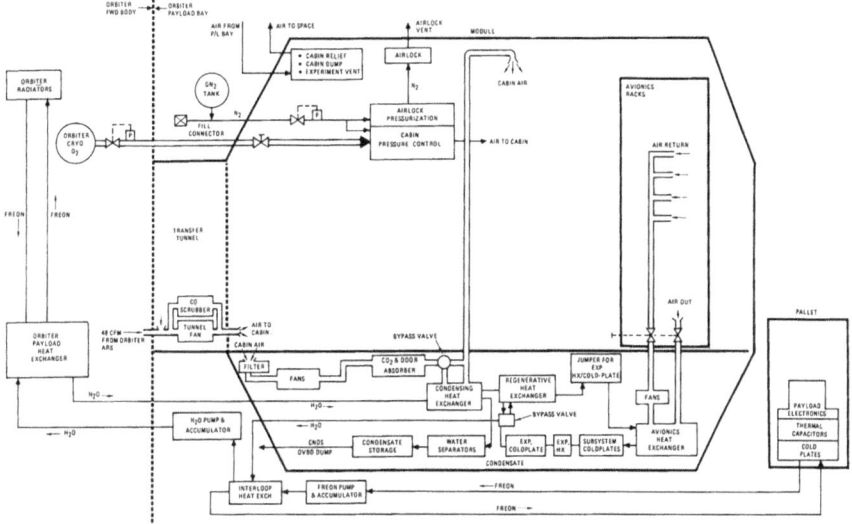

Schematics of the Spacelab ECLSS.

Power was provided by the Spacelab main bus which, in turn would normally be powered via one of the Orbiter's payload main buses. The system was to supply both direct current at 30 volts and 3-phase 400 Hz alternating current at 112/120 volts for the

---

[17] As ascent and re-entry (particularly the latter) imposed an intense heat load on the Orbit and the contents of the payload bay, it was necessary to equip Spacelab with a means of maintaining its internal temperature in an acceptable range in order to avoid damaging the avionics or spoiling the experiments.

[18] The initial configurations envisioned both the pressurized module and the pallet-only arrangement being pivoted out of the payload bay in order to allow surface mounted radiators adequate area and view angles. Other configurations had the module fixed in the bay and equipped with deployable radiators. Eventually, the design of the Orbiter's heat rejection system matured to a point where it had enough radiator surface and capability to deal with the thermal load generated by the pressurized module.

subsystems of the laboratory, and direct current at 28 volts and 3-phase 400 Hz alternating current at 115 volts for the experiments by way of a dedicated distribution network. The Orbiter would provide emergency direct current for Spacelab's critical environmental control subsystem sensors and valves, and its fire and smoke detection and suppression equipment, water lines, emergency lighting, intercom system, and caution and warning panel. If the power delivered to the module was degraded, the experiments would draw current from their own essential bus.[19]

To assure a safe working environment, caution and warning inputs were received by the Orbiter. This system could cater for a maximum of 36 safing commands[20] for the module's subsystems and 22 for its experiments. These included dealing with the most serious emergency that can occur on a manned spacecraft: fire/smoke and rapid depressurization. In either case, the caution and warning system of the Orbiter would sound an emergency tone simultaneously in the Orbiter and the Spacelab, in parallel with illuminating master alarms.

The complexity of the module and its experiments was managed primarily by the Command and Data Management System (CDMS). This included three identical IBM AP-101-SL computers. One was assigned to Spacelab experiments. It took care of activating, controlling, and monitoring payloads, and experiment data acquisition and handling. A second computer controlled and managed data for essential Spacelab experiment support services, such as power distribution and equipment cooling. The third computer served as a backup. The Data Processing Assembly (DPA) contained the software to process telemetry data, command data, and timing signals between the Orbiter, Spacelab subsystems, and experiments. It integrated software specific to the subsystems and the experiments, each with its operating system and applications. The software was stored in a mass memory unit tape recorder.

While the pressurized module was taking shape in Italy, the unpressurized pallet system was being developed by Hawker-Siddeley in Stevenage, near London in the UK. In order to maximize the usable volume of the Orbiter payload bay, a U-shaped aluminum frame and panel structure were chosen with provisions for four mountings on the payload bay longerons and one on the keel. The basic structure was made of five parallel U-shaped frames (four primary and one subframe) axially connected by a keel member, four longerons, and two sill members. Aluminum honeycomb panels of differing sizes would form the inner and outer load-bearing skin and the carriage of lightweight Spacelab subsystems and experiments, while larger payloads would be mounted on standard hard point assemblies at the intersections of the frames and the longerons.

---

[19] During the design of the electrical system, it was thought that power to run the Orbiter's subsystems while on-orbit would be adequately provided by just two fuel cells, freeing the third one to fully power Spacelab. This was a good news, because the European aerospace industry lacked experience in the development and design of fuel cells.

[20] A safing command allows a malfunctioning system or piece of equipment to be safed by either shutting it down, activating a redundant system, or downgrading it to a lesser performance.

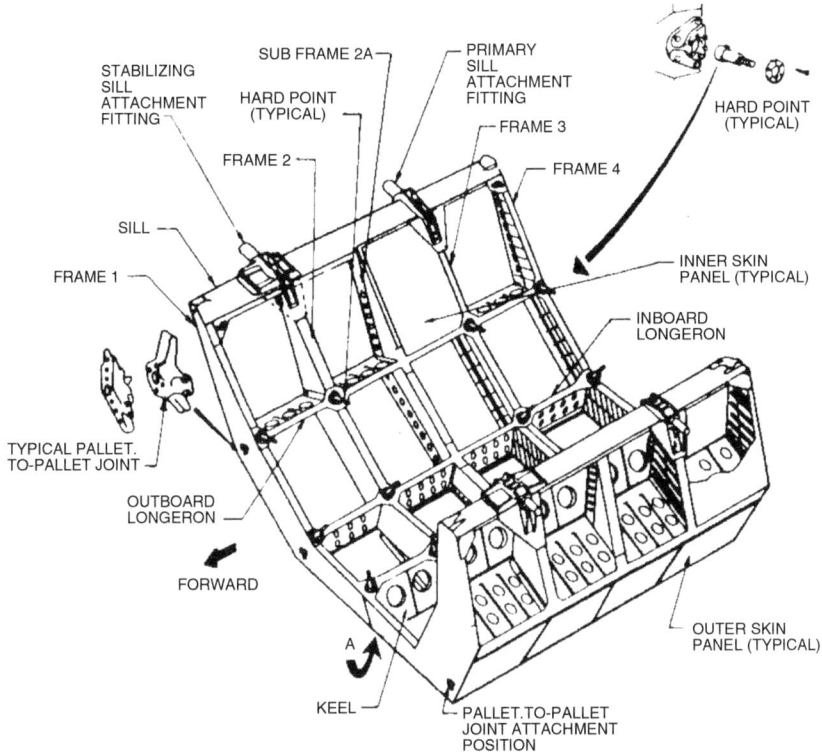

The Spacelab pallet structure.

Once again, the simplicity of such a structural configuration proved deceptive, as the designers pursued every possible weight saving measure. For instance, the face sheets of the sandwich panels[21] were very thin.[22] Local stiffening was not applied at hard points, nor in areas subjected to repeat installation. The structure was designed with the usual load factor of 1.4,[23] but to save weight they elected not to apply any reinforcement, even in areas that normally would be bolstered to improve durability. Fasteners were tightened

---

[21] A sandwich panel is a honeycomb core placed between two face sheets. This arrangement is equivalent to a solid I-beam but with the advantage of a lighter structure.

[22] The idea of replacing the honeycomb sandwich panels with laminate composite panels was soon rejected because of incompatibility of the coefficients of thermal expansion of the composite material and the underlying pallet aluminum structure.

[23] Simply put, this means the structure was sized to withstand an additional 40% of the maximum payload that the pallet was designed to house.

using extensiometers to apply the maximum possible torque and so derive every last ounce of strength and minimize the number of fasteners. These expedients delivered a considerable weight saving, but easiness in hardware handling and integration were both compromised. In fact, it was soon realized that fasteners would require to be replaced after each flight, that handling on the ground would need to be done very carefully to avoid damaging the delicate pallet structure, and that more complicated structural analysis had to be made for every flight than had been initially planned. Nevertheless, the pallets would play a key role on some of the most exciting Shuttle missions flown, carrying a wide variety of payloads.

Spacelab pressurized modules and unpressurized pallets gave the Shuttle program flexibility in mission planning and configuration that NASA had probably not even thought of when it first began to consider sortie missions. The modularity gave the flexibility to fly missions with either only the pressurized module (in its short or long configuration) or only pallets (a maximum of five with up to three rigidly connected), or any mixture to accommodate a mission's needs. In one way or another, the large volume of the payload bay of the Shuttle could be exploited to the fullest in order to maximize the scientific return.

SMALL PAYLOAD CARRIERS

As early as mid-1970, when NASA had already begun to define the payload manifest for the first Shuttle flights, it was realized that none of the primary payloads would be able to fully exploit the mass and volume capabilities of the payload bay. It became apparent that there was an opportunity to fill this leftover potential by offering to the science and university communities the opportunity to develop small payloads that could be flown very cheaply.

Hence in October 1976, John Yardley, Associate Administrator for the Office of Space Flight at NASA headquarters, announced the Small Self Contained Payload Program at a convention of the International Astronautical Federation that was held in Anaheim, California. The objectives were to encourage the use of space, enhance education with hands-on space research opportunities, inexpensively test ideas which could later grow into major space experiments, and generate new activities unique to space. The key requirement was that the experiments had to be self-contained. They could not rely upon the Shuttle for power or data transmission, and crew interaction would be limited to switching the hardware on and off. They had to incorporate their own power source, thermal control, data recording, and master event sequencer.[24] In addition, they would have to conform to any orbital or pointing requirements set by the primary payload(s) in the bay. Nevertheless, if special conditions were necessary, NASA would strive to assign the experiment to a flight that could closely match the experimenter's needs.

---

[24] In fact, telemetry would not be available either. Apart from being switched on and off by the crew, the experiment would have to run completely without supervision.

Despite the limitations, the program was enthusiastically accepted and the agency began to receive orders. For a rather modest price, an experiment could be flown at fairly short notice and be reflown to gather additional data or to test a modification to the apparatus. Clearly, new technologies could be rapidly tested in space and applied to their field of interest. Paperwork was minimized by developing standard interfaces and generic documentation to reduce the requirements placed upon the user. The user had to provide a detailed description of the experiment, its objectives and its design, and prove it would comply fully with the safety requirements for the Shuttle, crew, and other payloads.

In early 1977 responsibility for developing the technical specifications and flight hardware was bestowed upon the Sounding Rocket Division at the Goddard Space Flight Center in Greenbelt, Maryland, which was logical because the self-contained payloads closely resembled those flown on sounding rockets. Based on their 20 years of experience, the Division produced a suite of units collectively known as Attached Shuttle Payload (ASP) carriers.

GET AWAY SPECIALS

The first ASP was the so-called Get Away Special (GAS). In general terms, a GAS was a pressurized aluminum cylinder (referred to as a canister) which fully housed a self-contained experiment. Two canister sizes were offered in order to suit apparatus of differing mass and volume. The larger canister had 5 cubic feet of useful capacity in an envelope 19.75 inches in diameter and 28.25 inches tall. It could accommodate up to 200 pounds of equipment. The smaller container had the same diameter, but a user envelope only half as tall and capable. The experiment was to be mounted on the internal surface of the canister's top plate, which had a pattern of threaded inserts to account for a variety of hardware configurations. However, in many cases apparatus was installed in a rack attached to the top plate to exploit the available volume to the fullest and enable several experiments to share a canister (and also share the cost). To provide lateral stabilization and prevent the rack from contacting the sidewall of the canister, an adjustable bumper on the bottom end of the canister retained the free end of the rack.

The bottom end plate of the canister was reserved for electrical connection ports and pressure sensors, these being necessary to interface the container with the Orbiter and enable the crew to activate/deactivate the contents. The bottom end plate and the sidewall were both thermally insulated and shielded from radiation. Protection on the top plate was optional, depending on the experiment.

The GAS canisters were designed to maximize flexibility and accommodate the greatest variety of experiments. The internal pressure could be varied from vacuum to almost terrestrial pressure, and the container could be filled with either dry air or nitrogen. For experiments needing direct exposure to the vacuum of space or to view out through a window, a lid was created for the top plate that could be opened and closed on-orbit. An antenna top plate was also developed for investigations based on radio amateur communications.

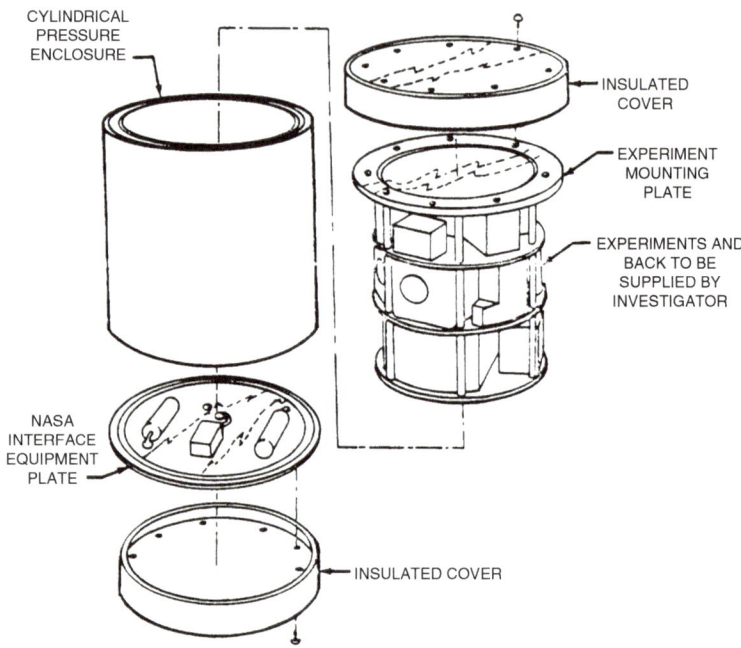

The Get Away Special container concept.

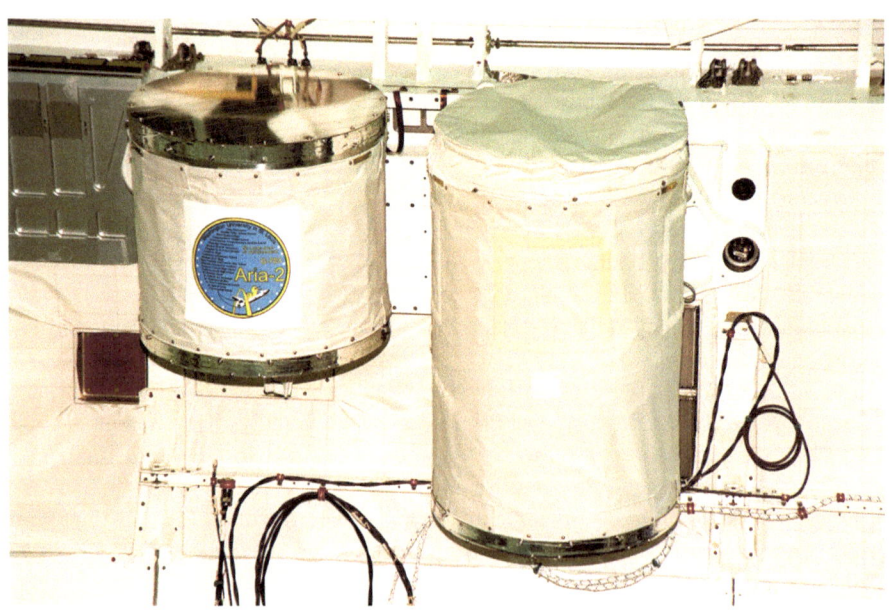

A pair of GAS containers in the payload bay.

The first canisters flew as early as STS-4, and the backlog of experiments waiting to be carried grew so fast that a so-called GAS bridge was rapidly designed and first flown on STS-61C. This was a truss structure spanning the full width of the payload bay and was capable of carrying between 5 and 12 canisters distributed along the two sides.

The GAS program was abruptly halted by the tragic loss of *Columbia* in February 2003, after about 200 canisters had been flown, it having been determined that from then on the Shuttle would be exclusively dedicated to completing the assembly of the International Space Station.

HITCHHIKER PROGRAM

Leveraging on the highly successful GAS program, in early 1984 the Office of Space Flight initiated the development of a new class of attached Shuttle payload carriers, named Hitchhikers. The primary objectives were to reduce flight lead times, increase opportunities for re-flights, reduce integration costs, and maximize the load factors of the Orbiter. Hitchhiker payloads were allowed access to the Shuttle's resources, such as power, real-time telemetry and ground command services in order to monitor and control the experiments, so long as this did not interfere with the primary payload(s). To address the program's objectives, two different platforms were developed, known as Hitchhiker-G and Hitchhiker-M. The former was developed by the Goddard Space Flight Center to carry up to six payloads weighing a total of up to 750 pounds. The latter was prepared by the Marshall Space Flight Center to carry up to three payloads with a total of 1,150 pounds.

The Hitchhiker-G (HH-G) platform consisted of aluminum experiment mounting plates that were secured to one or two GAS Adapter Beam Assemblies (ABA) which spanned the length between two consecutive payload bay frames. Experiments whose size or weight were not suitable for a GAS canister could be mounted on such plates, as indeed could GAS canisters themselves. Typically these platforms could support experiment packages of up to 150 pounds on 25×39-inch plates. They presented a grid pattern of holes for hardware installation using stainless steel bolts.

The HH-G platform could also accommodate the Shuttle Payload of Opportunity Carrier (SPOC), a system that was developed to support non-Hitchhiker experiments that were manifested as primary payload. The SPCO was designed to be modular and expandable, in accordance with payload requirements, to allow maximum efficiency in the use of Orbiter resources. The SPCO was a large 50×60-inch mounting plate to carry an avionics unit which had standard electrical and hardware interfaces between the Orbiter and up to six experiments. The unit also connected to a switch panel on the flight deck so that the crew could activate and deactivate the payload and impart control commands if there was a keyboard/display. The plate included facilities such as heaters and thermal blankets in addition to those of the experiment hardware. The SPOC could not be mounted directly on the GAS ABA; it required an additional adapter plate. As an option designed to maximize the payload mass, it was possible to mount an experiment directly on the ABA, eliminating the need for an experiment mounting plate. This method could accommodate payloads of up to 700 pounds, but it necessitated a detailed case-by-case structural and safety analysis and approval.

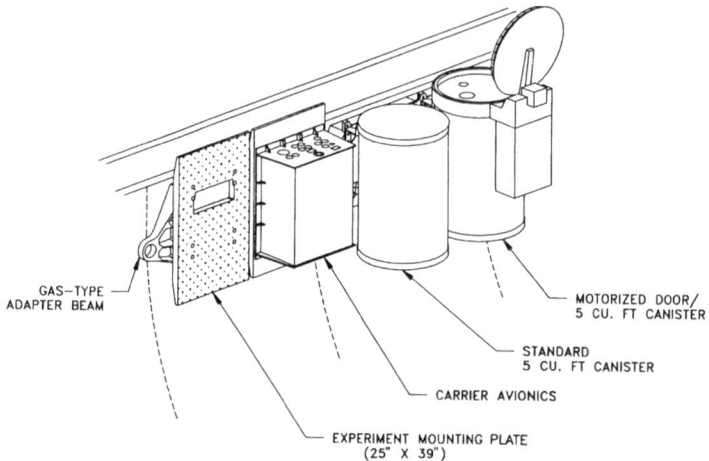

GAS–TYPE
ADAPTER BEAM

MOTORIZED DOOR/
5 CU. FT CANISTER

STANDARD
5 CU. FT CANISTER

CARRIER AVIONICS

EXPERIMENT MOUNTING PLATE
(25" X 39")

The carrier components for Hitchhiker-G.

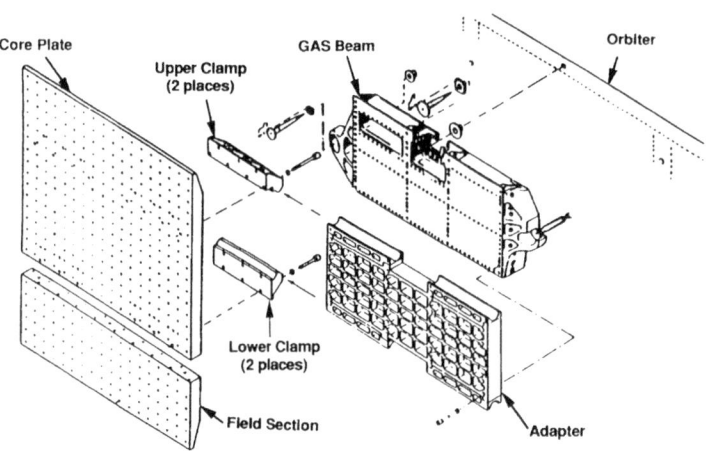

Core Plate

Upper Clamp
(2 places)

GAS Beam

Orbiter

Lower Clamp
(2 places)

Field Section

Adapter

The Hitchhiker-G SPOC plate structural assembly.

The larger Hitchhiker-M (HH-M) platform was a carrier based on a truss structure called the Hitchhiker Bridge Assembly (HHBA). It consisted of an upper and lower support structure that spanned the full width of the Orbiter's payload bay. The former contained the Mission Peculiar Equipment (MPE) consisting of a series of structural elements designed to accommodate the experiments, and it was fitted onto the lower structure once the Shuttle was on the launch pad. Once again experiments and even GAS canisters were attached to aluminum plates arranged on the top and sides of the upper structure. An SPCO avionics unit could be added, if required. Special brackets were used to mount the experiments on the plates because both the HHBA and the MPE were uninsulated and hence could experience significant temperature variations in space. If the experiments and mounting plates were temperature controlled, these brackets afforded thermal isolation and thermal expansion. As with the HH-G, direct mounting of experiments on the carrier structure was permitted, thus eliminating the requirement for mounting plates, but such experiments had to be designed to safely accommodate larger differential temperature changes between the equipment and the carrier.

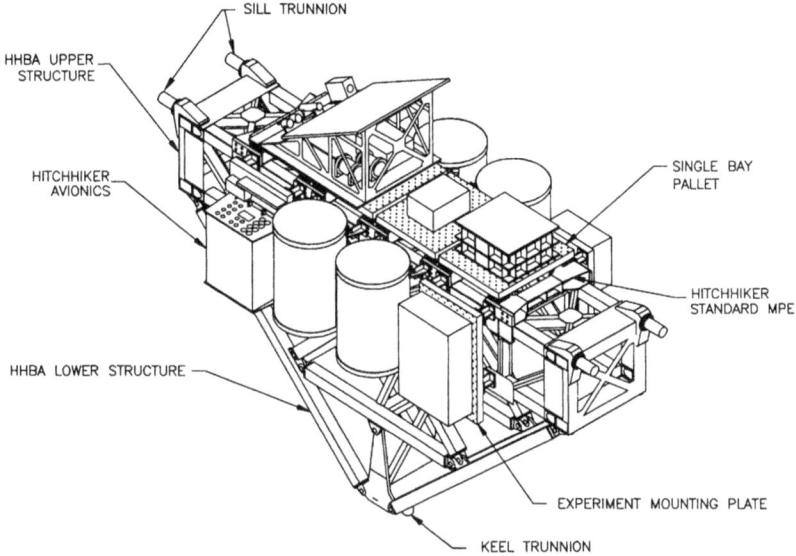

The Hitchhiker-M structural assembly.

The first Hitchhiker payloads were assigned to STS-61C in 1986. It carried three experiments of interest to NASA to study the distribution of particles in the Orbiter environment, test a new heat transfer system utilizing a capillary loop, and determine the effects of contamination and atomic oxygen on ultraviolet optics. The HH-M was introduced in April 1991 by STS-39. Hitchhikers were last flown on *Columbia*'s ill-fated STS-107 mission in 2003.

Although the Hitchhiker and GAS programs may appear rather similar, they were actually very different. The GAS program was simple, designed to allow experiments to be flown in canisters of two possible standard sizes, with no involvement from the crew except for activation and deactivation and without drawing upon the resources of the Orbiter. The Hitchhiker platforms instead allowed investigators the possibility of choosing between using GAS canisters or mounting plates for larger experiments whose equipment could not fit into a GAS container. Furthermore, and irrespective of an experiment's mounting place, resources were provided to allow monitoring and control either by the crew or by the investigator's own ground control center, or both.

Clearly the Hitchhiker program created a much needed bridge between the simple GAS-based experiments and the much larger and more sophisticated Spacelab-style investigations. They allowed the Shuttle to offer a broad range of flight opportunities for science and technology research while at the same time exploiting to the fullest the capabilities of the Orbiter.

7

Spacelab Stories

"AN UNQUALIFIED SUCCESS"

We could easily fill thousands of pages simply summarizing the scientific work and hundreds of experiments carried out in 30 years of Shuttle flight operations, which is clearly beyond the scope of this book. In this chapter, therefore, we will look at some of the science conducted on some Spacelab missions as a meaningful sample of how the Shuttle was able to morph into an orbital research facility and contribute to man-tended science on-orbit.

Spacelab's maiden flight was assigned to STS-9 by *Columbia* and included a mix of one long pressurized module and one unpressurized pallet, the latter located at the very rear of the payload bay. But this was not the first time that a Spacelab pallet was flown in space, because a simplified version carried small experimental packages on STS-2 and STS-3. These were the engineering test articles being used to validate the structural design of the pallet and to enable NASA engineers to acquaint themselves with ground handling and integration with the Orbiter. Although these items were not designed to be space-worthy, ESA accepted a NASA proposal to recondition them to carry a small selection of experiments and apparatus on the orbital flight tests. STS-9 was the first time that a fully space-worthy Spacelab was flown, so it is reasonable to call it Spacelab's first mission.

For this inaugural flight, a complement of 73 experiments was spread out across all of the Spacelab segments, carefully selected to put the payload bay laboratory through its paces and assess its merits. Together they were to perform investigations in the realms of astronomy, solar physics, Earth observations, technology, materials and life sciences. But the science was not the primary objective. STS-9 was to be a fully fledged test to verify the thousands of structural, mechanical, and electronic parts of the Spacelab, to validate its systems, their performance, their compatibility with the experiments and the Orbiter, and to measure its environment in the payload bay. In addition to the standard operational instrumentation of the Orbiter, some 200 sensors had been installed in the laboratory and *Columbia* as part of the Verification Flight Instrumentation (VFI) to validate the module's environmental control system, structure, command and data management, electrical power

© Springer International Publishing AG 2017
D. Sivolella, *The Space Shuttle Program*, Springer Praxis Books,
DOI 10.1007/978-3-319-54946-0_7

distribution, habitability, and compatibility with the space environment, including verifying that contamination would not degrade sensitive optical surfaces and other experiments.

The Spacelab-1 module and pallet ready to be installed in the payload bay of *Columbia* at the Kennedy Space Center. To the right is the pallet housing those experiments that required exposure to space.

Launched on November 28, 1983, STS-9 immediately began to set one first after the other. It was the first Shuttle to use an orbit inclined at 57° to the equator, the first to carry a payload of 33,070 pounds, and with a planned duration of 9 days it was to be the longest mission to date.

Within 3.5 hours, the crew of six had activated and entered Spacelab. They ran it continuously for the next 231 hours. While performing experiments and component testing, *Columbia* and Spacelab underwent more than 200 attitude changes designed to put the duo into the most demanding possible conditions for thermal and structural stress. For instance, from flight day two to flight day four the so-called cold test was undertaken by continuously facing the payload bay to deep space. The opposite hot test was carried out on flight day eight by pointing Spacelab to the Sun and flying an orbit that was continuously exposed to sunlight without passing through the Earth's shadow. During (and in between) these two tests, *Columbia*'s attitude was frequently altered to accommodate observation requirements of the experiments on the external pallet, since they lacked their own gimbaling mechanism. This offered the chance to assess the Shuttle as a platform for future operations involving astronomy and Earth observations.

As the days went by, the astronauts reported that Spacelab was a comfortable and functional environment in which to work, not at all unlike a ground-based laboratory. The onboard systems and experiments performed so well that by flight day four an extension of 1 day was already being discussed, with approval being granted 2 days later. The seals were so good that a lower-than-expected leakage rate had reduced the inflow of oxygen from the Orbiter to Spacelab. Carbon dioxide removal was better than initially expected; in fact, the lithium hydroxide canisters in the Orbiter's cabin were able to pick up the load without requiring any support from those available in the laboratory. Furthermore, the power consumption was lower than expected and as a precaution there was extra cryogenic hydrogen and oxygen for the fuel cells. All of these factors justified extending the mission.

Spacelab's thermal environment proved to be within predicted range, as both the active and passive control systems ran without any recorded anomalies.[1] In fact, the astronauts reported that the atmosphere in the module was fresh and the temperature cool and comfortable. The relative humidity levels remained below 40% throughout, and there was no condensation. The scientific airlock functioned flawlessly. There were 22 opening and closing sequences of the inner and outer hatches and six cycles of depressurization and repressurization. The experiment table was operated 23 times in support of two experiments and the hot and cold tests. Post-flight data verified that the mechanical and thermal stresses transferred between the Orbiter and the module were within design assumptions for launch, landing, and all the attitudes employed in flight.

Although the science work was secondary for this mission, the post-flight science report included an impressive list of achievements. For instance, the Fluid-Physics Module (FPM) rack discovered a new reformation mechanism for free-floating liquid zones, and unexpectedly revealed a slow demixing of oil and water in microgravity. The behavior of fluids in closed containers was investigated, and this helped improve understanding of fluid sloshing in spacecraft carrying large quantities of propellant or coolant. The rack also demonstrated for the first time the existence of the Marangoni convection in space.

This was also the first opportunity to carry out extensive investigations into life sciences. For example, the importance of the eye as a source of orientation in zero-gravity and the coupling between the eye and the vestibular system were studied. It was determined that in weightlessness the eye takes over a major role in the body's orientation system. Observations were made to investigate changes in spinal reflexes and postural behavior in weightlessness. And an experiment in mass determination in space showed that the crew adapted more readily to their environment than had been predicted.

A variety of spectrometers yielded sensitive and accurate measurements of minor constituents that nevertheless play a significant role in the chemistry and dynamics of the Earth's atmosphere. The discovery of carbon dioxide, water and methane in the mesosphere, a precise altitude profile for ozone, and a quantification of the amount of deuterium in the thermosphere were all significant achievements. The remote sensing instruments

---

[1] It is worth recalling that the active control system made use of water and Freon loops and the passive one adopted multi-layered insulation blankets, painted white, plastering the entire outer surfaces of both the module and the pallet.

measured atmospheric composition on a global scale with unprecedented accuracy, and a core set of reference measurements of the atmosphere were obtained.

In astronomy and solar physics, more than 90% of the planned observations and objectives were achieved. These included the high-resolution measurement of X-ray lines from stellar sources, an apparent far-ultraviolet background, and measurements of the solar spectrum and the solar constant (which, contrary to its name, varies). In space plasma physics, some 88% of the objectives were attained, including studies of how the electrical charge on a vehicle is neutralized by magnetoplasma dynamics and neutral gas plumes, observation of beam-plasma discharge phenomena, generation of energetic return electrons during beam firings, combined experiments using multiple instrumentations, and imaging of airglow layers in the upper atmosphere.

In the light of these achievements, on February 29, 1984, Samuel Keller, NASA's Deputy Associate Administrator for Space Science and Applications, reported, "The Spacelab-1 mission is judged a success. Demonstration of Spacelab as an orbiting laboratory is an unqualified success." This success was the result of careful mission planning and a competent crew.

As a matter of fact, planning for the first Spacelab flight began even before the first piece of metal was cut for the flight hardware. Shuttle astronaut Joe Allen, who was involved in the program, remembers, "It occurred to us that maybe we ought to [simulate] doing research in a volume the size of a Spacelab, in conditions that were not unlike living aboard the Shuttle…We set out to do this in a simulation called ASSESS…We used NASA research airplanes that were stationed out at the Ames Research Center, we confined research crew members to quarters similar to those quarters aboard what would later become the Shuttle, and we actually carried out real research. It had nothing to do with zero gravity, but it did have to do with confining research scientists in teams working for a period of about a week, eating and sleeping where you're working."

Standing for Airborne Science Spacelab Equipment System Simulation, ASSESS was a NASA-ESA joint program aimed at flying a Convair 990 aircraft whose cabin was fitted to reproduce as closely as possible the working environment that Spacelab would provide. Spacelab experiment racks and real experiments were installed on the aircraft and flown in two campaigns.

The first one, ASSESS I, started on June 2, 1975, and five 6-hour flights were made over a period of 5 days. ASSESS II, made seven flights between May 15–25, 1977. To make the simulation as close as possible to a real space mission, when they were not flying the two NASA and two ESA "astronauts" spent the rest of their day either onboard or in living quarters directly connected to the aircraft. Irrespective of the science undertaken by the ASSESS program,[2] its real purpose was to improve understanding of how a Spacelab flight should be planned and organised. The flights also confirmed the feasibility of carrying out a broad array of diverse experiments by astronauts who were not the principal investigators. In fact, they provided insight into how astronauts could interact with the principal investigators on the ground during a flight. They also helped in establishing guidelines for experiment development, crew training, and Spacelab operations. To ESA,

---

[2] In fact, to make the flights even more meaningful and the simulation more realistic, all experiments had to be real and part of genuine scientific investigations.

this program was a veritable playground to exercise management and integration schemes in the framework of manned space flight.

The second element that afforded Spacelab-1 its outstanding success was STS-9's crew of six, the largest to date. Commanded by the legendary moonwalker John W. Young, the make up of the team established another first. To understand why, it is necessary to recall that the first people to fly in space were jet pilots, and in the case of the Americans they were test pilots with thousands of flying hours logged. It was thought that only highly disciplined people accustomed to "pushing the envelope" of dangerous high performance flying machines would be able to handle the unknowns of the emerging experimental field of human space flight. In April 1959, September 1962, and October 1963, NASA recruited three groups of astronauts. All candidates were required to have at least a college or advanced degree, but the primary requisite was experience of military flight test of jet fighters. For the October 1963 group, the requirement was relaxed to experience of flying military jet fighters, thereby opening the way for operational pilots.

In June 1965 a brand new category of astronauts were recruited. Although four of the six had served in the military, they had been selected only on the merit of their academic background and research experience. In fact, they all had the title of doctor preceding their names. Unsurprisingly, they were labelled as the first of the scientist-astronauts. In October 1967 a second group was selected on the basis of academic merit, this time with 11 members. Although the Apollo program was still a few years away from landing on the Moon, post-Apollo plans were already envisaging Earth-orbiting space stations devoted largely to scientific research in space. These would be mainly operated by scientist-astronauts who would apply their research skills and academic knowledge to maximize the performance of the experiments developed by principal investigators on Earth.

By the last splashdown of an Apollo capsule on July 24, 1975, only four scientist-astronauts had flown in space; Harrison H. Schmitt walked on the Moon on the final Apollo lunar flight and Joseph P. Kerwin, Owen K. Garriott and Edward G. Gibson served aboard the Skylab space station.

The Shuttle program brought hope that the other scientist-astronauts would finally get assignments as mission specialists. This started with the first operational mission, STS-5. Unlike the Apollo capsule that was designed to carry three astronauts,[3] the Shuttle could easily accommodate up to seven crew.[4] With two pilots designated as drivers,[5] there were seats for missions specialists who would not only take care of the Orbiter systems but also perform nominal and/or contingency spacewalks, deploy and recover satellites, operate the robotic arm, conduct experiments, and so on. With the coming of Spacelab and the

---

[3] A modification intended to rescue a crew stranded aboard Skylab would have enabled the Apollo capsule to launch with two astronauts and return with five.

[4] In fact, several times a Shuttle flew with a crew of eight astronauts. For contingencies such the rescue of a stranded Shuttle crew, it could land with up to eleven people.

[5] Some people referred to them as "the dumb guys in the front" because all they were required to do was to launch, keep the Orbiter running in space, and fly it home, not too dissimilar to a bus driver. For this reason and taking into consideration the projected annual flight rate, it was expected that Shuttle pilots would fly very often.

projected use of the Shuttle for an array of scientific missions, NASA decided that yet another class of spacefarer ought to be allowed on board.

As astronaut Paul Weitz explains, "At the time it was felt that the best way to get the most 'bang for the buck,' as far as experiments (we call everything experiments, whether it is evaluation, data-gathering or on-site analysis, whatever) was to ideally – Let's say you're going to take up this new astronomical sensor for spectrographic or radio emissions or what have you, and that some guy had spent from the time he was a graduate student right through his 10-year tenure at a university and developed this detector. The most reasonable thing to do was to take that person up with it, then let them operate it." The concept of payload specialist was born. While they would have the privilege of flying in space, their role on a mission would be purely to carry out experiments and research. They would operate devices they themselves had designed or on behalf of a project for which they received appropriate training by the principal investigator. As Weitz continues, "That person wouldn't be a NASA employee. We wouldn't have to keep them. All we'd do is train them to not touch anything, any switch or any control in the Orbiter."

On any given flight, mission specialists would assist in carrying out the scientific work based on their academic knowledge and training. The payload specialists would have to carry out the bulk of the experimental and research activities. "So that was kind of the basic approach to payload specialists," Weitz concluded. "Each time we flew that package, or some version of it, then that person would go fly again."

Given their limited responsibilities, payload specialists would be given only a few hundred hours of training, just sufficient for them to learn how to live and cooperate with the professionals in space.

However, career astronauts were unimpressed when NASA began to advertise the advantages of payload specialists to attract the interest of industry and the scientific community. The main issue was the unknown of how a payload specialist with only basic training would behave when faced with the realities of space. As Henry W. Hartsfield, who commanded two missions which included payload specialists, says, "You had to determine their personality. It was hard to do in a short period of time. Were they going to be stable? If you had a problem on-orbit, am I going to have to babysit this person, or are they going to be able to respond to an emergency situation and take care of themselves like the crew has to? Because it could be detrimental. If you had a big problem, you could wind up having a person that wasn't used to that kind of condition endanger the rest of the crew because you have to attend to them… You're getting a person that you don't have a lot of experience with. Everybody you fly with, from a crew standpoint, you've worked with for several years. Everybody knows their personalities and their quirks. They had received a thorough psychiatric evaluation before they even got selected. We put them in hazardous situations. Those that weren't pilots that fly [as passengers] in the T-38s, you get to watch them. If you fly for long enough, you're going to have some contingencies, and you see how they react to that. I mean, you're building the database that this is a good, reliable person and you can count on them, because you've got to count on each other. Everybody's got a job to do, especially in contingency. Everybody's got something that is their responsibility to carry out. And you've got to have that confidence to make the flight successful." To further highlight the seriousness of the issue, Hartsfield says, "Early on when we were flying payload specialists, we had one that became obsessed with the hatch. 'You mean all

I've got to do is turn that handle and the hatch opens and all the air goes out?' It was kind of scary. Why did he keep asking about that? It turned out it was innocent, but at the time you don't know. We had some discussions, so we began to lock the hatch, carried a lock. Once we got on-orbit, we locked it, because you're not going to open it on-orbit."

Another reason for discontent, especially among the mission specialists, was that they had spent years waiting for an assignment, and trained very hard, yet a payload specialist could be added to a flight at short notice, reducing the number of seats for mission specialists. As Jerry L. Ross recalls, "We were giving away seats, is the way we kind of saw it, to nonprofessional astronauts when we thought that the astronauts could do the [payload] jobs if properly trained." This issue was aggravated when the space agency decided that as part of selling commercial satellite launch services, its customers could designate one of their employees to fly on the deployment mission, making what was essentially only a joyride in space. As Hartsfield explains, "It was marketing…It was sort of like a payola kind of thing or a marketing gimmick. People would get a Shuttle ride just because their company was putting [up a satellite]…In fact, we even made the deal that if they weren't able to fly with [their payload], we'd get them a flight on another mission. It was kind of a ticket to fly anytime. I wasn't alone in the Astronaut Office. A lot of us didn't like that at all."

Except for Spacelab and several other missions, in the initial years of operations a number of joyriding payload specialists rode an Orbiter into space and spent most of their time enjoying the view out of the window. Their experimental tasks were rather minimal, often offering themselves as volunteers (or otherwise agreeing to serve) as guinea pigs for life sciences studies. After the loss of *Challenger*, NASA continued to fly payload specialist but much less frequently and with a different approach.

It is natural to sympathize with the career astronauts, but the flying of payload specialists ought to be analysed in the light of another point of view. Astronaut John M. Fabian offers this reflection, "I felt like the Shuttle was giving us the capability to expand the flight environment to non-career people, and we ought to take advantage of that."

The delicate design of the Shuttle meant it could not withstand high accelerations during launch and re-entry for which the Mercury, Gemini and Apollo capsules were designed. In fact, the maximum acceleration underwent by the Orbiter during launch was only 3g's and for re-entry a mere 1.5g's. This contrasts with the 8g's endured by Apollo astronauts at both launch and re-entry. While an Apollo astronaut had to be in impeccable health and undergo stressful physical training, Shuttle astronauts would no longer be required to possess the health of an athlete; only to be in good physical condition. Because ordinary healthy citizens would be able to fly on the Shuttle, this would open up a new era of manned space exploration. At least that was the plan.

Two of the STS-9 crew were indeed payload specialists, both selected by a panel composed of the principal investigators whose experiments they would be operating. Aged 35, Dr. Byron K. Lichtenberg was the youngest member of the crew. An MIT biomedical researcher, he went into the history books of human space flight for being the first person to operate in space an experiment that he helped to develop and that he would analyze after the flight.[6] In fact, even prior to being selected to fly, he was in a research group which

---

[6] Up to STS-9, astronauts had operated and performed experiments as a result of the training they received, but not because they took part in the development process.

designed a Spacelab experiment to better understand how the otoliths, the vestibular organ of the inner ear, adapts to weightlessness. During the flight, Lichtenberg ran several other tests, both on himself and fellow crew members. The second payload specialist was Ulf Merbold, who represented another first for the Shuttle program. A doctor of physics specialized in materials, he was the first non-American to fly aboard a US spacecraft.[7] His inclusion in the crew resulted from an agreement with ESA that a European researcher would fly on the first Spacelab.

To carry out such a complex mission, the crew were divided into two shifts. Each shift tended to the Orbiter and Spacelab for 12 hours. This was another first. While one shift was busy in Spacelab, the other could enjoy well-deserved time off and rest in the Orbiter cabin, which for the first time was fitted with three bunk-type sleeping stations that were stacked vertically. Each bunk came complete with an individual light, communications station, fan, sound suppression blanket, pillow and sheets with microgravity restraints. On a mission where there would be some astronauts working while others were resting, the bunks offered a private and isolated space for sleeping. Later missions, in particular those on which the crew worked around the clock, made extensive use of these amenities. If they could not be manifested owing to apparatus occupying the middeck, the astronauts would attach sleeping bags to any convenient section of cabin wall.

"THE CHEAPEST EXPERIMENT THAT HAS EVER GONE INTO SPACE"

The memorandum of understanding signed by ESA and NASA on September 24, 1973, said that two qualification flights would be necessary to declare Spacelab ready for operations. As initially intended by the flight manifest, Spacelab-1 would test the pressurised module and one pallet configuration, while Spacelab-2 would put the all-pallet configuration through its paces. However, several issues with the Orbiter fleet, payload changes and resultant reshuffling of the flight manifest, and some problems with the supporting hardware for Spacelab-2 meant that Spacelab-3 was rescheduled ahead of the pallet-only flight.

On April 29, 1985, STS-51B entered a 215×219-mile orbit at an inclination of $57°$ to start the 7-day Spacelab-3 mission. For this first operational Spacelab flight, the payload bay of *Challenger* carried the same pressurised module as was employed almost 18 months earlier. This time, however, the scientific focus was narrower. The pressurized module had hardware for 10 investigations in materials and life sciences, fluid mechanics, and astronomy. There was other equipment in the Orbiter cabin for three further investigations in life sciences and one in astronomy. There were also two instruments for atmospheric and astronomical observations in the payload bay on an MPESS platform.

On Earth, materials processing and fluid dynamics are inevitably affected by the presence of gravity, and this often has a detrimental effect on the processing of pure materials or studies of the intimate nature of fluidic phenomena. For example, it is gravity that causes convection and sedimentation during the fabrication of industrial crystals, and such

---

[7]It might be objected that Soviet cosmonauts flew in an American spacecraft during the Apollo-Soyuz Test Project flight, but Merbold was the first non-American to fly on an American spacecraft from launch to landing, as opposed to being temporarily invited aboard on-orbit.

distortions and impurities limit their quality and performance for electronic devices. The weightlessness environment of space eliminates convection and buoyancy, making it possible to produce purer, and perhaps new substances that can have real commercial applications on Earth. Subtle and weak diffusion processes during materials processing that are normally disrupted by gravity can be observed in all their complexity, and the insights thereby obtained applied to improve processing on Earth. Likewise, fluid dynamic studies in microgravity can disclose weak forces at play during drop formation and lead to a better understanding of atmospheric weather both on Earth and on other planets. In order to preserve the necessary microgravity environment, *Challenger* kept its tail constantly pointed towards Earth and aimed the starboard wing in the direction of travel. This established a low-drag gravity gradient attitude.[8] To produce an even cleaner microgravity environment, the most sensitive experiments were clustered together near the center of gravity of the Orbiter, which was the most stable part of the vehicle.

In addition to the seven humans on *Challenger*, there were four squirrel monkeys and 24 rats. The rats were loaded into cages 18 hours prior to lift-off. To get access to this facility far down in the payload bay, two KSC technicians used for the first time the Module Vertical Access Kit (MVAK), a system of ropes, pulleys, and platforms to transfer payloads from the middeck into the Spacelab module while it was in the vertical orientation on the launch pad.[9]

As part of the life sciences research, the mammals were studied to seek clues to human physiology and behavior in space. In this case, the primary objective of the four projects with the monkeys and rats was to observe a large number of animals in space in a uniquely designed and independently controlled housing facility. Scientists had the opportunity to assess the efficacy of the facility and make recommendations for future flights on which more extensive studies would be carried out with animals.

For Don L. Lind, being assigned as a mission specialist was a dream come true after a 19-year wait since his selection as an astronaut in 1966. On STS-51B he was responsible for supervising the proper functioning of Spacelab and its experiments. "I was the Laboratory Director…When the mission was being formalized, I thought, 'Now, there are several of those experiments that I'll be responsible for that will run in an automatic mode. I will have to check [on] them once an hour, but they may run for 2.5 hours in an automatic mode. So I'm going to have some free time in space.'" Rather than simply use such free

---

[8] By human standards, orbital space can be regarded as a pure vacuum. But it is not empty, it is occupied by dust grains, wandering molecules and atomic species. Despite the extremely low density of this environment, a vehicle traveling at orbital speed suffers a tiny but constant aerodynamic drag which, over time, alters its orbit. It also disturbs the microgravity environment, imparting a small but non-negligible acceleration which impairs the perfect balance between the pull of gravity and centrifugal force that is the very reason for microgravity research on a spacecraft. Maintaining the starboard wing in the direction of travel allowed *Challenger* to minimize the exposed frontal surface and hence drag, thereby preserving as much as possible the required microenvironment. The gravity gradient attitude will be explained in detail in Chapter 10.

[9] It was common for life sciences experiments, sensitive investigations, and delicate or perishable payloads to be loaded with the Shuttle already in the launch configuration, just hours prior to lift-off.

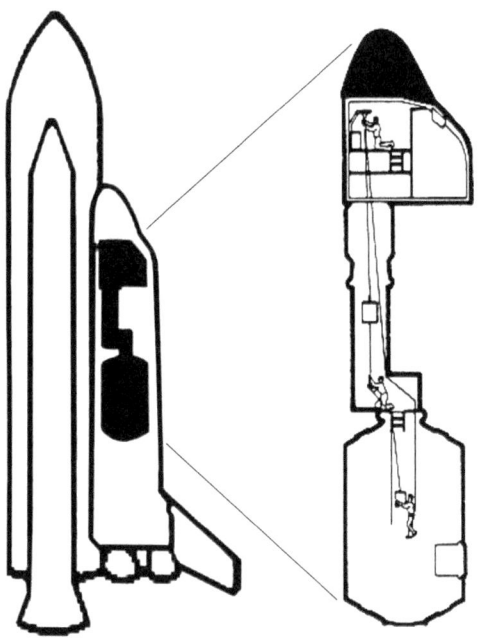

How to load a payload using the Module Vertical Access Kit.

time to enjoy the view outside, Lind knew he would be able to do something of benefit to the scientific community. The magnificent play of light of the auroras had been rarely photographed from space and the quality of the pictures was not very good. "Before our mission, the aurora had been photographed [only] by some slow scan photometers, which gives you a blurred picture. [It is] like trying to take a picture of a waterfall. [You] just [get] a white blur. Owen Garriott in Skylab had taken, I think, six or nine pictures of the aurora that were way off on the horizon." Together with a friend from the time he was in Alaska for his post-doctoral study, Lind set out to plan an experiment to photograph auroras in high resolution. "The first thing we wanted were high time-resolution pictures of the aurora with a TV camera. So we started looking around. What TV system could we get that would be sensitive enough in such a low light level? It turned out that the TV camera that was already on the Shuttle was as good as any TV camera we could've bought in the world. But we had to take off the color wheel and photograph in black and white, instead of color. So we got one of the cameras modified. Because now we would be photographing only in black and white, we wanted to take some still photographs in color to document the color the auroral light, since that would identify what particles were emitting the light. So we started to look around for an appropriate camera and camera lens. It turned out that the camera that we already had on board, and the lens we already had, were again as good as we could have gotten anywhere. NASA only had to buy three rolls of special sensitive color film. So this experiment cost NASA $36. It is the cheapest experiment that has ever gone into space." This experiment yielded outstanding results. "We found out that there is a different component to the mechanism that creates the aurora, involving microwaves that was not understood before. So the theorists had to add one more element in the equation for the creation of the auroral light…We claimed that we could do more science per dollar per pound than anybody else in the space program."

"THE MOST IMPORTANT SCIENTIFIC MISSION"

With the successes of Spacelab-1 and Spacelab-3 the made-in-Europe laboratory was almost open for business; only the all-pallet configuration had to be demonstrated. If successful, Spacelab would enable NASA to undertake any scientific program that it wished.[10]

Assigned to *Challenger* on STS-51F, Spacelab-2's launch date was scheduled for 4:30 pm EDT on July 12, 1985, with an ascent designed to put the Orbiter into a 208-mile high orbit at an inclination of 49.5°. With 6 seconds remaining, each of the three main engine on the Orbiter started its own ignition sequence. Three seconds later the launch sequencer initiated the second launch pad abort of the Shuttle program, in this case due to the shutdown of a coolant valve on all three engines.

At 5:00 pm on July 29, *Challenger* and its seven crew lifted off. But 5 minutes 43 seconds into the ascent the chances of accomplishing the mission suddenly took a turn for the worse. Two sensors on the center engine had detected higher than usual temperatures in the fuel high pressure turbopump, and when the limit of 950°C was exceeded the engine controller automatically initiated an in-flight shutdown and then an Abort To Orbit (ATO). ATO is one of the four so-called intact launch abort modes designed to allow a Shuttle to safely react to the loss of one or more engines during ascent.[11] In order for an ATO to be declared, the vehicle must have an altitude and speed that provide enough energy to reach an orbit at 105 miles. This is a lower than planned orbit, but sufficient to keep the Orbiter in space for at least one day, to allow Mission Control to determine whether the mission was viable, although perhaps with certain limitations.

While *Challenger* kept climbing on its two good engines, the orbital maneuvering engines were fired to control the center of gravity and to reduce the overall weight to improve performance. At 8 minutes 13 seconds, a temperature sensor on main engine number three[12] began to report the same issue as had occurred 3 minutes earlier, and the engine controller was about to enact the shutdown procedure. At this point, it was questionable whether Spacelab-2 would make it to orbit at all.

---

[10] It was NASA Administrator Dr. Fletcher who insisted that at least half of the projected Spacelab missions should be flown in the all-pallet configuration, in order to satisfy that part of the scientific community which felt a manned laboratory would not be necessary on board the Shuttle. It must be said however, that the idea of flying all-pallet missions had been outlined from the beginning. Nevertheless, the "rule" of flying such missions for at least half of the Spacelab program appears rather arbitrary.

[11] Having suffered a launch pad abort and an ATO abort, STS-51F hold the records as the only crew to have experienced two aborts on the same mission.

[12] This is the right engine as seen looking forward.

Only the smartest engineers are hired as flight controllers in Mission Control, and one of them recognised the absurdity of the situation. One engine shutdown may well happen, but two for the same cause was extremely unlikely. The crew were therefore told to inhibit the engine shutdown and to continue with the ATO ascent. This quick thinking saved the mission. Later analysis of the telemetry found that the turbopumps on the two "ailing" engines were far short of a temperature red limit. In fact, material recrystallization had caused the wiring within the thermocouples that measured the engine temperatures to become brittle and issue false data. The thermocouples were redesigned to avoid a recurrence.

Once it was verified that the mission could proceed, *Challenger* raised its ATO orbit to 170.7 × 169.8 miles. On opening the two large payload bay doors, Spacelab-2 could finally taste the emptiness of space. The bay was filled with 11 strange-looking telescopes and detectors carried aboard a Spacelab that consisted of one pallet and a double-pallet train.[13] In fact, the configuration for Spacelab-2 had been the object of discussion between ESA and NASA for several years. ESA had initially suggested having two trains, each of two pallets. But NASA had in mind to fly a single large experiment in the payload bay and therefore proposed a train of three pallets. Later, based on the mission research manifest and a projected weight of 13,098 pounds, the issue was settled by selecting the one pallet plus two-pallet train configuration, since the three-pallet train could carry only 10,638 pounds.

When a Spacelab included a pressurized module, the subsystem equipment was in the core segment of that module; for the all-pallet configuration it was located in the "igloo" developed by SABCA of Belgium.[14] This compact pressurized compartment provided a dry air environment at normal Earth temperature. Mounted vertically,[15] the primary structure consisted of a cylindrical shell made of aluminum forged rings, closed at the bottom end, and locally strengthened. External to this were fittings to fasten the igloo to the pallet, for thermal control insulation, and for ground handling. A pair of feedthrough plates accommodated utility lines and a pressure relief valve. The interior had mountings for subsystem equipment. The top end was covered by a flange to support a hinge-fastened[16] secondary structure on which other subsystems could be mounted. The secondary structure was then covered using a cylindrical shell of aluminum alloy installed in the style of an inverted garbage can and mechanically joined to the mounting flange to provide an airtight seal and allow the equipment to operate in a pressurized environment. In the early

---

[13] A pallet train consisted of two or more pallets physically connected to each other in order to create a long pallet to carry experiments and apparatus that was too large for a single pallet. On the same flight, therefore, you could have single pallets, each with their own experiments, and one pallet train for bigger experiments.

[14] Only the subsystems necessary for the proper working of the pallet-mounted instruments would be housed in the igloo; e.g. data and command handling and computers. Subsystems such as the environmental control life support system or fire control would obviously not be carried on such missions.

[15] Originally, the igloo was meant to be mounted horizontally.

[16] This expedient allowed easy access to the equipment fitted at the bottom of the secondary structure, and to equipment mounted inside the primary structure.

stages of development, this airtight seal proved to be a major source of concern, as there were doubts about its ability to maintain a pressurized environment without its own supply of gas. In fact, this would have greatly simplified the design and operation of the igloo. One alternative was to over pressurize the igloo for lift-off. While this would allow longer missions, NASA rejected it for safety reasons. Eventually, it was demonstrated that the seals between the base and cover were sufficient to maintain the pressure at a satisfactory level for up to 12 days. SABCA also developed a utility support structures for the fluid lines and wiring harness between the module and an attached pallet, or between separately mounted pallets. Depending on the size of the gap, these cantilevered support mounts varied in length between 6 and 30 inches.

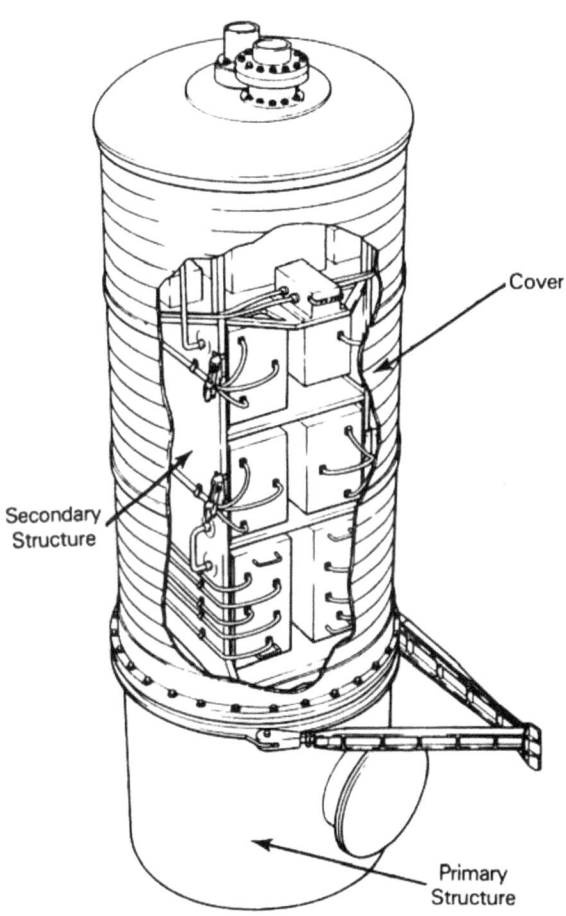

The main structural components of the "igloo" for Spacelab.

The first pallet of the Spacelab-2 mission carried the Instrument Pointing System (IPS), a versatile system to aim scientific instruments that required greater pointing accuracy and stability than that which could be provided by the Orbiter. It proved to be the most difficult Spacelab element to develop. In fact, together with the required pointing accuracy of plus or minus an arc-second, the specifications also addressed quiescent stability, disturbance errors and stability rates for line-of-sight and the roll axis. To achieve these requirements demanded three bearing/drive units (to move and stop the pointing system in three axes), a payload/gimbal separation mechanism (to off-load the payload weight during the ascent and re-entry phases of the mission), a replaceable extension column (to place the pointing system in vertical position in the payload bay), a supporting structure (to mount the pointing system on the pallet), an optical sensor package (to provide pointing references to known stars and the Sun), a payload clamp assembly (to carry the weight of the IPS payload during ascent and re-entry), and a thermal control system (for temperature control). Power and data would have to be made available to the drive units and routed across the gimbal system to and from the instruments on the IPS.

The gimbal drive mechanisms were complicated devices that incorporated direct current torquers, resolvers, main and auxiliary shafts, a load bypass mechanism, ball bearings, and provisions for passing heavy power cables and a large number of signal wires through the drive assemblies to the experiments. The drive systems required to provide a reasonable rate of slewing, but with accurate braking to secure the desired pointing and stability precision.

The requirements were so demanding that Dornier, the German company tasked by ESA to supply the IPS, was forced to change the design even as the first unit was nearing completion. Problems in creating a mechanical system and control software capable of satisfying the stringent pointing requirements meant that the entire project needed a radical rethink. Meanwhile, it was decided to fly Spacelab-3 ahead of the original schedule.

What ESA and Dornier eventually provided to NASA was a 3-axis gimbal system on a gimbal support structure, and a control system based on the inertial reference of a 3-axis gyro package and operated by a gimbal-mounted minicomputer. The gimbal system was the basic structural hardware. It incorporated three bearing/drive units, a payload/gimbal separation mechanism, a replaceable extension column, a support structure, a thermal control system, and an emergency jettisoning device. The gimbal structure was a yoke plus an inner and outer gimbal to which the payload would be attached via a payload-mounted integration ring.

It is interesting that at that time, most concepts for pointing systems employed a yoke handling the instrument at its center of gravity. However the IPS requirements for a 3-axis system with a pointing accuracy of 1 arc-second, capable of positioning a 4,273-pound instrument with a diameter of up to 6.56 feet and a length of up to 13.12 feet required switching to an end-mounted approach which was no longer connected to the center of gravity. A circular mounting frame provided the physical connection between the integration ring, on which the instrument(s) would be assembled, and the gimbal mechanism. For launch and re-entry, the two assemblies (the gimbal system and instrument package) would be disconnected from each other and clamped to the surrounding pallet structure so they would not impart loads on each other. In space, they would be pulled together, locked, and readied for work. This expedient enabled the gimbal to be sized to handle only the

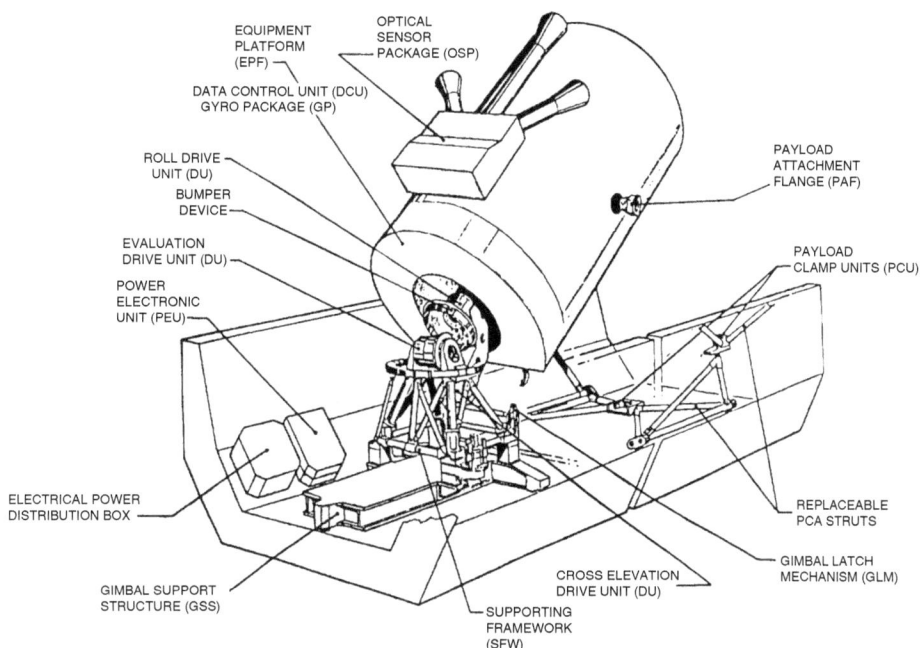

The main components of the Spacelab Instrument Pointing System.

momentum of the instrument(s). Although this approach created severe problems for ground testing and mission simulations, it provided a relatively lightweight pointing system.

The IPS was controlled through the Spacelab subsystem computer, a data display unit, and one keyboard. It could be operated automatically, via ground commands, or by the crew in either a pressurized Spacelab module (if present) or the aft flight deck of the Orbiter (if not). This flexibility of manual control, extended periods of pointing at a single object, slow scan mapping, and so on, catered for the requirements of each investigation. Precise pointing was achieved using an optical sensor package of one boresighted and two skewed fixed-head star tracker. Each star tracker operated as an independent unit and was used to null out the IPS inertial platform attitude and rate errors resulting from gyro drift and scale factors. For solar observations, the boresight tracker was equipped with a four-quadrant sunlight beam-splitter that would produce a negative image.

For the Spacelab-2 field trial, the IPS was extensively operated by maneuvering a suite of four instruments, three of which were devoted to studying the atmosphere of the Sun. The Solar Magnetic and Velocity Field Measurement System/Solar Optical Universal Polarimeter (SOUP) observed the strength, structure, and evolution of the magnetic fields in the atmosphere to help to determine the relationship between these fields and other solar features. The Coronal Helium Abundance Spacelab Experiment (CHASE) determined the abundance of helium. The High Resolution Telescope and Spectrograph (HRTS) investigated the chromosphere, the corona, and the transition zone between them. And the Solar Ultraviolet Spectral Irradiance Monitor (SUSIM) investigated both long-term and short-term variations in the total ultraviolet flux from the Sun.

The IPS and its four instruments shared *Challenger*'s payload bay with a number of other instruments that were to study plasma physics, high-energy astrophysics, and infra-red astronomy.

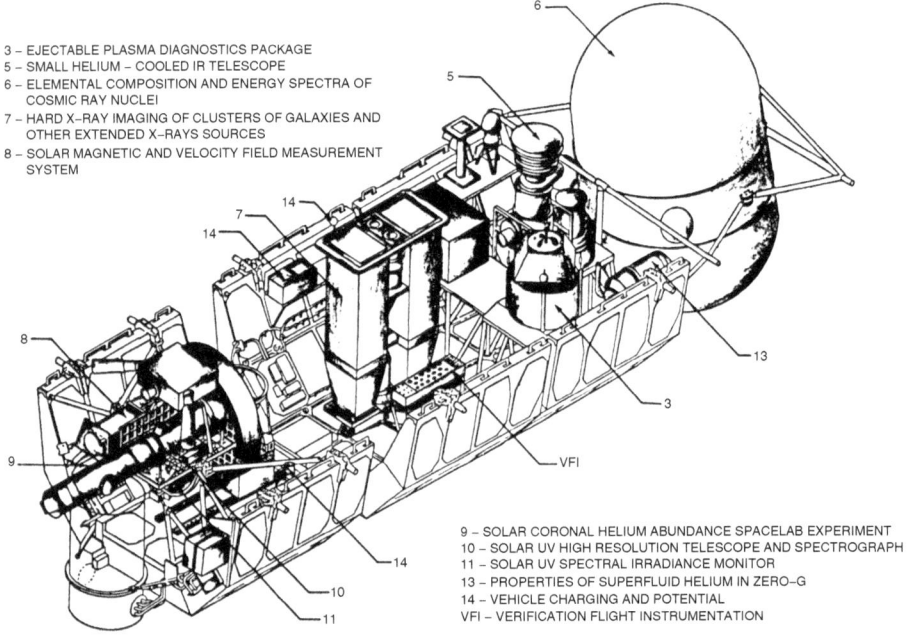

3 – EJECTABLE PLASMA DIAGNOSTICS PACKAGE
5 – SMALL HELIUM – COOLED IR TELESCOPE
6 – ELEMENTAL COMPOSITION AND ENERGY SPECTRA OF
    COSMIC RAY NUCLEI
7 – HARD X-RAY IMAGING OF CLUSTERS OF GALAXIES AND
    OTHER EXTENDED X-RAYS SOURCES
8 – SOLAR MAGNETIC AND VELOCITY FIELD MEASUREMENT
    SYSTEM

9 – SOLAR CORONAL HELIUM ABUNDANCE SPACELAB EXPERIMENT
10 – SOLAR UV HIGH RESOLUTION TELESCOPE AND SPECTROGRAPH
11 – SOLAR UV SPECTRAL IRRADIANCE MONITOR
13 – PROPERTIES OF SUPERFLUID HELIUM IN ZERO-G
14 – VEHICLE CHARGING AND POTENTIAL
VFI – VERIFICATION FLIGHT INSTRUMENTATION

The STS-51F payload complement.

Known as the fourth state of matter, plasma is dominated by ionized atoms and molecules. The ionosphere is a vast bubble of plasma which forms the outer fringe of the atmosphere. Satellites in low orbit travel through it. As the ionosphere is similar to the plasma environment around stars and planets, the results obtained by studying our ionosphere can help to explain plasma processes elsewhere in the universe.

Spacelab-2 carried three experiments to investigate the Earth's ionosphere and the processes occurring within it. The first one was a re-flight of the Plasma Diagnostic Package (PDP) of STS-3. On that flight, the package took measurements while it was installed in the Shuttle payload bay and when on the manipulator arm. It successfully measured electromagnetic noise created by the Shuttle, and detected other electrical interactions between the Shuttle and the ionosphere. For its second trip, the PDP took further ionospheric measurements in the vicinity of the Shuttle and, for the first time, was released as a free-flyer to monitor how the plasma responded when *Challenger* maneuvered several hundred yards away. Using the Vehicle Charging and Potential Experiment the ionosphere's traits were revealed by firing beams of electrons. Probes on the pallet that carried the PDP observed the state of the plasma environment prior to, during, and after such disturbances to study the conditions that naturally disrupt the atmosphere.

Additional investigations were fulfilled by the Plasma Depletion Experiment for Ionospheric and Radio Astronomical Studies. This research employed the Shuttle as an active experimental probe to create artificial "holes" in the ionosphere by way of firing thrusters over specific ground sites, thereby releasing 2,800 pounds of exhaust vapors, primarily of water. These exhaust gases caused the plasma to exhibit specific behaviors that were detected by radar and other instruments at ground observatories over which the Orbiter was passing at the moment of the firings.

High energy astrophysics is another field well suited for space-borne instruments, as they can observe without the hindrance caused by the Earth's atmosphere. In fact, X-rays and cosmic rays can be absorbed by the atmosphere or give rise to showers of secondary particles that destroy any possibility of studying the information carried by this electromagnetic radiation. Experiments in this field are routinely flown on high-altitude balloons to improve their chances of capturing such radiation before it is lost in the dense lower atmosphere. But high-altitude balloons can remain aloft only for limited periods, and in any case they cannot ascend to altitudes sufficient for "raw" exposure to cosmic rays. The all-pallet Spacelab-2 presented an ideal opportunity to attempt such observations on an unprecedented scale.

On the first pallet of the train, the X-Ray Telescope (XRT) imaged and examined X-ray emissions from clusters of galaxies to investigate the mechanism that causes high-temperature emissions and to measure the masses of such clusters. At the very aft end of the payload bay, just behind the pallet train, was the egg-shaped Cosmic Ray Nuclei Experiment (CRNE) with detectors to analyze the composition of high-energy cosmic rays. Its independent installation in the payload bay was a matter of concern to ESA, lest NASA be seeking to develop an alternative to the pallet. In fact, this was not the case. The rationale for mounting CRNE on a purpose-built support structure was the need to maximize the size of the instrument to enhance its detection capabilities. CRNE was an impressive device that made use of lightweight plastic to reduce as much as possible its mass. It was the largest cosmic ray detector ever flown in space up to that time. At its heart were two detectors, each sensitive in a different range of particle energies. It extended the sampling to regions of the energy spectrum never before explored. As the particles entered the detectors, they released electrical pulses. The characteristics of the pulses enabled physicists to identify the particles at energies almost 100 times greater than those previously possible. The scientists were able to garner valuable data about the nature of the particle sources, the method of particle acceleration, and the properties of the interstellar medium through which the particles had traveled.

Although infrared astronomy is possible on Earth with substantial limitations, it is best conducted in the vacuum of space. There are "windows" which permit infrared radiation in certain wavelength ranges to penetrate the atmosphere, but much of it is absorbed by water vapor and other gases in the troposphere (which is why telescopes are placed on mountain tops). Above the atmosphere, the entire infrared spectrum is visible.

By the time of Spacelab-2, the astronomical community had wetted its appetite with the 10-month mission of the Infrared Astronomical Satellite (IRAS) in 1983. It made four surveys of 95% of the sky and cataloged a quarter of a million sources that produced discoveries such as rings of dust in the solar system, new comets, new star-forming regions, and improved understanding of our galaxy. The Infrared Telescope (IRT) on the third pallet was to take a second look at some of the IRAS sources, plug gaps in the IRAS coverage, and add to the catalog. It was also to extend knowledge of the infrared sky to shorter wavelengths.

Any infrared telescope must be protected from the heat that is emitted by its own equipment, as otherwise this will entirely mask the faint radiation that the instrument is intended to detect. To eliminate this background noise, infrared detectors and their surrounding equipment and structure are cooled to temperatures a few degrees above absolute zero. Helium is the coolant of choice for such telescopes, and this was so for the IRT. Spacelab-2 presented an ideal opportunity to investigate the characteristics of superfluid helium in space in preparation for using it in space applications. In fact, although the superfluid helium on IRAS had maintained its temperature uniform to a few thousandths of a degree, astronomers expected future instruments to require their temperatures to be stable to a few billionths of a degree. So understanding precisely how superfluid helium behaves in microgravity, in particular its movement patterns and temperature fluctuations, was of paramount importance for the next generation of infrared space-based telescopes. Mounted on the same pallet as IRT, the Properties of Superfluid Helium In Zero-Gravity experiment was a barrel-shaped container which held 100 liters of superfluid helium, two fluid experiments, and sensors to monitor the behavior of the helium.

One such experiment was an open-structure vessel filled with superfluid helium and 133 liquid-vapor phase sensors and 12 semiconductors. The liquid-vapor phase sensors were to determine the location of the helium, and the semiconductors were to gauge temperature variations within the liquid and any temperature fluctuations that were induced by Shuttle motions. The data would indicate whether fluid movements affect sensitive experiment systems and whether temperature changes can cause an uneven distribution of coolant. The second experiment consisted of five ring-shaped tracks that contained sufficient helium to form films between 0.5 and 2 micrometers thick. It studied the velocity of capillary waves (feeble acoustic waves) on the surface of the superfluid helium films. Given the extreme sensibility of this experiment to external disturbances, *Challenger* was placed into a gravity gradient attitude with its tail towards Earth and while the experiment was underway the crew minimized their activities.

Perhaps not surprisingly, the IPS caused some frustrations during the first half of the flight when a series of software-related glitches crippled the gimbaled platform. First, the package had trouble tracking the Sun, then it was fooled by the background illumination level. On the third day, communications issues arose between the optical sensor package and the Spacelab computer. Another error was found on the seventh day, this time involving the interaction between the IPS roll angle commands and the computer mass memory data. In short, the IPS was struggling to find the Sun. When this problem was overcome by uploading software patches, the IPS refused to stay in "fine tracking" operation mode. This was also fixed by new software code. However, the IPS continued to be temperamental, taking a long time (4 to 5 minutes) to acquire a target. Persistent efforts on Earth and on-orbit eventually fixed all these operational problems, with the result that the IPS performed satisfactorily for the second half of the flight.

The four IPS-mounted telescopes were able to accomplish most of their planned observations with the desired pointing accuracy and stability. Data and image quality varied from very good to spectacular. It was also noticed that disturbances from crew movements were greater than predicted. Although the mass of a human is small in relation to the Orbiter, crew activities within the cabin were nevertheless sufficient to disturb the IPS and generate pointing errors. On the other hand, it was observed that these errors were rapidly damped out. Because the pointing accuracy was better than the design specification, there was optimism that the system would be able to support its future assignments.

Each experiment began with clearly defined objectives and designated periods of operation, but the lower than intended orbital altitude forced a quick rewriting of the flight plan. Also, daily analysis of the data which was being gathered often prompted the scientists to call for revisions to the activity plan. Solar observations, for example, were scheduled orbit-by-orbit on the basis of the data attained during prior orbits and the changing conditions of the Sun. If anomalies spoiled the chance to do a particular experiment, scientists shifted their focus to another topic in order to keep taking good data in the limited time available. As always on a space mission, time was a precious resource.

The changes were received via the onboard teleprinter, which for the first time in the Shuttle program ran out of paper because of the number of flight plan revisions and the IPS troubleshooting activities. The crew did a superb job of keeping up with the dynamic plans, operating the pallet-mounted instruments via computer keyboards and screens on the aft flight deck, on which they could see in real time the data being gathered. To ensure the greatest scientific return, the crew were split into two shifts, each of which had at least one mission specialist and payload specialist for efficient operation of both the Orbiter and the payloads. The hardware to test the performance of the new Spacelab configuration and verify its compatibility with both the Orbiter and the payload, confirmed that the pallet-only mode was indeed an opportunity for scientists to have sophisticated experiments conducted in the payload bay. And as we shall see, a number of missions would be flown with the all-pallet Spacelab.

There were a great many operational and scientific highlights in this third flight of Spacelab. Almost 13,000 commands were transmitted to the Orbiter, exceeding any previous Shuttle flight by 50%. Some 1.25 trillion bits of data were transmitted to the ground, requiring 230 miles of magnetic tape to store. Despite the capricious IPS and complications with the SOUP experiment, more images (and at greater resolution) of the Sun from space were obtained than during the 171 days of observations made by the three crews of Skylab in the mid-1970s. Although SOUP did not start operating until late in the flight, and achieved only 16 hours of the planned 50 hours of data, it was returned with 12,800 pictures on film. It was able to take high-resolution images every 2 seconds for durations of up to 40 minutes. The high-resolution telescope used all 4,150 images available in its film magazines, taking data on 20 orbits as compared to a mere 16 minutes of sounding rocket observation time achieved using this type of instrument over the previous 10 years. The crew were able to observe the rise and fall of spicules on a precise schedule, and to conduct a systematic search of the solar disk for coronal bullets.

The helium experiment obtained 70% of the data needed to record the abundance of this element in the Sun, but owing of the lower than intended orbit the quality of the data was not as good as it would have been if the Shuttle had reached the planned altitude. Another frustration was that the IRT, cooled by cryogenic helium, had most of its wavelength capabilities saturated throughout the mission because of thermal emission by the Orbiter. This raised serious concerns about the feasibility of infrared astronomy from the Shuttle. Four of the eight OMS engine burns were performed for the plasma depletion experiments, despite the consumption of propellant while using the OMS to limp into orbit. As expected, these burns were effective in opening holes in the ionosphere for ground-based observations. The PDP experiment was extremely successful and achieved at least 70% of its mission objectives. The superfluid helium experiment accomplished at least 80% of its goals. The CRNE instrument received about 100 impacts per second, producing a bonanza of data on cosmic rays for post-flight analysis.

When STS-51F touched down at Edwards Air Force Base in California at 12:45 pm PDT, August 6, 1985, Jesse Moore, NASA's Associate Administrator for Space Transportation, told reporters that, despite its problems, Spacelab-2 had "returned a wealth of information. In fact, this may be the most important scientific mission that the Shuttle has flown."

"THE WHOLE CREW WAS PUT TO THE TEST"

The first operational use of the IPS came 5 years later with STS-35. *Columbia* was launched on December 2, 1990, for a 9-day mission. It was the first Spacelab devoted to a single discipline and, as the name ASTRO-1 implies, this was astrophysics.

The IPS carried a suite of three instruments: the Hopkins Ultraviolet Telescope (HUT), the Wisconsin Ultraviolet Photo-Polarimeter Experiment (WUPPE), and the Ultraviolet Imaging Telescope (UIT). A fourth instrument, the Broad Band X-Ray Telescope (BBXRT) was on a separate pointing system secured to a support structure installed aft of the two-pallet Spacelab train. They were to examine objects such as quasars, active galactic nuclei and normal galaxies at different ultraviolet and X-ray wavelengths. Though individually designed to address specific questions, when they were used in concert their capabilities and scientific yields were greatly enhanced. In fact, the BBXRT could make joint observations with the IPS by following the same alignment as the ultraviolet telescopes.

Again, the IPS showed its capricious character. As mission commander Vance D. Brand, says, "The whole crew was put to the test on this mission. There weren't very many anomalies with the spacecraft itself, but this telescope system, and particularly the mount that held and pointed the ultraviolet telescopes, acted up on us. It had been tested on the ground thoroughly, but there are things about being in space that can't be simulated on the ground. For example, the mount itself is rather weak structurally, and you cannot do an end-to-end test on the ground with it pointing at stars because it is not strong enough. In weightlessness, it is plenty strong enough to do its job. On-orbit we found there were some things about the automatic pointing system that did not work right. There was a very commendable effort between the STS-35 crew and the Johnson and Marshall centers to coordinate manual maneuvers and workarounds so that we could get most of the data. It took about a day to get this all working. In a typical viewing sequence the telescope would be pointed by first turning the Orbiter to an attitude, and then having one of the four astronomers onboard manually point it. The fine pointing was a coordinated exercise between the astronomers and Marshall scientists. Anyhow, the coordinated contingency operations saved the mission."

The situation was further complicated by the loss of both IPS control units early on. As STS-35 mission specialist Jeffrey A. Hoffman says, "There are two Spacelab computers which you use to control the telescopes. They are different from those that control the Shuttle. After a few hours of operation, we smelled this burning. One of the computers had basically overheated and burned up. We operated for several days, and then the second computer burned up. Now we had no Spacelab computers left. It was actually the control unit. (The computer itself was out in the payload bay. It was the interior control unit.) It turned out later that it hadn't been cleaned properly, and it had spent so much time on the ground that it had accumulated a bunch of lint. Should have been cleaned, but wasn't. The lint impeded the airflow and it didn't get proper cooling. I'll tell you. It's a very uncomfortable feeling to wake up in the morning and smell smoke in a spacecraft. It's not fun. Anyway that's the way it was. Then there's a question: Do we just abort the mission and come home? No. We still had a limited ability to command, because we had a hand controller. They figured out a way where the ground could send up most of the commands that we would have sent from the onboard computers. Then we would do the final positioning with basically the hand controller, which is something that astronomers traditionally have done…There was less for us to do than there would have been if we'd had to operate the entire system. The whole thing worked out."

Despite the challenges, which meant repeated revisions of the observing schedule, ASTRO-1 managed to provide a substantial scientific return. In fact, many celestial objects were viewed for the first time in extremely short ultraviolet wavelengths and the first extensive studies were made of ultraviolet polarization. The targets spanned solar system objects, the interstellar medium, stars, star clusters, individual nebulae, galaxies, and distant quasars.

In 1993 NASA solicited proposals from the astronomical community to take part in a re-flight of the ASTRO package. Broad involvement was an important feature of this mission. Although each instrument had been developed by a team of scientists and engineers at a particular university or government facility, "guest investigators" had been invited to use the telescopes for their own observations. After scientific and technical peer review, NASA selected 10 proposals for inclusion in the program for the mission.

ASTRO-2 would be the last chance for the IPS to demonstrate that it could indeed serve as the stable and trustworthy platform that it was meant to be. In fact, a special team was put together by mission management at the Marshall Space Flight Center to extensively modify and test the IPS software and make other improvements to ensure that it would work correctly. For instance, the Image Motion Compensation System (IMCS) was developed to detect unwanted motion in the instruments and adjust the mirrors of the telescopes to eliminate jitter caused by crew movements and thruster firings during long observations. The ASTRO-2 package was the primary payload of STS-67, which was to attempt a 16-day flight in order to maximize the time available for astronomical observing. It comprised the same trio of ultraviolet telescopes on the IPS as previously, but this time the BBXRT was not carried.

It was decided to launch *Endeavour* on March 2, 1995, and to improve the quality of the observations, particularly of faint objects, it would set off at night. The time of launch was also selected to produce a trajectory which would intersect the so-called South Atlantic Anomaly mainly on the day side of the orbit. In this region of space, the Earth's magnetic

ASTRO-1 in *Columbia*'s payload bay. Notice the Spacelab igloo at the front of the pallet train. Just aft of the igloo, the experiment supporting platform of the IPS is visible in its stowed configuration. Just aft of the pallet train is the BBXRT telescope on its own purpose-built pallet.

field is weaker (which is why it is an "anomaly"). The radiation of the van Allen belts dips closer to the surface, causing a lot more background noise in space-borne electronic detectors such as those of the ASTRO package. In addition, scattered light and residual ultraviolet airglow can spoil the measurements by these detectors. This optical noise is obviously strongest when on the day side of an orbit. Launching at night permitted observations of the faintest (and often highest priority) astronomical targets to be made during orbital darkness and well clear of the South Atlantic Anomaly. The brighter targets, being less sensitive to interference, would be observed in orbital daylight.

The STS-67 flight plan called for 20 hours of checkout and then the observations starting immediately after that and continuing throughout the mission, with only brief pauses for activities such as water dumps[17] and Shuttle tests. As the mission report states,

---

[17] These would dump overboard both urine and excessive water produced by the three fuel cells available for power generation. During these events, the doors over the three telescopes were closed to prevent contamination of the sensitive optics.

The Instrument Pointing System being maneuvered on-orbit.

"The ASTRO-2 laboratory and its systems, including the IPS, performed in an outstanding manner from activation on flight day 1 to deactivation on flight day 15." All principal investigator teams reported that the science far exceeded their pre-flight expectations. The UIT team said that all of the planned targets were studied, the HUT team said more than 100 targets were observed, and the WUPPE group said they had obtained over three times as much data as ASTRO-1. Other accomplished objectives were the first successful ultraviolet observations of the Moon and detecting helium in the intergalactic medium. The reported ended, "This extremely successful scientific endeavor produced a wealth of data that will occupy the ultraviolet astronomy community for some time to come."

Towards the end of the flight, ASTRO-2's mission manager Robert Jayroe said, "In my estimation, the IMCS and IPS teams have done everything but make their hardware stand up and do a tap dance." There could not have been a better finale for the IPS and the ASTRO program.

"WE CALLED IT THE IGNOROSPHERE"

The all-pallet configuration was also employed by the Atmospheric Laboratory for Applications and Science (ATLAS) program, which was one of the several research activities undertaken within the framework of what NASA referred to as its Mission to Planet Earth.

In 1989 a presidential initiative called the US Global Change Research Program (USGCRP) set out to delve more deeply into issues such as global warming, climate change, and ozone depletion. In 1990 the initiative was ratified by Congress with the Global Change Research Act that called for "a comprehensive and integrated United States research program which will assist the Nation and the world to understand, assess, predict and respond to human-induced and natural processes of global change."

To participate in this research, NASA set up the Mission to Planet Earth program to undertake long-term worldwide measurements of the interactions of large systems, such as the atmosphere, oceans and land masses. Space-based observations would be coordinated with those by aircraft and ground sites. Such data would enable scientists to distinguish between natural environmental processes and the impact of humans.

A few Shuttle missions were flown under the banner of the Planet Earth initiative. The first major contribution was the deployment of the Upper Atmosphere Research Satellite (UARS) on STS-48 by mission specialist Mark N. Brown on September 15, 1991. This was an orbiting platform composed of a module with ten experiments coupled to a Multi-mission Modular Spacecraft (MMS) bus. A hydrazine propulsion module was used for orbital adjustments and the Modular Attitude Control System[18] provided the fine attitude control for long-duration measurements of the atmosphere. Paint, blankets, coatings, and electrical heaters were a passive but effective means of thermal control.

As mission commander John O. Creighton says, "It looked at several different things. One was high altitude winds around the world. It also looked at the chemistry of the atmosphere, and at the energy balance – how much energy came in from the Sun and how much was reflected back from the Earth [into space]." To study ozone depletion in more detail and to better understand other aspects of the planet's fragile atmosphere, the science program had been designed as a single experiment with nine component instruments to study the chemical, dynamic and energy processes of the upper atmosphere. A tenth instrument, though not strictly part of the UARS mission, was to measure the energy output of the Sun.

As Creighton says, "One of the things they saw is the wind patterns and what-not, that the chlorofluorocarbons that originate in the industrialized northern hemisphere were

---

[18] This was the same Modular Attitude Control System module that originally flew on Solar Max. Following the servicing and repair mission of STS-41C (see Chapter 5) the module was returned home, refurbished, and fitted for use by UARS.

migrating down over the Antarctic. That was a direct correlation to what was causing the destruction of the ozone layer in the Antarctic spring. It would release all of those things that were trapped in the lower atmosphere and then it would spiral up because of the circular wind patterns, up into the ozone and then create that hole. So it was kind of exciting to say that what people had long suspected was proven; that that was the cause of it."

Shortly before the mission, Mount Pinatubo in the Philippines had erupted a vast volcanic plume. When UARS started operating, it measured the composition of the stratosphere. The results were spectacular. As Creighton says, "It had circled around the Earth, right around the equator, close to the latitude of Mount Pinatubo, and lo and behold, there was destruction of the ozone layer right around the equator because of the eruption of the volcano, and nobody had expected that. So that was kind of a surprise."[19]

Preparing to release the large UARS payload.

---

[19] The satellite was meant to operate for 5 years. It fact, with six instruments still fully operational, it was decommissioned in 2005 owing to funding cuts. On September 24, 2011, it re-entered the atmosphere, drawing a lot of media attention for several days because preliminary projections indicated that some parts might crash on populated areas. Ultimately, the debris of the 15,432-pound satellite plunged harmlessly into a remote part of the Pacific Ocean.

In general terms, the ATLAS program undertaken by the Shuttle was to provide a comprehensive and systematic set of data which would help to establish benchmarks for atmospheric conditions and the influence of the Sun. The specific objectives were to determine how the atmosphere reacted to natural and human-induced atmospheric variations in order to be able to better distinguish between anthropogenic effects and factors beyond our control. As no single set of atmospheric and solar measurements made by a single mission lasting 7 to 10 days would be sufficient to characterize the ever-changing atmosphere, and because the Sun varies over an 11-year cycle, NASA envisaged the ATLAS program as a series of ten missions that would be flown every 12 to 18 months to monitor both seasonal variations and the longer term solar effects. Each flight would build on the results of its predecessors, and indeed with missions such as the UARS.

The Shuttle offered a tremendous payload capability, the ability to position highly calibrated experiments in an orbital location best suited for gathering data at different times of the day, and the possibility of calibrating the instruments prior to each flight and returning them to Earth for reverification, upgrading, and reuse. The instruments for ATLAS would also provide higher quality measurements than those by a number of other satellites on-orbit at that time, and offer an invaluable opportunity to relate the different sets of data to one another. And ATLAS data would permit the rate at which exposure to ultraviolet radiation and contaminants such as atomic oxygen was causing the calibrations of satellite instruments to drift.[20]

ATLAS-1 was launched on March 24, 1992, as the primary payload for STS-45 with seven crew members. As mission commander Charles F. Bolden explains, "The focus of the flight…was principally to help us understand the middle part of Earth's atmosphere, the mesosphere, because that's where everything gets mixed. We understand Earth relatively well, and we understand space relatively well…What we didn't understand at all back then was the middle atmosphere. In fact, we called it the 'ignorosphere,' the portion of the atmosphere that was ignored by most scientists and about which we knew very little…And that's where a lot of the weather is made, a lot of pollution comes about, and we just wanted to understand exactly what's going on there. How does God do this thing in what we call the middle atmosphere?"

Atlantis had a two-pallet train plus an igloo in its payload bay as the platform for an array of a dozen instruments. Most of them had already flown on Spacelab-1 and Spacelab-3, thereby demonstrating the capability to return sophisticated instruments to Earth for refurbishment, upgrade and reuse, together with resulting cost savings. In all, there were 14 investigations into how the Sun interacted with the atmosphere. In particular, the scientists wanted to probe the chemistry, physics and movement of the middle and upper parts of the atmosphere by measuring the energy from the Sun and the distribution of trace chemicals within the atmosphere. In addition, links between the magnetic fields and plasma that lie between the Sun and Earth were scrutinised to broaden the study.

To maximise the scientific return from the 8 days planned for the flight, the crew were divided into two shifts. As Bolden says, "The advantage of working around the clock for

---

[20] If the calibrations of satellite instruments drifted, long term trends in their data could be misinterpreted. The frequent reverification of the ATLAS instruments would provide a basis for identifying the degradation of satellite instruments.

us was that we had an opportunity to look at daylight and darkness in both hemispheres… daylight and darkness in both north and south and east and west. You got daylight in one part of the world that you wouldn't have been able to get had you gone to bed when it was daylight in your normal launch time." Kathryn D. Sullivan explains another reason for the two-shift pattern, "There was not enough automated interface to operate all the experiments from the ground if you put the whole crew to sleep at the same time."

The 186-mile orbit inclined at 57° to the equator was chosen to yield the greatest scientific return. This allowed *Atlantis* to collect data from most of the atmosphere at orbital sunrise and sunset at latitudes from the tropics to the auroral regions, spanning rainforests, deserts, oceans and land masses.

Most of the experiments and investigations were automatically run by Spacelab's computer, which sent data to scientists at Spacelab Mission Operations Control at the Marshall Space Flight Center in Huntsville, Alabama. But the crew still played a key role in ensuring the smooth running of the instruments. They also carried out a series of manual observations. There was a space plasma physics experiment called Space Experiments with Particle Accelerators (SEPAC). As Sullivan recalls, "It was really cool. We were all pretty jazzed about this. What this thing was, it was basically an electron gun…in a sense, a large capacitor in the payload bay that would build up a charge to a certain level and then release a bolt of electrons. The idea was to orientate the Orbiter so that the aperture of the instrument would inject these electrons roughly along the magnetic field line, down towards the atmosphere near the polar regions. You can think of it as a dose-response experiment. In medicine or other experiments, you put in a dose and you see what happens, then you dial the next dose up or down and see how the response varies. The idea in this case was to try to better understand the physics behind auroral phenomena by injecting a known dose of electron energy into the upper atmosphere. Then a camera out in the payload bay and one in the cabin measured the brightness of the auroral-type glow that this dose of electrons induced. If I know I put in this many kilovolts of energy and I measured that luminosity, then maybe I can start to get a clearer understanding of how the energy of the incoming solar particles couples into the atmosphere and creates auroral luminescence. It was pretty exciting."

Performing the SEPAC observations was a complex task, as Sullivan notes, "The complexity in the cabin was we had a handheld camera with photomultipliers and a number of different filters as the adjunct to SEPAC, and you wanted to get auroral photography, Earth limb photography, and shots of these patches that we'd fired the electrons into. That became a pretty elaborate ensemble of dark shades so you could shield the window, not have any glare, get yourself all dark-adapted, stack up all the filters and diffraction gratings, get these photo observations. It was photography back then, so we're talking film and film changes…complex photographic assembly and disassembly." SEPAC had been included on Spacelab-1 but it had failed early in the flight, so there were high expectations for this re-flight. "SEPAC was going to fire, finally, a day or so into the flight…The SEPAC looked like…out in the payload bay it would remind you of a house paint canister. You could see a little bit obliquely into the top of it. I felt I was in a science fiction movie… There's an oscillating blue blob accumulating vaguely in and around the can, as if some luminescent blue creature is about to ooze out of this can. It's getting a little brighter and larger. Then suddenly this parcel of blue energy leaps away from the Orbiter, just jumps out of the can. You could see it starting to curve away. You could see the curvature of the magnetic field line. You could just see it begin to spiral along. All this material you drilled into your head in college physics, you're now seeing in front of your eyes: the curvature of the magnetic field lines and the electron gyroradius as this thing spirals around it. 'Wow, fabulous!' We joked around a lot, and I still joke sometimes, that I'm sure we are the only Space Shuttle crew to ever fire photon torpedoes because that's what this was. It was like firing photon torpedoes down towards the atmosphere."

But unlikely the *Enterprise* of *Star Trek* or *Millennium Falcon* of *Star Wars* that seemingly have unlimited reserves, *Atlantis*'s supply of electrons proved to be very limited. In fact, at the third firing, a fuse installed to protect the Orbiter and the other experiments from a power surge in the SEPAC blew up, curtailing the experiment for the second time.

When *Atlantis* landed at KSC on April 2, it successfully wrapped up the first deep study of Earth's atmosphere conducted by a Shuttle. The ATLAS-1 data established a voluminous baseline of atmospheric and solar data against which to measure future global changes.

The next ATLAS mission would investigate subtle variations in solar activity and in atmospheric composition since the first flight, paying particular attention to the mechanism which caused the depletion of the ozone layer to identify how to mitigate its harmful effects on human life and food crops. Installed as the primary payload for the STS-56 mission by *Discovery*, ATLAS-2 consisted of a single Spacelab pallet for six experiments that were being reflown, plus a seventh experiment contained in two canisters that were mounted on the wall of the payload bay.

The mission was launched on April 8, 1993, and the five astronauts ran two 12-hour shifts in order to operate the payloads continuously. Only twice did they have to interrupt their work. This was to deploy and later to retrieve the SPARTAN 201 free-flyer which was equipped to discover how the Sun generates the solar wind and to observe aspects of the corona. A wave off due to adverse weather allowed an extra day of orbital activities, and five of the seven experiments were powered back up to make further observations. When the mission concluded after 9 days, ATLAS-2 was declared a success.

Although NASA had hoped to fly 10 missions for the ATLAS program, funding constraints, the reduced flight rate, and the increasing demand to accommodate other payloads terminated the atmospheric research agenda prematurely. The final mission was on STS-66, which was launched on November 3, 1994. It carried the same suite of instruments as its predecessors and the 11-day mission continued the investigation of the processes influencing the ozone layer.

On flight day two, ESA astronaut Jean-François Clervoy used the robotic arm to hoist the Shuttle Pallet Satellite (SPAS) from *Atlantis*'s payload bay and deploy it. In this case the free-flyer carried the Cryogenic Infrared Spectrometers and Telescopes for the Atmosphere (CRISTA) on behalf of the University of Wuppertal in Germany, and the Middle Atmosphere High Resolution Spectrograph Investigation (MAHRSI) for the Naval Research Laboratory in Washington DC.

STS-66 was another spectacular success for the ATLAS program. For instance, the data indicated that the Antarctic ozone hole was a self-contained region, and that there had been a substantial increase of Freon-22 in the stratosphere and therefore in stratospheric chlorine. A notable discovery was the lack of a direct link between the Antarctic ozone hole and ozone depletion at mid-latitudes, implying that atmospheric processes were still poorly understood. CRISTA made infrared measurements both inside and outside of the south polar vortex,[21] and observed the edge of the Antarctic ozone hole. A total of 51,000 vertical scans from the upper atmosphere down almost to the ground gave a three-dimensional profile of the distribution and movement of ten trace gases involved in ozone chemistry. CRISTA also produced the first global map of atomic oxygen, believed to help to cool the Earth. The MAHRSI instrument gave an unprecedented look at the ozone-destroying hydroxyl molecule.

Atlantis's landing at Edwards Air Force Base on November 22 brought to an end the ATLAS program. As one representative said, "The decisions we make to protect the health of our environment need to be based on accurate scientific knowledge, and that is what the ATLAS missions were all about…Results will be shared by scientists all over the world in their studies of the Sun and Earth's changing atmosphere."

One accomplishment by ATLAS-1 was to vertically profile the concentrations of chemicals which result from the breakdown of industrial chlorofluorocarbons (CFC). This was the most direct confirmation that CFCs were responsible for the increase of chlorine in the atmosphere. By measuring the constituents of the middle atmosphere at high northern latitudes in daylight, ATLAS-2 assisted scientists to understand the behavior of the atmosphere after a winter which had seen unprecedentedly low ozone levels. Indications were that the total ozone decreased by 10% at mid-latitudes in the northern hemisphere between ATLAS-1 and ATLAS-2. Later on, ATLAS-3 made detailed measurements of the northern hemisphere in late autumn, allowing scientists to study critical (yet poorly understood) processes as the atmosphere adjusted from a relatively quiet summer to an active winter. Investigators also studied the chemical processes in and near the Antarctic ozone hole, which usually peaks in early October. Observations of both areas at this time of year provided a valuable comparison with data obtained in the spring by the previous two flights.

"KIND OF BUSY, BUT VERY COMFORTABLE"

As *Challenger* lifted off from launch pad 39B at noon on October 30, 1985, to begin its ninth space voyage, STS-61A was already making history in manned space flight. The demands of the flight plan had obliged NASA to assign a record crew of eight, the largest yet for a single mission.[22] Mission commander Henry W. Hartsfield and pilot Steven R. Nagel were joined by mission specialists Bonnie J. Dunbar, James F. Buchli and Guion S. Bluford. While these NASA crew took care of the systems of the Orbiter and the Spacelab, payload specialists Reinard Furrer, Ernst Messerschmid and Wubbo Ockels would undertake the majority of the scientific research. The crew arrangement for ascent and re-entry called for four people on both the flight deck and the middeck. In space, they ran the two-shift pattern which by now was common for Spacelab flights. It was, as Dunbar

---

[21] Generally speaking, a polar vortex is a large-scale, low-pressure zone close to one or other of the Earth's poles. It is located between the middle and upper troposphere and extends into the stratosphere. It is in the polar vortex that ozone depletion occurs, in particular that at the South Pole.

[22] This record of eight crew members at launch still stands, and is likely to do so for a long time.

recalls, "kind of busy, but very comfortable." The middeck was fitted with sleep stations that were occupied in the usual hot-bunking fashion. For this flight, however, as on an old oceangoing vessel, the commander had the luxury of his own bunk. As Hartsfield notes, "Since we had eight people, we had four bunks. I was given one. They decided rather than mess around, the commander ought to have one that's all his, because if we get into a contingency, he might need to rest at a different time or something. So the other three bunks were shared between the seven crew members."

On this occasion the Spacelab was being flown primarily for Germany, so it was known as Spacelab-D1. This represented another historic first, because the German Ministry of Research and Technology had leased a Shuttle and Spacelab for US$150 million to carry out a battery of some 70 experiments that had been supplied by eight European countries and the USA. Payload specialists Furrer and Messerschmid were from West Germany, and Ockels was from The Netherlands. As Dunbar recalls, this was an important mission for Germany, "They were also trying to inspire their youth, their population, to look to the future – that there was still, they felt, a heavy burden coming out of World War II, in terms of optimism, and overcoming that. So this was also a symbolic flight, not just a science flight. There was a lot of youth interest in it. There was a lot of Bundestag, which is basically their political Congress, interest as well."

For the first time, mission control functions would be shared with an international partner. As Dunbar recalls, "The Mission Control Center in Houston was responsible for managing the total mission, and all the Space Shuttle and Spacelab systems. The German Space Operations Center (GSOC) in Oberpfaffenhofen was responsible for payload operations. The payload crew worked with the principal investigators and the CICs[23] in the GSOC and the flight crew worked with the Capcoms in Houston. We spoke with both control centers during the flight. Back in the lab, we didn't say, 'Houston, *Challenger*,' we said, 'Munich' or [sometimes] 'München' in the German phraseology. We would talk to their engineers when we were operating the payloads, or we would talk to their researchers if they were enabled. If we wanted to talk about Spacelab systems or the Shuttle systems then we'd talk back to Houston."

With several nationalities flying, there were some cultural issues to be sorted out. The most important one concerned which language would be spoken. As the mission was funded by Germany and most of the experiments were prepared by the scientific institutions of that country, Hartsfield remembers, "They wanted to use the German language and talk to the ground crews in Germany and speak German. I opposed that for safety reasons. We can't have one part of the crew unable to understand what the other is getting ready to do. It was made clear up front, that the operational language would be English." Dunbar agreed, "We didn't want to get in a situation where there was a discussion on an experiment that interfaced with a system, and a switch was thrown that we weren't aware of…Since I needed to be responsible for the Spacelab systems, I needed to know what was being discussed…So it wasn't a control issue, it was a safety issue. We had to ensure that everybody understood everyone else all the time." However, Hartsfield accepted that at particular times the non-English speaking crew would prefer to use their native language.

---

[23] The Crew Interface Coordinator was the equivalent in the GSOC of a Capcom in Mission Control (the astronaut who acts as liaison between the ground and the crew on-orbit).

"We finally cut a deal that in special cases, where there was real urgency, that we could have another language used, but before an action was taken it had to be translated into English so that the commander or my other shift operator lead and the payload crew could understand it. So we cut a good deal, but it took us a while to get there."

The experiments were contained in eight facilities, six by West Germany and two by the European Space Agency. The investigations sought to investigate aspects of microgravity.[24] Assisted by the NASA crew, the three payload specialists operated single and double racks in Spacelab packed with experiments for materials and life sciences.

The materials investigations were done using the Materials Science Double Rack (MSDR) containing experiments in fluid physics, the solidification of metallic melts, and the growth of single crystals. In another facility, the Process Chamber provided optical diagnosis equipment for experiments in fluid physics and the solidification of transparent melts. The MEDEA[25] rack was a sort of second-generation MSDR with apparatus for the melting and solidification of metallic and semiconductor materials using very precise thermostats.

The Biorack was primarily for life sciences investigations into cellular functions and developmental processes. The Vestibular Sled worked with a rack of apparatus to deliver linear accelerations to study the human vestibular (balance) organs. Both of these racks were developed by ESA. So was the System Rack, which had thermostats both for its own botany experiments and for cooling the furnaces in the other racks.

On a truss across the payload bay aft of the Spacelab pressurized module was the Materials Experiment Assembly (MEA) containing five NASA experiments using a number of furnaces and an acoustic positioning apparatus. There was also the Navex experiment. This was the only investigation unrelated to microgravity, instead being to observe the behavior of caesium atomic clocks in space, including measurement of relativistic time dilation, and the monitoring of two-way synchronisation between the clocks onboard and their counterparts on the ground to an accuracy of better than 10 nanoseconds. One-way distance measurements and position determination better than 42.3 feet were also undertaken, along with investigating how the ionosphere affects such measurements. The results from Navex would find application in the design of atomic clocks for satellites scheduled to be used to determine positions on Earth for navigation purposes.[26]

To minimize disturbances to the microgravity environment, *Challenger* adopted the low-drag gravity gradient attitude that was introduced for Spacelab-2. As Nagel says, "On the Spacelab mission, the first one, they wanted as good a microgravity as they could get, and they felt that turning all the reaction control jets off was the way to do that. But what you've got to do, you have to put it in an attitude where it will be stable…We would get it

---

[24] The contents of a spacecraft, regardless of whether they are living beings or equipment, in principle are subjected to weightlessness as the balance between pull of gravity and centrifugal force would be perfectly nil. In practice, however, astronauts' movements and equipment operations disturb the perfect balance of these forces. For this reason, it has become customary to speak of "microgravity" rather than weightlessness when referring to space flight.

[25] It is the German acronym for "Material Science Double Rack for Individual Experiments and Dedicated Apparatus."

[26] In particular the constellation of satellites of the Global Positioning System (GPS).

in this attitude, which was nose at the Earth and the right wing pretty well forward…so we'd do that, get it all stable and turn off the jets, and it would just stay there." In this attitude, *Challenger* was also subjected to minimum aerodynamic drag. It flew like that for 8 to 10 hours per day, and spent the remaining time in the so-called minus ZLV, with the payload bay facing downward.

Despite this being the fourth Spacelab flight, and the third using the pressurized module, STS-61A had some lessons to teach about experiment design and human factors. As Dunbar observes, "You ought to train like you fly, and fly like you train. In a couple of the experiments they simply didn't have the training hardware ready, so when we trained in the lab we had like wooden blocks. The procedure would say, 'Throw this switch,' and we would pretend throwing it. Well when we got into space there is a piece of hardware that we've never seen before. In one case, we threw a switch, thought things were on, but it turned out it was almost a two-position switch with a second detent we didn't get past and the experiment was never activated. But there was also no light feedback. That's another rule. You must have some kind of feedback system to let the crew know that it's on, just like when you turn the light switch on, the light comes on. When there's power to the experiment, there ought to be a little green light someplace. So that was a challenge. That was a disappointment, to come back and find out that it was never activated."

This experiment was not the only one to lack basic human factors design criteria. As Dunbar notes, "There was no integrated philosophy on panels and switches when we flew the flight, because they had a lot of different vendors bringing in experiment racks, and because for them it was international. I mean, they had fluid physics from Italy. They had a glovebox from ESTEC [European Space Technology Center] in the Netherlands. So the rack lighting, these trouble lights and stuff, were not integrated. When you looked down the lab, you'd see this large array of different colored lights, green ones, amber (they called them yellow) and red. But they didn't always have the same function. On one particular rack, it was one red light's okay, two is caution, and three is stop. Well, when you do what we call a cockpit scan, and you see a red light [that means 'good'] it would be better if everything that was working was green, you know. Or if something was all red, then you knew it was stopped. That was always a challenge. Especially in the zero-g environment where you're not always up straight. You have to read the nomenclature on the light and the switch in order to make sure everything's in the appropriate position. So that [lack of uniformity] added a little bit of time overhead."

Challenger landed after 7 days of activities which kept the crew so busy that they hardly had any time to enjoy the view outside or the experience of being in space. As Dunbar points out, "You can go full speed for seven days. We would typically try to catch up in pre-sleep, and you really shouldn't be doing that. But again, we looked at time as money. I didn't personally want to come home and think, 'Well, gee, I got to look out the window for an hour, but I've got to come back and tell this engineer or this scientist that I didn't have time to do something for them.' So we felt a certain amount of pride in accomplishing everything we were supposed to do."

In addition to its scientific results, the success of STS-61A was it proved to be an excellent exercise in international cooperation. NASA learned important lessons that it would apply many years later in developing and operating the International Space Station. In particular, the cultural differences smoothed out by STS-61A helped when working with international partners to the extent of sharing responsibilities with other control centers.

8

Space Industries

"THE PROSPECT OF DOING SOMETHING USEFUL FOR HUMANITY"

It is almost a standard question in a job interview to ask the prospective employee what their expectations are, if they are hired. In fact, during his interview with the McDonnell Douglas Astronautics Company in St. Louis, Missouri, 29-year-old aerospace engineer Charles D. Walker had a crystal clear view of the type of career he was seeking. His answer was simple, "Technical work, design development for a few years, opportunity to move into management and oh, by the way, along the way, if anything I'm working on has the opportunity to fly into space, I would like the opportunity to approach NASA to go fly with it."

Walker had graduated in 1971 from Purdue University, Indiana, with a degree in aeronautical and astronautical engineering, and since then he had endeavored to work not merely in the space industry but in particular for those projects that would also offer a commercial aspect. In the early 1970s, as the last Moon samples were being collected by the crew of Apollo 17, NASA and the aerospace industry were already gearing up to develop the Shuttle. Unlike Apollo, which focused only on brief visits to the lunar surface, the Shuttle was meant to become all sorts of things, one of which was to serve as a platform for commercial exploitation of the peculiar characteristics of the orbital environment, such as the apparent absence of gravity. In fact, it was felt that new materials and substances could be made by processes that were impossible to exploit (or even to reproduce) on Earth due to the ever present effects of gravity. Furthermore, the chances were that in the Shuttle era, space would no longer be the realm of a privileged elite of professional astronauts, it would also be made available to scientists and engineers who would conduct their own experiments and research.

McDonnell Douglas was one of the companies willing to study how to transform the space industry from one exclusively dedicated to exploration to one which would deliver profits to innovative commercial ventures. As Walker says, "They knew the prospect of utilizing space and the unique environment of space, and the company wanted to find a way to…find interesting and fruitful ways to invest private capital to produce a profit for

© Springer International Publishing AG 2017
D. Sivolella, *The Space Shuttle Program*, Springer Praxis Books,
DOI 10.1007/978-3-319-54946-0_8

the stockholders…They thought they could do that by looking at what could be done in the low gravity of orbital space through the very near-term flight and operations of Space Shuttle." His far-reaching career ambitions, therefore, did not raise any concerns, since they matched corporate aspirations. For the moment though, the principal task of the company was to supply parts of the Orbiter to prime contractor North American Rockwell.

Walker began as a test engineer for the development of the Orbital Maneuvering System for the Orbiter. This "sounded very interesting," Walker reflects. "I mean…it was space systems and engineering!" Less than a year later, he joined a small group of engineers in the Materials Processing Group who had a small company budget to develop a means of producing pharmaceutical drugs in weightlessness. The interest in this particular market stemmed from a collaboration between McDonnell Douglas Astronautics and the NASA Marshall Space Flight Center which began in 1975. The partnership sought to identify types of industrial processes or materials which could be used to manufacture real commercial products in space for eventual sale back on Earth. It was soon realized that pharmaceutical purification would yield a significant financial return in one of the most profitable and richest industries on the planet. As Walker says, "If you can purify medical materials such as hormones and enzymes, which make up the basic components of treatments for a variety of diseases, you can purify materials to become the therapeutic treatment for diseases, and you can do that purification to a degree that is impossible here on Earth."

Drug constituents of higher purity yield more effective medicines that reduce the required dosage and hence enable a pharmaceutical company to earn more money for each production batch of a given drug. On Earth, gravity interferes with any chemical separation process. That effectively puts an upper limit on the degree of purification that it is possible to achieve. By eliminating this obstacle, the weightless environment permits a chemical separation process to attain a superior level of refinement. In its studies, McDonnell Douglas had determined that it should readily be able to attain a four-fold or five-fold increase in purification. It was also recognized that operating an apparatus in space would yield greater production rates; up to five hundred times the volume that was achievable in the same time on Earth. The commercialization of low Earth orbit promised lots of work for the Shuttle.

In reviewing the various chemical separation processes used in the preparation of drug ingredients, McDonnell Douglas concluded that the electrokinetic phenomenon called electrophoresis would be the most suitable for a device that would operate in space. Electrophoresis is the motion of dispersed particles within a fluid, relative to the fluid itself, under the influence of a spatially uniform electric field. It was first observed in 1807 by Ferdinand Frederic Reuss of Moscow State University when he noticed that clay particles dispersed in water would migrate under the influence of a constant electric field. The cause of the phenomenon has to do with the presence of a charged interface between the particles and the surrounding fluid. The applications for biochemistry are noteworthy.

For example, as Walker says, "Blood is a complex mixture of proteins and cells, and every protein molecule of a particular kind has a resident charge, an electrical charge to it, very small, but different from every other chemical type of protein or a cellular body within that mixture. Apply an electric field and they will all move as a group toward the attracting electrical pole, and they'll move at different rates, and so if you expose that sample to an electric field for a period of time, when you shut the field off, you will have the groups of compounds all separated from one another. So whereas you had an original mixture, at the end of that process you have purified into individually obtainable groups each component of that original mixture. That is the nature of electrophoresis."

McDonnell Douglas set out to find a partner to establish a profitable joint venture. After surveying the market, the aerospace giant signed an agreement with the Ortho Pharmaceutical Company of the Johnson & Johnson Companies Group. As Walker points out, "Johnson & Johnson agreed to invest their own monies, private monies, in product development of one particular hormone, maybe two, that they thought had a commercial market and would benefit from space processing." The agreement was signed in 1977 and prompted McDonnell Douglas to create a prototype for a ground-based apparatus to refine their knowledge and understanding of the fluid-mechanical equations of electrophoresis as a step towards a space-worthy system for continuous flow electrophoresis processing for pharmaceutical purification. Ortho would provide the financial backing for the developmental work.

As would be revealed some years later, Ortho were seeking space production of highly purified erythropoietin. This hormone is released by the adrenal gland, and its effect is to stimulate bone marrow to produce red blood cells. A lack (or shortage) of erythropoietin causes the condition known as anemia, in which the yield of red blood cells is insufficient to sustain a healthy level of oxygen in the body tissues. Anemia is also recognized as a side effect of some forms of blood cancer. The manufacture of erythropoietin was already taking place but in limited quantities and with a moderate level of purification, restricting its effectiveness on patients. As a substantial portion the population in the United Stated was affected in one way or another by anemia, it was evident the cheap mass production of highly purified erythropoietin would mean a significant profit for the Ortho Pharmaceutical Company.

An early proposal to acquire an experimental continuous flow electrophoresis unit which was operating in Europe was rejected because it was too small, and also too expensive. Therefore, Walker and his team elected to design from scratch their own experimental apparatus. To further complicate the already challenging task, there was the need for aerospace engineers to enter the newly emerging field of biotechnology. As Walker says, "The word 'biotechnology' was really not in general use then, it was just coming into use in the technical community in the late 1970s or early 1980s. So there were plenty of challenges in terms of materials, in terms of the electrochemical substances and in terms of keeping the cells viable. When we were using living cells as either the tissue-culture source for the biochemical materials, or the enzymes and hormones that were produced by the living cells in culture in the laboratory, we had to keep them viable…for periods of, oh, up to six or seven days." As Walker notes, "Biological substances require very delicate chemical balances in the fluids in which they're immersed…They need a very balanced temperature and pH to maintain those [balances] in the isolated and resource-constrained environment…of a Space Shuttle cabin." Although developers of Spacelab experiments were facing the same issues, it is shameful that there was no exchange of information. Each community addressed their own challenges by themselves.

Although working on a judiciously planned development program with a limited budget, the Materials Processing Group was confident of success, and its personnel routinely put in 60-hour weeks at the expense of their social lives.

By the end of the 1970s the McDonnell Douglas Astronautics Division had begun informal discussions with the NASA Marshall Space Flight Center on the possibility of utilizing the Shuttle to serve the commercial interests of a private company. The Astronaut Office of the Johnson Space Center assigned astronaut Don L. Lind as a mentor for Walker's team, and Lind assisted in developing a space-worthy device for astronauts to operate. As Walker says, "He very quickly came up to St. Louis, to our facility, and began working with us. We would exchange information, give briefings. We'd show him designs and prototypes of equipment, and he'd help us understand the basics such as how an astronaut would want to see procedures described for both training processes as well as for in-flight. We began to learn directly from Don what kind of switch designations, what kind of instrumentation, what kind of procedures and processes were mandated, basically, by the Astronaut Office, in terms of controls to the device, the design of the instruments, as well as the design of the procedures to operate the device."

In January 1980 the informal discussions gave rise to a joint endeavor agreement that entitled McDonnell Douglas to fly a proof-of-concept experiment on six Shuttle missions. In line with the NASA acronym-philia custom, the machine was designated the Continuous Flow Electrophoresis System (CFES). As Walker explains, "It was, as attorneys say, a quid pro quo arrangement…One side provides these resources, the other side provides these other resources to accomplish an overall objective without funds being exchanged." NASA would provide transportation without compensation. McDonnell Douglas, in turn, would develop and build the CFES with its own money. But the agreement allowed NASA to make use of the apparatus for its own research needs.

As Walker explains, "We'd been working with the Marshall Materials Processing Lab folks for a period of years. They were interested in continuing to do research, in particular electrophoretic separation research. Their interest was, from an academic standpoint, to do this research in space. So they would like to use the electrophoresis device to do research in the microgravity environment and the agreement embodied that. About one-third of the time of the electrophoresis device operation on-orbit for those tests and proof-of-development flights, those six flights, there would be NASA samples on board that we would inject into the apparatus, separate, and collect. And because the device had a clear acrylic front cover so that you could actually see the separation process when there was a coloration or a dye in the specific sample stream being separated, we would take photographs. So NASA, in the form of the Marshall Space Flight Center Materials Lab folks, had the opportunity to use a device that was produced at the expense of the private sector for private-sector research, to undertake NASA research for up to a third of the time on-orbit in exchange for the opportunity to have it there aboard Shuttle. Their only expense was the preparation of samples, their collection, and the post-flight analysis in their own laboratories."

Owing to its large size, the agreement envisaged the CFES being carried inside a pressurized Spacelab module. As Walker notes, "We were talking about pretty good-size equipment. I mean, even in concept form, this thing was going to weigh several hundred pounds. It was going to be at least the size of a file cabinet with four or five drawers, with probably another module or two for electronic controls and storing the biologic materials.

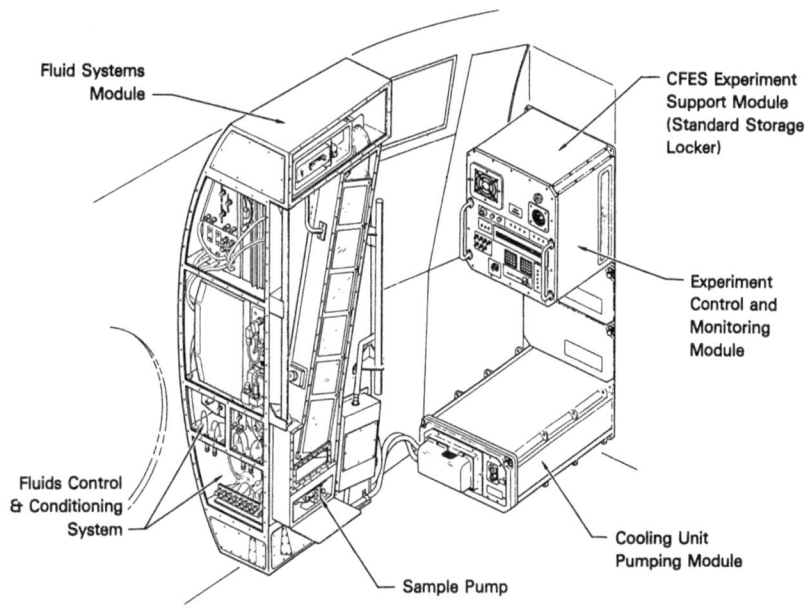

The main components of the Continuous Flow Electrophoresis System.

So we knew it was going to be big, and Spacelab, again, was the obvious place for this kind of thing, mounted in a rack."

Even prior to the joint endeavor agreement, it was determined that the third flight of Spacelab would be the first suitable opportunity to fly the CFES. This mission was scheduled for late 1982 or early in the following year. But the continuing slippage of the first Spacelab and its follow-ons prompted McDonnell Douglas to ask NASA to bring forward the first test of the CFES. All in all, even though the company did not have to pay to fly the experiment, it was accountable to its shareholders, eager to see profit emerge from this space endeavor. The pharmaceutical industry was even more impatient. The first test flight therefore had to take place as soon as possible.

It was Glynn Lunney of the Space Transportation System Group at JSC that came up with the idea of flying the apparatus on the middeck. As Walker remembers, "He knew the design of the systems and areas within the Space Shuttle Orbiter very well, and his thinking was, there's a galley on the port side in the middeck that is about the same size as these electrophoresis program folks are telling me they're going to need to fly. Why don't we on a few flights, we're not talking about many flights here, take the galley out, put in this electrophoresis materials processing device." As soon as it was confirmed that this switch was feasible, Walker and his team ran the numbers and made the necessary modifications to allow the CFES to be safely and correctly quartered in the middeck.

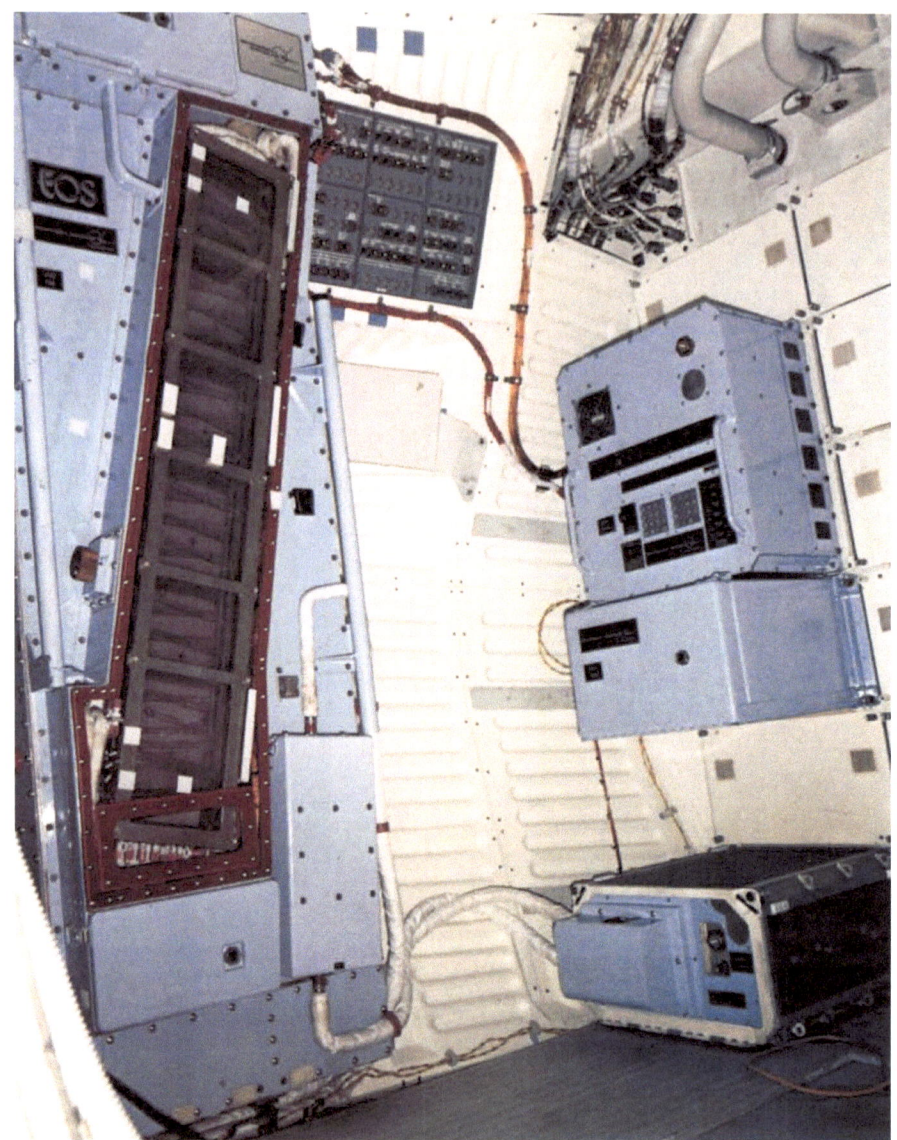

A view of the CFES fitted in place of the galley on the Orbiter's middeck.

As 1980 drew to a close, and with *Columbia*'s maiden flight imminent, the CFES was manifested for STS-4. It would become the first of a vast number of experiments that would be routinely carried out by the astronauts on the middeck right through to the very last flight three decades later.

With the CFES now officially manifested, Don Lind handed over his work to the ground support team and flight crew of STS-4. Mission commander T.K. Mattingly made his pilot, Henry Hartsfield, responsible for operating the device. As the primary objectives of the mission were to continue testing the Orbiter's systems and operating capabilities, the CFES was not a priority for crew training. For almost 2 years up to a few days before launch, Walker paid short visits to JSC to train Hartsfield on how to operate the CFES using a purpose-built simulator of the experimental apparatus and its electronics.

On June 28, 1982, the CFES came to life in the weightlessness of space. To save on development costs and time, it was not capable of sending any telemetry to Earth. This meant that Walker and his colleagues in the support rooms at Mission Control could only rely on the crew reporting back to them strange but meaningful readouts from the control apparatus, plus any interesting observations they might make about how the device was working. At this point in the Shuttle program the vehicle was in communication only when it had a line of sight to a ground station, and during those brief periods of contact up to three people in the support rooms would carefully listen to the astronauts, scribbling down every scrap of information on the experiment. As Walker explains, "Sometimes the pass was maybe so short, and there were so many things going on otherwise with the vehicle or with other primary payloads that the crew would just give a quick verbal burst of information." To make sure that nothing was missed, they requested and obtained a copy of the audiotapes after each pass for review, seeking details that might not have been caught the first time. Based on this feedback, Walker and his group could generate a new set of instructions for the crew to carry out by altering the settings and configuration. "There was the opportunity to make…some physical changes, changing valves, changing flow rates, but there was also the opportunity to change pump speeds and electric field, that kind of thing." For the McDonnell Douglas engineers, communicating the changes to the crew involved an indirect procedure. "We couldn't talk to the crew directly from the back room. We had to go through the Payloads Officer in the front room, who then took [our request] to the Flight Director, who then okayed the Capcom to ask such-and-such a question of the crew." Nevertheless, this far-from-ideal procedure was the best that the space agency could offer at that early point in the Shuttle program

When the CFES experiment was deactivated on the fourth flight day, Walker and his team had learned a lot of lessons. First and foremost, they had established that the CFES could indeed operate in space in accordance with their mathematical models. This was a real achievement. As Walker enthusiastically explains, "This was a fluid process with pumps pushing gallons of fluid through a very thin chamber – I mean, literally 3 millimeters thick and 16 centimeters wide and 120 centimeters in length – and doing so without bubbles forming, because bubbles would block the process. So it had to be bubble-free. It had to be ultraclean because of the biological nature of it." In fact, the device operated without bubbles for the entire duration of the mission, and the temperature and electric field were fully controllable. At the same time, the ground handling procedures confirmed the device, when already fitted on the Orbiter on the launch pad, could be sterilized, serviced with the electrochemically conductive buffering compound, and kept biologically clean and bubble free for several days in the run up to launch. This readiness capability must not be underestimated, as for any rocket the final few days preceding blast off are always hectic with a great many final checks and activities underway. Having the experiment ready to go gave NASA and the astronauts one less thing to concern themselves with, and for sure helped increase their confidence in the professionalism of the McDonnell Douglas engineers.

Although designed to operate with biological materials, for the first test flight the CFES used only a simulant just to validate the process and the overall working of the apparatus. Walker explains, "We used small styrene beads, as a matter of fact, which was a simulant that was proposed by the Materials Processing Lab folks at Marshall. These microscopic styrene beads were dyed with different colors. They were mixed together, and in the chamber they would separate into the different color streams. On STS-4, [Hartsfield] could readily see and photograph the results of the separation, and then collect the samples for return to the ground." The simulant test confirmed not only that the device was able to operate as intended, it also verified that it could separate the desired compounds from the undesired ones. "We wanted to know, for the practical purpose of being able to collect pure enzyme X with an inherent electric charge of Z volts, predict what tube – and we're talking about little exit tubes out of that array of 198 exit tubes, each maybe one half of a millimeter in diameter – predict which one of those tubes it will come out of, so that you can collect X in pure form, versus all the other compounds, all the other liquids, of no use to the pharmaceutical company. So you wanted to be able to predict where it was going to come out…We were very practical in terms of our objectives…We had proven that we could predict adequately for production processing what we needed to know."

Shortly after the flight, Walker and his manager Jim Rose visited Lunney to brief him on the CFES results and to discuss further development of the joint agreement. Little did Walker know that he was about to see his life-long dream become a reality. As Walker says, "Jim told me, 'I just want to tell you, as we walk into this meeting if I get an indication from Glynn that he's happy with the results, too, from the NASA side, I'm going to ask for a payload specialist opportunity. Are you okay with that?'" It was music to Walker's ears, because his perseverance in forging a career path that would see him fly above the Kármán line[1] was about to pay off.

Lunney was indeed very satisfied with the first test flight, particularly with how McDonnell Douglas was honoring the joint agreement. Rose seized the moment. He started by explaining to Lunney that despite the good job Hartsfield did in operating the device, an astronaut could spend only a small amount of time on it because of the need to deal with more important duties. But the device would need constant care to provide the best possible results. What McDonnell Douglas needed was a person that would be exclusively focused on the apparatus. In other words, it required a payload specialist. Walker continues, "As I remember it, Glynn chewed on his cigar a little bit…then said something like. 'Well, we've been wanting to move into this payload specialist thing, so if you've got somebody that is qualified, somebody that can meet all the astronaut selection criteria, put in the application. Let's do it…Do you have somebody in mind?' At this point Rose indicated to me and said, 'You're looking at him!'"

---

[1] The Kármán "line" at an altitude of 62 miles represents the boundary between the atmosphere and space. Simply put, whoever crosses this line earns the title of astronaut.

Evidently Lunney was not as surprised as he might have been, because Walker had applied when NASA had set out to recruit astronauts for the Shuttle. Paperwork was sent to NASA headquarters for review, and approved. Walker thus became the first private Shuttle payload specialist. To that point, all payload specialists to have been selected for flights derived from the academic and research scenes and were the type of people that would participate in flights such as the Spacelab missions. Walker was not an academic, he did not belong to the industrial research community, and he was not a scientist interested in performing research in space. He was an engineer in a private company seeking to create opportunities for commercial manufacturing in space. It was hoped by all concerned that he would be the first of a new generation of spacefarers who would use the Shuttle for business activities.

Assigned to STS-41D, which was to be the maiden flight of *Discovery*, Walker started indoctrination into the complex world of Shuttle training. Due to his coaching of the STS-4 crew to operate the CFES apparatus, it was a world with which he was already quite familiar. Although Walker's status as a private payload specialist meant that he received only a basic training which focused on the systems of the Orbiter, he participated in a number of simulations with his crewmates. What was omitted, as he says, were the "big-expense items, the items like emergency training. I didn't get the emergency water training. I didn't get the survival training in deserts or the jungles."

However, mission commander, Henry Hartsfield, with whom Walker had worked previously, pushed Walker's training a little bit further than what NASA had in mind by allowing him to learn simple tasks like water dumps or turning on and off a given system, and more complex ones such as operating the robotic arm. It was Hartsfield's opinion that this would help Walker to bond better and more rapidly with the rest of the crew, and that it would be helpful to have someone perform such housekeeping tasks to allow the professional astronauts to spend more time on their various mission tasks. When NASA rejected this additional training of a person who was intended to fly only once, essentially on the basis of its cost, Walker reverted to the basic training regime. As he remembers, "I was still invited by the crew to look over their shoulders at some individual training that they would receive on such as the RMS or onboard systems, so that I was at least familiar, not only in textbook, but with the hands-on operation of these systems. The whole point of the training in that regard was for me to be familiar." Of course, along with Shuttle training, Walker had also to take care of his own training in St. Louis. And as if this was not enough, he also supported the other CFES flights that were taking place on STS-6, STS-7, and STS-8 while he was in training for his own flight.

On August 30, 1984, Walker's dream of flying into space was finally realized. As the engines of the Orbiter shut down, he felt immediately that gravity was no longer pulling on him. As a payload specialist, or in other words a passenger, his flight plan required him to remain strapped in his seat until a professional astronaut, in this case Judy Resnik, came to assist him. This proved to be a sensible procedure, because he soon developed the symptoms of Space Adaptation Syndrome (SAS), NASA's name for space sickness. For the next 3 days, Walker suffered nausea, vomiting, sweaty palms, and so on. Not fully understood even today, SAS affects at least 50% of space fliers and arises from the manner in which the body attempts to cope with the various physiological changes that occur in weightlessness.

Despite the discomfort of SAS, Walker set about starting the CFES operations on the following day. He ran it until the penultimate flight day, when it was deactivated. In supervising the experiment he slept next to it at night because, as he points out, "It was supposed to operate 24 hours a day for three or four days of the flight, and the electronics had a beeper. The beeper was disabled at night, so I'd plug in a headset to alert me through my ear. I had to be real close by, so that I'd be alerted, and if I had to get up I wouldn't disturb anybody else, hopefully, when I took my little flashlight to try to fix whatever the problem might be, if that were the case." During the day, he would consistently work next to the machine to ensure it was running smoothly, and in the case of glitches he would figure out how to fix the problem either by himself or in consultation with his support team at Mission Control. With only a single TDRS satellite on-orbit, communications were still rather limited and whenever he wanted to talk to Earth he had to ask permission of the crew.

When *Discovery* landed at Edwards Air Force Base in California on September 5, Walker had every reason to feel very satisfied with himself and the CFES. But less than 24 hours later his enthusiasm was slightly tempered when it was discovered that the purified biological samples had been contaminated by bacteria that had infected the raw material during its pre-flight preparation. From a technical point of view, of course, the test had been successful because it further validated the CFES procedures and operation. But it was a total loss in terms of returning useful product. In fact, the samples could not even be used for an animal trial, an essential step in verifying that drugs produced in space could actually work on humans and clear the way for Ortho Pharmaceutical to approve their production on an industrial scale. As Walker notes, "Our partner in what we hoped would soon be a commercial project, was Johnson & Johnson's Ortho Pharmaceutical [Division] and they intended to use some of these [highly purified] samples for first-level animal testing. It would have been many tens of months later, after much more testing, that they first could have gotten Food and Drug Administration approval to use this material for a pharmaceutical product."

Another important lesson learned was that the apparatus needed a lot of care and direct personal attention to keep it under control and within the design conditions for production of pure samples. Encouraged, McDonnell Douglas asked to fly a payload specialist on the next available opportunity for the CFES. As Walker recalls, "We expressed that to NASA management, and the management apparently said, 'Well, if we do that, we'll train another one of your engineers…go through that process. Or, Walker seems to have made it back okay, and none of the crew really got ticked off at him on this flight, so maybe he can fly again, if you want to do that.'" This was an unexpected but very welcome offer. In fact, payload specialists were meant to fly just once. There had never been plans for a career payload specialist. Walker knew this very well. In fact, the primary purpose of his flight had been to better understand the operation of the apparatus, learn as much as possible from its way of working, make the necessary improvements, then fly the CFES routinely in a semi-automated mode in which an astronaut would need only to turn it on and off.

Six months after STS-41D, on April 12, 1985, Walker was strapped into his seat on *Discovery*'s middeck to fly in space for a second time with an improved CFES. "I was to be operating some relatively new equipment…a change to our electrophoresis apparatus in terms of some of the electromechanical subsystems within it, and some of the processes. We had changed a number of processes to improve the biological sterility, to avoid the problem we had, and some other anticipated possible problems that we had discerned over the previous few months of experience and working with it on the ground." Among the issues Walker hinted at, was an annoying tendency of the buffer liquid[2] to bubble, because that impaired the purification process.

Talking about his second experience in space, on STS-51D, Walker says, "It went pretty well. There was the usual little snags here and there of some software with a glitch. As I was monitoring the progress, I would reprogram the controlling software and adjust the temperature or flow rate or the electrical power settings. I was kind of manually – semi-manually – flying it, so to speak, through its processes, adjusting it, fine-tuning its processes." The only significant issue was the formation of bubbles. This was attributed to the subsystems rather than to the electrophoresis process itself. The contingency spacewalk by Jeff Hoffman and Dave Griggs[3] also caused some CFES issues. The device had been designed with a 14.7 psi reference pressure, which was nominal for the Orbiter on-orbit. The day prior to the extravehicular activity, the cabin pressure was lowered to 10.2 psi to help the spacewalkers prepare for the low-pressure of their spacesuits. As Walker says, "That was really going to play hob with my electrophoresis process. So I had to shut down my process and kind of secure the device before cabin depressurization, and then on the other side when we boosted the pressure up again after the EVA was done, to reconfigure my device and turn it back on. I had some extra procedures and work to do while I was helping them get ready for their EVA."

The gaps in communication that had plagued the CFES project since STS-4 were partially solved by recording voice data on the on-board recorder. Prior to retiring to bed, and with the permission of the mission commander, Walker would spend a few minutes reading a prepared summary of the day into the voice recorder, recapping his results, the problem that he encountered, and requesting assistance and guidance. The recording would be download as part of the telemetry, and overnight his colleagues in the support rooms at Mission Control would prepare new instructions that Walker would receive the following morning. Although slow, this procedure "worked very well."

After STS-51D, the agreement between McDonnell Douglas and NASA allowed one more flight opportunity for the CFES. But the company was looking ahead to an extension of the agreement that would provide further flights in order to advance the project. In phase two, the know-how developed from the six middeck CFES flights would be applied to create an automated system to be carried in the payload bay on at least two flights. The only astronaut involvement would be to turn the device on and off. Successful completion of such tests would clear the way for McDonnell Douglas to produce pharmaceutical-grade materials for sale. NASA would receive part of the operating profits.

---

[2] The buffer liquid is the carrier of the substances that are to be separated via electrophoresis.

[3] Refer to Chapter 5 for details of the first contingency spacewalk of the Shuttle program.

The design for the new apparatus, the Electrophoresis Operations in Space (EOS), was completed in mid-1985 and McDonnell Douglas set to work building it. Walker says it was an "across-the-cargo-bay system" weighing 5,000 pounds. "It was [equal to] 24 electrophoresis chambers; I think there were six of them. It had 24 times the capability of the individual electrophoresis chamber that I was flying in the middeck, and advanced electronics, and advanced monitoring and control systems as well as all the necessary support structure, insulation, and volumetric capacity for the fluids, so that it could run out in the open cargo bay and run through a 7-day mission almost continuously, producing large quantities of purified pharmaceutical-grade material."

Carriage of EOS-1 would not mean an end to activities on the middeck though, because McDonnell Douglas was eager to apply the electrophoresis process to other materials and extend their client list for independent sources of revenue. The wisdom of this strategy was shown in September 1985, when Ortho Pharmaceutical withdrew its financial and commercial backing. This came about because advances in the field of genetics had made possible a breakthrough in another method of mass producing drugs or their components. Although the purity of these products was not as good as could be attained in space, the volumes were high and it was a marketing opportunity too attractive to ignore.

McDonnell Douglas also had other prospects. It opened contracts with 3M Riker, a competitor of Ortho Pharmaceutical, and initiated talks with other rivals in Europe and Japan. The pharmaceutical industry was not the only target. As Walker explains, "The electrophoresis process is the electrophoresis process, and you can separate one of tens of thousands of types of hormones or enzymes – proteins of all kinds – in solution. You can separate dozens if not hundreds of types of living cellular bodies in suspension in the device. We were actually even at the time doing our own internal research at McDonnell Douglas and consulting with some companies on doing wild things such as separating solid suspended materials like rocky ores using this device, which could never be done with this process on Earth because sedimentation would make it completely just laughable to even talk about it. But when there wasn't any such thing as sedimentation in a place where gravity is not a concern, then it becomes possible to use the electrical charge characteristics of any small, microscopic, or even very tiny particles suspended in a fluid to separate it from any other type of material, and so you could separate any kind of thing. So we were talking to others about kind of wild projects." However, as he notes, none of this ever "got into the laboratory test phase on Earth."

In line with the company's optimistic near-future plans, for the final flight of the original NASA agreement the CFES that was assigned to STS-61B was upgraded to preproduction test level to qualify the production processes, settings, and software for EOS-1. Once again, Walker would take care of the device. In the span of 18 months, Walker would fly three times, which was more often than professional astronauts. In effect, he had become a career payload specialist, and McDonnell Douglas was eager to have more of its employees fly missions. In fact STS-61B's crew not only had to get to know Walker but also his backup, Robert Wood, who was one of the computer gurus in the project. With many flights to come and the prospect of having a facility like EOS-1 on the future space station, McDonnell Douglas and Walker himself were dreaming of establishing a cadre of commercial astronauts to periodically fly in space on company business. For his first flight, Wood was scheduled to fly with EOS-1 on STS-61M in the summer of 1986. But for the moment he had to shadow Walker and be ready to step in if, for any reason, Walker could not fly.[4]

The STS-61B mission started on November 27, 1985. As his primary assignment, Walker spent 175 hours purifying a large single sample of 1.1 litres of hormone that was representative of the material which EOS-1 would process. The results helped to validate the apparatus. Then he spent the rest of his time tinkering with the device to explore its operating limits. In particular, he tested it using different combinations of sample and buffer fluid concentrations in order to determine the maximum levels that the process could work with.

Owing to the mechanical complexity of the device, Walker also had spare pumps and other items that he could insert in the event of a mechanical failure. In particular, he had a de-aeration pump for a recurrence of the bubble problem. And sure enough, shortly into the test, bubbles started to form. For short flights, this would not threaten the mission objectives, but the CFES was about to start production operations and as Walker says, "That could really become a major problem in the long term, many tens of hours of continuous operations. When you need a stable electrophoresis process, bubbles could cause instabilities." He decided to install the pump. This was not an easy task, because it involved getting direct access to the interior of the apparatus. "I literally took the side panel off the flow chamber apparatus. It was probably 4 feet in height and 1.5 feet wide, and bolted onto the side of the flow apparatus with 20 or 30 quick-release bolts. And there was a seal to prevent any liquid that might be leaking inside from getting out into the cabin. But I needed to get in there to interconnect this de-aeration apparatus." With the pump successfully installed, the device ran bubble-free for the remainder of the flight. At the end, Walker reversed the assembly process to remove the pump and rebuild the side panel that he had removed to make room for the pump.

In line with the need to run the CFES as if it were operating already in production mode, at least twice per day Walker sampled the products, not only to ascertain their concentration but also their cleanness. Mindful of the contamination which impaired the first flight, he inserted a small quantity of liquid into a compact test kit. Bacterial contamination would be shown by visual evidence of a growing bacterial colony. "I was literally going to be able to see, in flight, if I had contaminated materials or if the process had become contaminated. We never saw that. So I was verifying…that we were running without the problems that had developed on 41D."

STS-61B was unquestionably the most successful flight of the CFES, as indeed it should have been, given all the development work and prior in-flight testing. Back in St. Louis, laboratory tests confirmed that the apparatus had separated and purified the material as required. Everything was looking good for phase two.

---

[4] From the first US manned flight into space through to the early years of Shuttle program, it was standard practice to have a backup crew for each mission to guarantee that a given mission would be able to be flown on time in the event of a member of the primary crew becoming unavailable. With this in mind, NASA convinced McDonnell Douglas that they too should have a backup for Walker, as it could have been difficult for a mission specialist to work on the CFES if Walker was unable to participate.

Unfortunately, the tragic loss of *Challenger* and its crew on January 26, 1986, had a profound negative influence on the future of the Shuttle for commercial processing. The surviving fleet of Orbiters were grounded for almost 3 years while a variety of issues were fixed. Then on September 29, 1988, STS-26, the return-to-flight mission, renewed flight operations. However, not only had the hardware and procedures of the Shuttle been revised, so, too, had its mandate. From now on, it would be restricted to carrying out scientific research, to the assembly of the space station, and to deploying big telescopes and deep space probes. It would no longer carry commercial satellites into space and there would be no room for activities such as the McDonnell Douglas electrophoresis project. With no other means of transportation, the project was unable to proceed and had to be abandoned.

Reflecting on those incredible days when he and his colleagues were developing the CFES, and he was flying in space, Walker muses, "We were all enamoured with the prospect of ready access to space for all interests in this nation, government, civil, and commercial. And the prospects that we all believed were going to flow from that were just going to be enormous. So it was a period of time in which there was a lot of enthusiasm for exploiting and applying technology, and commercial opportunities in low Earth orbit via the Space Shuttle. All of us had, in the back of our minds, at least, the prospect of doing something useful for the nation and for all of humanity in the medical arena. All of us had in mind that the objective was to provide useful products for the medical research community and treatments for diseases."

McDonnell Douglas was a visionary company which had understood that space is not merely a playground for science but also a fertile place for a new economy based on products created directly in space. The company was far ahead of its time, since industry in general was not yet ready to recognize the space environment as a source of economic growth. As the second decade of the 21st century now draws to a close, we are starting to see the emergence of an industry that views space as a territory for new enterprises and economic opportunities.

"I WANTED TO BE PART OF THE ACTION"

During development of the Space Shuttle Orbiter the middeck was conceived simply to support the crew in mundane activities such as meal preparation, resting, personal hygiene, housekeeping, exercising, and so on.

Most of the crew equipment and their provisions were stowed on the middeck in lightweight lockers made of epoxy-coated or polyimide-coated Kevlar honeycomb panels joined at the corners with aluminum channels to provide for structural rigidity and bearing of the launch and re-entry loads. Typical dimensions were 11 x 18 x 21 inches, each providing about 2 cubic feet of volume for up to 66 pounds of contents. Designed for modularity, the lockers were interchangeable. They used spring-loaded captive bolts as attaching fasteners to the vehicle structure. In this way, they could be installed wherever there was room, and also removed and installed elsewhere by the crew in flight. There were some lockers on the flight deck to hold flight manuals or trash, but 95% of the storage was on the middeck. The usual configuration consisted of plastering the middeck forward wall with up to 33 lockers. A niche with room for 9 additional containers could be found on the back wall, to the starboard side of the airlock module. At times, lockers were replaced with the apparatus for experiments.

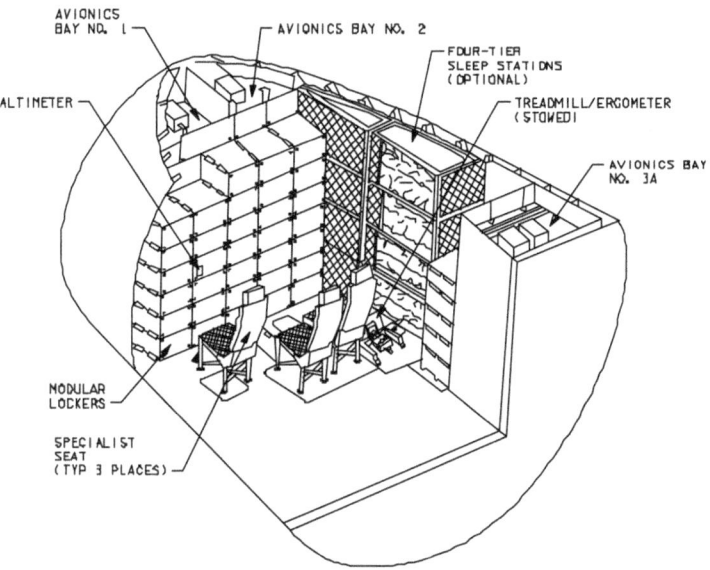

Middeck forward modular lockers.

The locker side which was directly accessible by astronauts had a door hinged to one of the long edges. In this way, the door prevented the contents from inadvertently leaving the container. Once on-orbit, the door was released and opened a full 180° by rotating inboard two quarter-turn self-aligning ball latches that provided for structural retention of the door against the locker structure for launch and re-entry. In space, the door was held closed by double magnetic latches that afforded easy and fast opening and closing. The contents of the lockers were often stashed on insertable trays which could be adapted to accommodate a wide variety of soft goods, loose equipment, and food. They came in four different sizes: single trays (two of which could fit inside a locker), double trays, half-length single trays (four of which could fit inside a locker) and half-length double trays. Foam inserts helped to retain the trays in position inside the locker in weightlessness. Some trays had straps, snaps, and mesh retention nets to retain their contents in place.

The forward bulkhead included a pattern of holes for the attachment of straps and brackets via pip pins to restrain additional equipment or mount experiments on-orbit. In the event that a locker door could not be closed or latched owing to misalignment problems, it would be removed and replaced with turnbuckles to maintain structural integrity during re-entry.

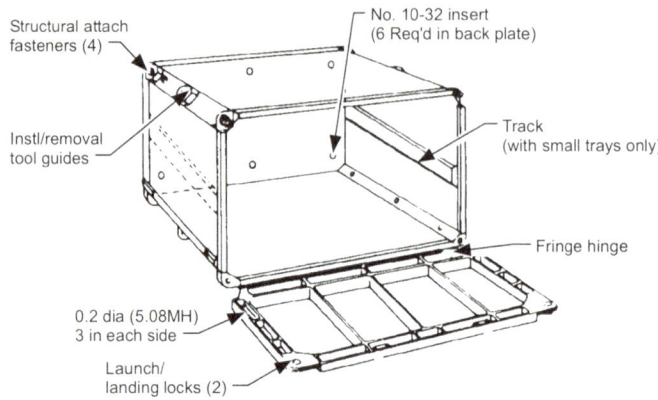

A schematic of a modular locker.

While on-orbit, lockers were often used as mounting platforms for experiments.

Later, the Middeck Accommodation Rack (MAR) was added to the complement of stowage for supplemental small payloads and experiments. It would be housed just forward of the side hatch, aft of the galley. A MAR could provide about 15 cubic feet of volume for a maximum payload of about 340 pounds, and it could draw power to run powered experiments. An active thermal control system that used either a water-to-air heat exchanger or water circulating through cold plates was able to dissipate up to 1,000 watts

Lockers could easily be replaced with apparatus for experiments. Other experiments could fit inside a locker and present a custom-built interface on the door to allow an astronaut to undertake the investigation.

of thermal load. The Lightweight MAR (LWMAR) was introduced for unpowered payloads. This provided approximately the same storage capacity, but its lighter structure of carbon-fibre composite material enabled it to hold up to about 390 pounds of payload.

In retrospect it is rather surprising that little thought was initially devoted to using the middeck as a rudimentary laboratory for science and technology investigations. But NASA soon realized this potentiality when the early flights demonstrated that the astronauts were at ease in handling and doing simple experiments within its confines. In addition, the middeck was also recognized as a cost-effective way of performing "crew-tended" scientific and commercial microgravity research in a manner that best used the available crew time. The success of those early flights was such that there was soon a backlog of requests for middeck investigations. This situation obliged the agency to slip an ever increasing number of experiments to later flights, to the clear disappointment of the eager science community. In fact, despite the large number of lockers, most of them carried cameras, clothing, food, tools, and various other objects that were required by the crew in carrying out the mission. Only a few could be set aside for experiment apparatus. This situation required to be resolved as a matter of urgency.

In the last decade, we have grown accustomed to witnessing privately owned and commercial non-governmental enterprises staking out various claims in the realm of human space flight. In all likelihood, by the end of 2018 astronauts heading for the International Space Station will catch a ride on a capsule privately developed in the USA rather than a venerable Russian Soyuz capsule. Plans are already being drawn and contracts penned for

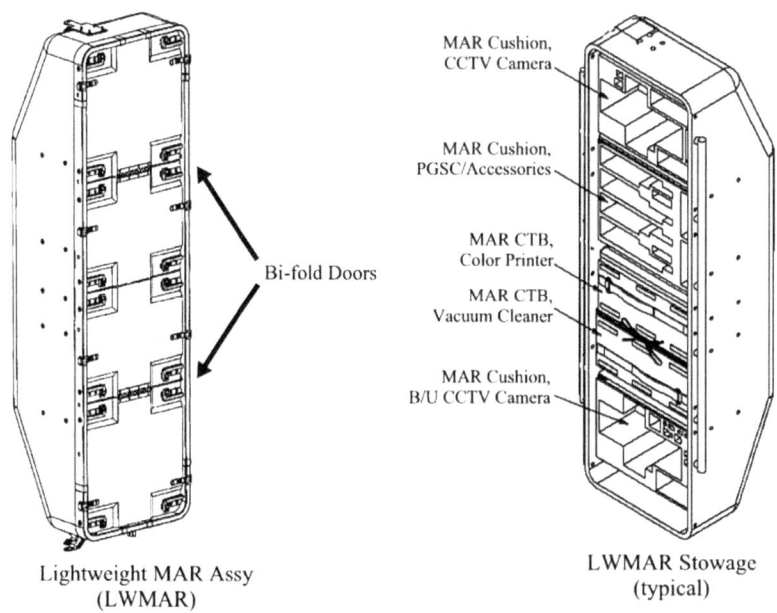

MAR Cushion,
CCTV Camera

MAR Cushion,
PGSC/Accessories

MAR CTB,
Color Printer

MAR CTB,
Vacuum Cleaner

MAR Cushion,
B/U CCTV Camera

Bi-fold Doors

Lightweight MAR Assy
(LWMAR)

LWMAR Stowage
(typical)

A schematic of the Lightweight Middeck Accommodations Rack.

commercial space stations, for private mission to the Moon and even Mars by wealthy and visionary billionaires.

In the early 1980s it was almost preposterous to advocate that human space flight could be accomplished with private money rather than taxpayer funding. This did not deter aerospace engineer and entrepreneur Robert A. Citron from founding the Space Development Corporation (SDC) in Seattle, Washington. In his own words, the goal "was to become the first privately financed company in the world to support human space flight." In fact SDC was intended to be the holding company for seven space ventures by which Citron planned to cover different commercial space markets such as space manufacturing and space tourism. "I was excited by a flying Space Shuttle fleet and I wanted to be part of the action."

As luck would have it, this was just the right time because NASA was seeking a means of relieving the middeck experiment backlog. Citron soon realized that what the agency needed was a way to augment the usable habitable volume of the Orbiter. His solution took the shape of a pressurized module of modest dimensions that would be installed at the forward end of the payload bay and connect to the middeck using a Spacelab-type tunnel. Inside its volume, a great number of middeck lockers could be mounted in order to greatly increase the total number of experiments which could be attended by astronauts. The module could also house racks of the type designed for the space station, to test technologies and experiments that were to be installed on the station. By May 1985 Citron had applied to the US Patent Office for a patent on this concept and this was granted in September 1989.

Meanwhile, SDC established SPACEHAB Incorporated as the first company in the SDC portfolio. This was to focus on the development of a research and logistics module for the Shuttle.[5] Citron and his employees at SPACEHAB Inc., embarked on an advertising campaign aimed at raising awareness amid the commercial space payload developers and raising private sector financing. Between 1985 and 1986 they gathered several million dollars in private equity investment. With these funds, they authorized Alenia Spazio in Turin, Italy, to begin engineering studies of Phase-A for the pressurized facility. The Italian company had vital experience in the design and fabrication of pressurized modules for human space flight, having built Spacelab. In addition, Citron received encouragement when Rockwell International[6] confirmed that the proposed module would be able to be carried in the payload bay.

NASA was also keen on the concept because it matched their desire to develop commercial space activities. In fact, by 1985 the agency had already supported the founding of seventeen Centers for the Commercial Development of Space (CCDS) right across the country, each focused on one or more of eight industry-driven space-based high-technology investigations in materials processing, biotechnology, remote sensing, communications, automation and robotics, propulsion, space structures, and space power. Each center received annual funding of up to $1 million, and in return, NASA borrowed their scientific and technical expertise for activities in cooperation with private or industrial partners. A vital facet of these centers was the additional financial and in-kind contributions being received from industrial affiliates, state and other government agencies that typically exceeded the NASA funding level. So from its inception the CCDS program served as an incubator for future commercial space industries by facilitating new commercial space ventures designed to spawn research and development that promised enormous social and economic benefit for all parties involved. The module being presented by Citron promised exciting opportunities for these NASA-sponsored commercial space activities.

In December 1985 NASA signed a memorandum of understanding with Citron's company by which NASA would cooperate in the development of the project. In the spring of 1986, negotiations for a Space System Development Agreement (SSDA) got underway to outline the conditions under which a SPACEHAB module would be permitted to fly aboard the Shuttle. By the fall of 1986, engineering studies of Phase-B were initiated with McDonnell Douglas Aerospace in Huntsville, Alabama, as the prime contractor and Alenia Spazio as the principal sub-contractor. In the meantime, Citron continued to engage in market research to assess the existing and likely future demand for such a Shuttle-based module, both at national and international levels.

In 1988 a new SSDA was signed to ensure that SPACEHAB modules would be on at least six flights, starting in 1991. The agreement also allowed commitment to construction contracts with McDonnell Douglas and Alenia Spazio. Private investors, investment groups, and a consortium of banks led by the Chase Manhattan Bank, were able to procure a total of $100 million in equity and debt financing to build and test the module. Lloyds of London supplied risk and program termination insurance.

---

[5] Although the name SPACEHAB is written in capital letters is not an acronym.

[6] As Rockwell International built all the Space Shuttle Orbiters, they were well suited to critically evaluate the Citron proposal.

The tragic loss of *Challenger* in January 1986 slowed work on the SPACEHAB development, and it was almost a victim of the new policy of the Shuttle no longer flying commercial payloads. However, the grounding of the fleet for almost 3 years greatly increased the backlog of middeck experiments and the new module was the only realistic way to satisfy this demand for experiments in the space environment.

In the light of President Ronald W. Reagan's National Space Policy of February 1989, NASA issued an industry-wide request for proposals for what it had decided to call a Commercial Middeck Augmentation Module. This would accommodate both the microgravity experiments developed by NASA-sponsored CCDS and also "crew-tended" experiments for advanced systems development, particularly those meant for the space station. NASA was well aware that Citron's company was offering exactly what they required, but federal regulations demanded the agency run a competition in which any American company could bid for the contract.

Unsurprisingly, there was only one bidder and in December 1990 SPACEHAB Inc. was awarded a $184 million contract in which NASA would lease at least two-thirds of the payload space for 8 missions, becoming an "anchor tenant" and thereby addressing the backlog of commercial experiments. The company could rent out the remaining capacity to other parties wishing to have their own experiments carried on the Shuttle.[7] Of course, all experiments had to pass NASA standards.

By the end of 1992 Alenia Spazio had delivered two SPACEHAB modules, plus a third one for ground trials. Meanwhile, a payload processing facility was built near the Kennedy Space Center in Florida to house the modules in between flights, as well to integrate and de-integrate experiments and provide training for experimenters and payload specialists alike.

SPACEHAB VERSUS SPACELAB

To a casual observer the SPACEHAB and Spacelab pressurized modules might look rather similar, but they were very different in detail. The SPACEHAB module had a volume of 1,000 cubic feet and was a cylinder 10 feet long and 13.5 feet in diameter that was truncated on top to offer a mounting platform for external experiments that would receive power and data through two feedthrough plates. If not required, these plates could be replaced with glass to facilitate investigations requiring viewports for Earth or deep space observations. Pressurized access to the Orbiter was provided by a tunnel adapter that connected to a transition section. Spacelab was designed to carry out complex experiments in large racks, but SPACEHAB was purposely designed to carry miscellaneous Shuttle middeck lockers and space station racks. In other words, SPACEHAB was to carry a lot of small experiments rather than a few large ones. It offered flexibility. In its original configuration, the module could fit 79 middeck-type lockers, or a double Spacelab-type rack and

---

[7]The difference between the SSDA signed in 1989 and the NASA contract ratified in 1990 was that the former allowed a SPACEHAB module to fly on the Shuttle as privately owned payload with commercial and private payload along with NASA own experiments. The latter formally specified NASA as a user of the module to fulfill its own needs and become an anchor tenant for the first eight flights.

57 lockers, or two double racks and 45 lockers. Experiments that did not fit into a locker or rack could be accommodated by mounting plates located on the walls and ceiling.

In contrast with Spacelab, the SPACEHAB module was designed to occupy only the first quarter of the payload bay, enabling other, potentially very large, items to be carried to maximize the objectives of a mission.[8] While Spacelab had been financed in Europe at governmental level, SPACEHAB was a privately funded commercial venture. This is what made it possible to develop a payload carrier for the Shuttle that offered flexible accommodations for experiments, simple interfaces, and thanks to a faster turnaround time, more frequent flight opportunities than could be provided by a Spacelab module.

In flight, the Orbiter supplied the power to feed the experiments and the module's environmental control subsystem. The design was integrated, so the heat rejection for both subsystems and experiments was by a link up with the vehicle's thermal control system. The fire suppression subsystem consisted of one handheld fire extinguisher, smoke sensors, fire suppression bottles, and other equipment. Detection and warning of an unsafe condition were provided by the Orbiter's caution and warning system, which worked together with the module to alert the crew of any subsystem failure. Upon detection of an alarm condition, a signal would be sent to activate the master alarm light in both the Orbiter and the module. Panels on the flight deck allowed the crew to control and monitor subsystem operations, including a dedicated display for interaction with the module software.

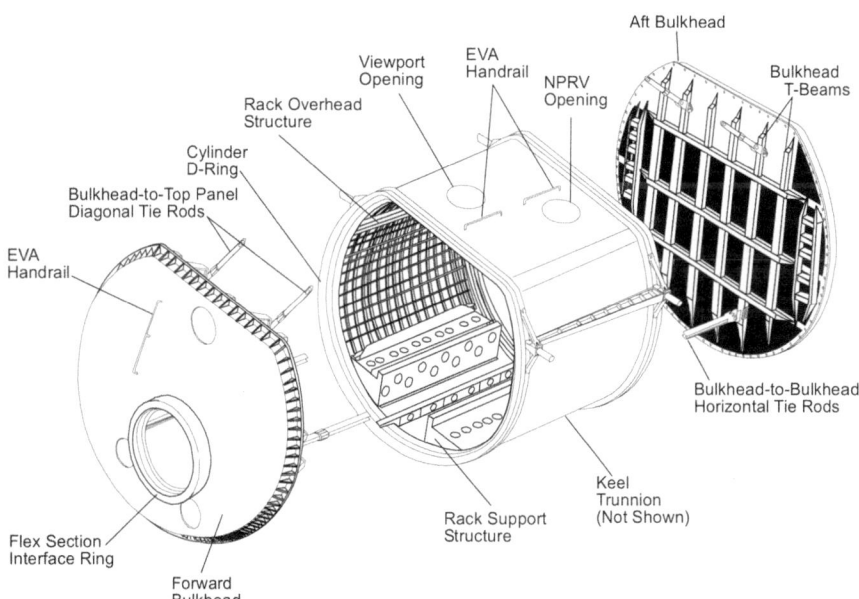

The main structural elements of a SPACEHAB module.

---

[8] The dimensions of a Spacelab laboratory meant it required a dedicated flight with no sharing opportunities with other payloads.

A SPACEHAB module (far right) is being loaded into the payload bay, along with a full complement of additional payloads. The flat ceiling that characterized this type of module is clearly visible.

A SPACEHAB module is undergoing ground processing in preparation for an upcoming mission. Clearly visible is the interior being filled with middeck-type lockers and space station-type racks.

SPACEHAB MISSIONS

SPACEHAB's maiden flight began on June 21, 1993, as one of the primary payloads of *Endeavour* on the STS-57 mission. The 21 experiments addressed a wide range of investigations ranging from improving drugs to feeding plants, from cell splitting to intergalactic particles, from the first soldering experiment in space by US astronauts to high-temperature melting of metals. In line with the goal of using the module to foster the development of commercial space payloads, a dozen of the experiments were sponsored by some of the NASA Centers for the Commercial Development of Space and one by the agency's Langley Research Center in Virginia. In addition, the Environmental Control and Life Support System (ECLSS) Flight Experiment (EFE) was to test components of the water recycling system intended for the space station.

As described in the mission post-flight report, SPACEHAB's first flight was an unqualified success because it achieved more than 90% of the planned objectives. The structure survived the launch undamaged, and there was no loss of pressure when the module was isolated from the Orbiter to enable astronauts to carry out a spacewalk. The avionics subsystems (fire detection and suppression, command and data, crew communications, and displays and controls) performed properly. So did the electrical subsystem, which consumed less power than predicted. Interestingly, this affected the environmental control subsystem as, at the beginning, it kept the module at a cooler-than-expected temperature. For this reason, two unscheduled in-flight maintenance procedures were performed to adjust a cold water bypass valve manually. This intervention succeeded in raising the internal temperature. Once at this optimal setting, the subsystem performed well for the rest of the mission. The successful first flight of the SPACAHAB module was also a clear demonstration that the concept worked, and that commercial space payloads could be relatively easily and rapidly manifested onto a Shuttle flight. At last, the backlog of middeck experiments could be tackled apace.

SPACEHAB modules were flown on 18 Shuttle missions, using four different configurations. The inaugural flight used what came to be known as the Research Single Module (RSM). As the name implies, this was dedicated to experiments for science and technology research. The Shuttle-Mir program made substantial use of the SPACEHAB module, this time not for research but rather as a logistics carrier to transport apparatus for the Russian space station. This Logistic Single Module (LSM) was first flown on STS-76. The flight by STS-79 to Mir saw the debut of a third configuration, the Logistics Double Module (LDM) which consisted of two LSMs joined together by an intermediate adapter for increased capacity. The fourth configuration, a double module dedicated to research and therefore known as the Research Double Module (RDM), was introduced by STS-107, which proved to be *Columbia*'s last mission. SPACEHAB flew twice more, on STS-116 and STS-118 using the LSM configuration with supplies for the International Space Station.

All subsequent logistics flights to the International Space Station employed the Italian-built Multi-Purpose Logistics Modules, which not only had a much greater capacity but could be temporarily berthed on the station for easier transfer of bulky cargo.

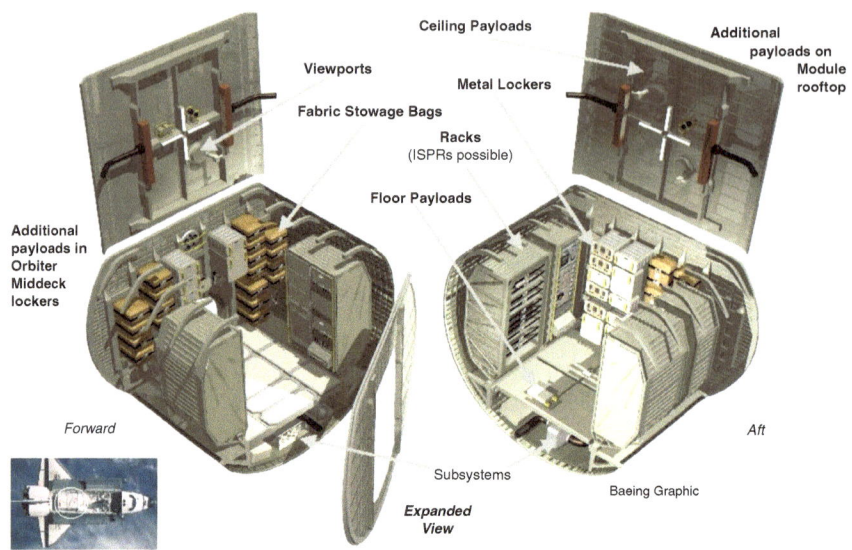

Ceiling Payloads

Viewports

Metal Lockers

Fabric Stowage Bags

Racks
(ISPRs possible)

Floor Payloads

Additional
payloads on
Module
rooftop

Additional
payloads in
Orbiter
Middeck
lockers

Forward

Aft

Subsystems

Baeing Graphic

*Expanded
View*

An exploded view of the Research Double Module flown on STS-107.

A PIONEERING INITIATIVE: THE INDUSTRIAL SPACE FACILITY

Maxime Allen "Max" Faget is, and will always be, a prominent figure in the NASA history books. He gained his degree in mechanical engineering from Louisiana State University in 1943, spent 3 years as submariner in the US Navy, and then turned his extraordinary mind toward the sky. He began by working on the experimental X-15 hypersonic rocket plane, then joined the fledgling NASA as one of the 35 engineers of the Space Task Group, where he played a leading role in designing the Mercury capsule. He went on to make major contributions to Gemini and Apollo and then the Shuttle.

As one of the chief architects of the early American space program, nobody could have expected that Faget would pose a formidable threat to the agency which he had done so much to help win the space race. His new role started in 1981 when a trio of Houston University academics, eating at one of the local delicatessens, brainstormed commercial space activities.

Space architecture professor Guillermo Trotti, Larry Bell, and James Calaway, who had just returned to Houston after graduating from the University of Oxford in England, spent hours musing about a commercial space station for the production of goods of superior grade that would be sold on Earth. The weightlessness of the space environment eliminates the disturbances that gravity imposes on exquisite and fragile industrial processes, such as pharmaceuticals and computer microprocessors. NASA had studied the potential of space for manufacturing and the exploitation of lunar and asteroidal resources, but this was merely conceptual thinking. The only real materials science experiments had been carried out aboard Skylab in 1973.

Nowadays we are growing accustomed to the concept of space operations being undertaken by private companies, often owned by visionary billionaire entrepreneurs, but in the 1980s things were very different. As the group of Houston academics soon discovered, the main hurdle facing any such proposal was securing financial backing. It was around this time that Faget retired from NASA. Bell decided to contact him in the hope of getting his endorsement. If someone as respected as Faget supported the plan, surely this would indicate that its advocates had done their homework and that their project was based upon sound engineering. Faget was immediately hooked, and lost no time in converting the concept into engineering.

After some innovative thinking, Faget had the proverbial Eureka moment. As he recalls, "The basic idea was that we built something that could be launched on the Shuttle and had a pretty good amount of volume. Nothing had to be added to it. Bring it up to orbit, turn it loose, and it would work for 30 to 90 days, or maybe 120 days. You go up and rebuilt it, resupply it, leave it up on-orbit, and the thing would work. It was to be a man-tended facility with an internal volume that was kept pressurized. Man could enter it. They would live off the life support system on the Shuttle simply by transferring air [across], so we wouldn't have to put a lot of life support system in there. When you wanted to make it bigger, you'd just add another unit to it. We took the viewpoint that [if] you want to go into space, it's sort of like forming a beachhead anyplace. You don't go in there with the idea of building high, multi-story buildings, first class buildings. You go in there with putting down enough to support a landing, a beachhead, and you build from there. It's the way we explored the frontier, here. We had small cabins and we built up from there. After two or three launches of this thing, you'd get out onto it. Each one would be independent of the other. That had a pretty nice effect. If one of them failed, you'd just move into another one. They were independent and self-sufficient, but you could equip one with a life support system in it that would clean air for all the others. You could get up to maybe about six or eight of these things attached together for the equivalent of a space station."

The beauty of the proposal was that the expansion of the facilities on-orbit would be driven by the proven value of their production. Even the first one would be able to deliver output. This proposal was aptly named the Industrial Space Facility (ISF). To make it into orbit, a company called Space Industries Incorporated was established. It was not long before private funding began to appear. Thanks to the oilman and future US ambassador to Austria Roy M. Huffington, the early investors included James Elking, co-founder of the Vinson & Elkins law firm, and developer Walter Mischer. Later, NASA astronaut Joe Allen and the Westinghouse Electric Corporation joined the venture.

With his excellent contacts at NASA, Faget was able to secure an agreement by which the space agency would provide three free Shuttle rides having a total value of $750 million. This deal would enable the company to launch and make operational two ISFs. Subsequently, once Space Industries had begun making a profit by selling manufactured goods, they would be able to pay NASA for servicing the facilities on-orbit. This would greatly assist both parties. In particular, the cash saved by the three launches could be spent on the research and development of the ISF. In addition, the company would have extra time to establish and mature commercial relationship that would generate revenue. The space agency would be assured of a supplemental flow of income.

Because the ISF had to fit within the Orbiter's payload bay, it took the form of a Spacelab-type cylinder, 35 feet long and 14.5 feet wide. This would be the so-called "facility module" in which microgravity materials processing, research, and various other projects would occur. This flexibility was of paramount importance to the very survival of the entire endeavor.

In fact, Faget and his team were conscious of the McDonnell Douglas experience with electrophoresis processing. Joe Walker had shaken down and proved the CFES could yield the expected products in microgravity. But when the time came to initiate full-scale production, the main (and at that point only) supporter retreated in response to technological advances in bioengineering that undermined the case for production in space. McDonnell Douglas had then failed to attract interest in using this apparatus for other applications. Faget knew the ISF had to offer amenities to suit the needs of many customers. For this reason, it would have at least seven identical experimental space station-type racks. There would also be room for six modular containers, each filled with four Shuttle middeck locker trays. The environmental life support system apparatus was added to restore shirt-sleeve conditions during a Shuttle crew visit. An outer shell would give thermal protection and shield against micrometeoroids. A pair of 105-foot-long solar arrays with an area of 3,000 square feet would provide a total output of 28 kW, some 20 kW of which would be at the disposal of the production hardware. Development of the power plant had already started by a contract awarded to Lockheed Martin, which had experience with the OAST-1 solar array deployment mechanism tested by STS-41D.[9]

The facility was also being planned with a view to the future to ward off a CFES-style fate. Four berthing ports, one at each end and one on each side, would provide expandability, allowing additional modules to be connected. One of the two end ports would be used to connect a 5.9-foot-long and 14.5-foot-diameter auxiliary or support module that, basically, would be a systems module containing additional experiments and provisions for the manufacturing activities. It would also include a water vapor-based orbital maneuvering system which the ISF could autonomously use to enter a higher orbit after a Shuttle visit or return to a lower altitude to greet the next visitor. Although water is not one of the most powerful rocket propellants, its selection was nevertheless a clever one in this case because it would make use of the surplus water created by the fuel cells of the Orbiter. The module would also have a 100-foot-long retractable boom that had a 193-pound mass at its end to provide a gravity gradient attitude, thereby eliminating the need for an active control system that would tend to disturb the microgravity that was vital for the manufacturing processes. If necessary, the module could be changed for a fresh new one during a Shuttle visit. Apart from a dedicated flight to deploy the facility, the servicing flights could be scheduled as part of other missions, such as when carrying commercial satellites, in order to minimize the incremental costs to Space Industries Inc.

Despite the enthusiasm generated by the press and industry, the time was not yet right for a privately funded commercial space manufacturing facility, and by the mid-1980s there were still insufficient private players willing to book and pay for payload space on the ISF, so the company changed strategy. At that time, Courtney A. Stadd was the Director of the Office of Commercial Space Transportation within the Office of the Secretary of the US Department of Transportation (DoT). He was monitoring the situation, and recalls that Faget suggested "that the government act as an anchor tenant. In sum, what that entailed was NASA's agreement, up front, to purchase a share, if you will, or agree to procure, a

---

[9] Refer to Chapter 4 for details on the OAST-1 experiment.

part of this facility, much the same way that when a real estate firm is establishing a shopping mall they look for a prestigious department store to be the first tenant; the obvious hope being that they become a magnet for other tenants, and the result is, hopefully, a viable facility."

NASA had been willing to give free rides to the ISF, but awarding real money to a venture which had an uncertain return of profits was entirely different. In fact, the agency was worried that there might not be sufficient funding to support the ISF and also to build its own space station. In effect, the ISF was becoming a threat to NASA. As Stadd points out, "During the Reagan era, with its focus on market-led initiatives, commercialization, [the ISF initiative] was quite favourably received by the White House; very well received by those of us who were working in the commercial space policy arena at the Department of Commerce and the Department of Transportation. And there was a lot of sympathy for this concept at the Office of Management and Budget, particularly since it promised to save the taxpayer some monies in terms of developing a research facility on-orbit. We [in DoT], our allies at Commerce, people within the White House, and [in the] OMB, were successful in having the President specifically call out support for the ISF in an executive order, which, when you think about it, is rather extraordinary, for a President to specifically note an initiative in such detail."

As a result, NASA was ordered to make an immediate commitment to lease 70% of the ISF with an annual investment of $140 million for a period of 5 years. The ISF was definitely a threat. And of course, the ISF was designed with growth in mind in a similar manner to that of the space station. This was something that the detractors of the space station eagerly noted. As Faget says, "If this thing ever got up there and we started adding onto it, people would say 'Why do we need the Space Station? Why don't we just keep going this way?'"

Worse for NASA, the simplicity of the ISF held out the prospect of space-worthy hardware becoming available for launch much sooner than that for the space station, the Phase-A study for which Congress had barely granted approval. The agency was also well aware that several high places in Washington DC were skeptical about its ability to deliver the station within the proposed timeline and budget. In view of how long it took NASA to introduce the Shuttle, such doubts were easy to understand. As Stadd says, "The agency was probably feeling – indeed, was feeling exposed in terms of getting solid support for the Station." As a frightened animal fights for its survival, so NASA launched a war against what it perceived to be a predator. Stadd continues, "The agency really went out of its way to oppose the ISF, and they even went to the extent of going to the Academy of Sciences to get a report that would suggest that the facility, as structured, was not as viable as Mr. Faget and others were suggesting. The House Science Committee had been sufficiently exercised by NASA and its allies – the contractors – that they were also very concerned that the ISF was driven more by ideology, more by free-market Reagan Republican ideology, than by practicality, and were concerned that perhaps it was a misuse of taxpayer money and an unnecessary diversion from the efforts to support the NASA-proposed Space Station."

Compared to NASA, Space Industries Inc. was a small fish. By the end of 1989, the agency had derailed Max Faget's latest creation. As Stadd concludes, "I feel very strongly that if it had been supported, we would have had this facility in place years ago, and, frankly, I think it would have generated a robust research community that would have been eager for the larger capabilities of the International Space Station that we currently have on-orbit. So I think it was a most unfortunate turn of events."

ULTRA-VACUUM: THE WAKE SHIELD FACILITY

Despite the unfortunate demise of the ISF, there were other parties seeking to exploit the orbital environment to conduct experimental work with a view to establishing in-space manufacturing. In addition to microgravity, low orbit offers another condition that is impossible to reproduce satisfactorily on the ground, namely an extremely low atmospheric pressure or hard vacuum. On Earth, several industrial processes require to be run within a vacuum that is as pure as possible to yield products of the highest quality.

Chief among these is the manufacturing of silicon wafers, microchips and Micro Electro Mechanical Systems (MEMS). Microchips and MEMS devices sport features in the micron, almost nanometer size,[10] and sophisticated techniques are required to deposit materials on a silicon substrate (wafer) to build an exquisitely precise piece of hardware like a computer chip or an accelerometer and gyroscope of the type that a smartphone incorporates so that you can play video games.

In 1990, $56.8 billion was spent worldwide on making microelectronic devices, of which 40% were for computer applications, 18% for communications, and 15% for the military. By 1994 it was estimated that the market for semiconductor devices would be $109 billion. Although silicon is the most widely adopted semiconductor for microelectronics,[11] there are other materials with better predicted performances in terms of power consumption and operating speed. One good example is provided by devices based on gallium arsenide (GaAs). In 1990 these devices accounted for a meager 0.5% of the whole semiconductor market. However, it was predicted that by 1995 this share would grow to at least 2%, and perhaps more if there was an increase in the supply of high-quality GaAs. In fact, the stumbling block was that GaAs of the required quality could be only artificially created using a deposition technique called Molecular Beam Epitaxy (MBE) or epitaxial thin film growth. In general terms, this is a powerful tool for synthesizing new materials with prescribed characteristics and fabricating novel microelectronics devices.

Put simply, MBE requires shining one or more beams of atoms or molecules, e.g. arsenic and gallium, onto a pre-heated substrate that forms an atomic template, or pattern, upon which the atoms or molecules are deposited to make a thin film or layer which follows the crystal pattern of the substrate. Multiple layers can be grown by a single process, each with a perfect interface with the neighboring strata. Being such a delicate process, it can only work in ultra-vacuum conditions, where the presence of contaminants that can spoil the film is reduced to be point of being almost negligible. Obviously there are limits to the vacuum that can be achieved on the ground, and the deeper the vacuum the harder and more expensive it becomes to attain and maintain that condition.

The ideal vacuum for MBE would be even deeper than that in low Earth orbit. In fact, by the 1970s NASA had already outlined the concept of an "ultra-vacuum" that could be produced in the "wake" of an object traveling at orbital velocity. A suitably shaped object would push aside the few atoms and molecules at that altitude, leaving even fewer (if any) in its wake. It was estimated that this vacuum would be 1,000 to 10,000 times better than

---

[10] A micron is one millionth of a meter, while a nanometer is one billionth of a meter.

[11] A semiconductor behaves either as a conductor or as an electric insulator upon careful manipulation and exploitation of its atomic properties.

the best that could be achieved in a vacuum chamber on the ground. Theoretical models predicted that a concave disk traveling in low orbit might produce a wake pressures of about 10^{-14} torr. But the concept did not find a practical application until 1987 when the Space Vacuum Epitaxy Center (SVEC)[12] formed a consortium of industry, academia and government laboratories to create and analyze an ultra-vacuum in low Earth orbit and use it for growing epitaxial thin films. One of the main tasks would be the growth of gallium arsenide wafers. Some 2 years later, the consortium assigned to Space Industries Inc. the task of building a free-flyer that would be launched aboard a Shuttle, deployed on-orbit to perform its work, and then retrieved for return to Earth.

This project was named the Wake Shield Facility (WSF). It comprised the Shuttle Cross Bay Carrier (SCBC) and the free-flyer, with the approximately 9,000 pounds about equally divided between the two pieces of hardware.[13] Because the disk-like platform was carried in a horizontal configuration, it consumed 25% of the payload bay volume. The SCBC supported the free-flyer during ascent, re-entry, and on-orbit when the platform was not flying solo. It doubled as stand-alone communications system, acting as a radio bridge between the Orbiter and the platform, even when the two were tens of miles apart. There was room on the supporting structure for GAS-type canisters that could benefit from the data and power resources provided by the SCBC.

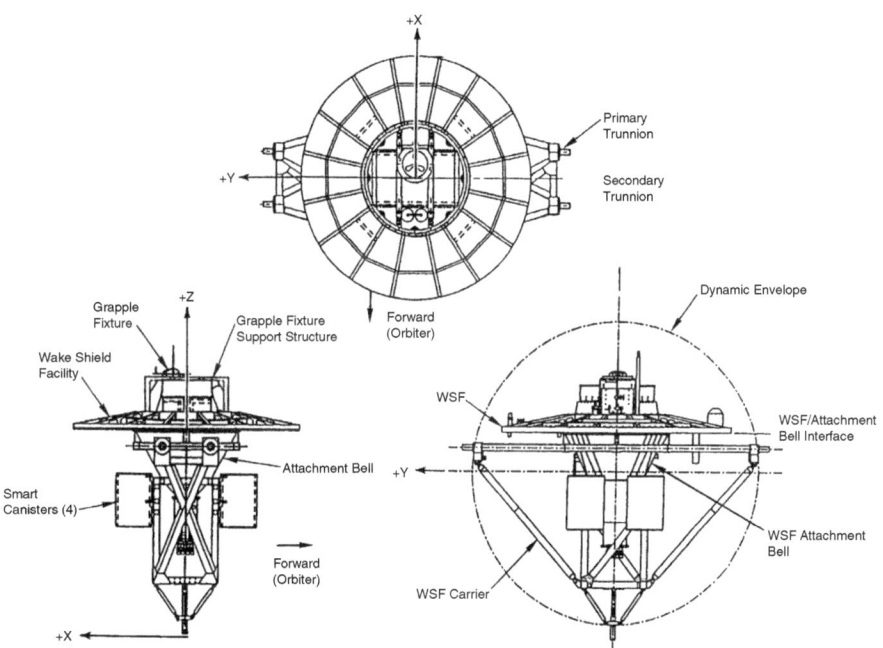

The main components of the Wake Shield Facility.

---

[12] Based in Houston, the SVEC was another NASA Center for the Commercial Development of Space.
[13] Although the free-flyer is often referred to as the WSF, strictly speaking the WSF was the combination of the SCBC and the free-flyer.

The free-flyer was a 12-foot-diameter welded stainless steel satellite which was equipped to function as an independent spacecraft. Propulsion for maneuvering away from the Shuttle was provided by a thruster fueled by cold nitrogen gas, and it would sense and maintain its orientation using a momentum-bias attitude control system, a horizon scanner, a 2-axis magnetometer, and a 3-axis magnetic torquer. Silver-zinc batteries with a capacity of 45 kW/h were adequate to power the growth of thin films, the process controllers, a sophisticated array of sensors, and various other on-board experiments.

The WSF was to fly with its disk in a vertical orientation, facing the direction of travel. This would push the few particles as occur at that altitude out of the way and create an exceptional vacuum in its wake. At the origin of the wake side was the so-called carousel, a cylindrical canister that rotated about the axis parallel to the plane of the platform. The carousel held seven GaAs substrates, each with its own heater. Only one substrate at a time could be exposed to the vacuum for processing by MBE. A thin film growth run would start by rotating the carousel until the desired substrate was exposed to space. From its position, the substrate was also looking directly at the source cell assembly. This was a cluster of eight canted molecular beam source cells that were held in place by an eight-rod strut support structure to maintain the center line of the assembly passing through the center of the sample that was locked in the active slot of the carousel. During processing, a molecular beam flux was created by heating one of the source cells. This was deposited on the exposed substrate to grow a thin film layer by layer.

The carousel and cell assembly made it possible to experiment with a number of different combinations of substrate and cell materials. For instance, cells containing gallium, arsenic, silicon, and triethylgallium were used to grow gallium arsenide and silicon-doped gallium arsenide membranes. The flux levels and growth rates were controlled by monitoring the data supplied by a mass spectrometer and total pressure gauges. Meanwhile the quality and uniformity of the films were monitored by a high-energy electron diffraction system.

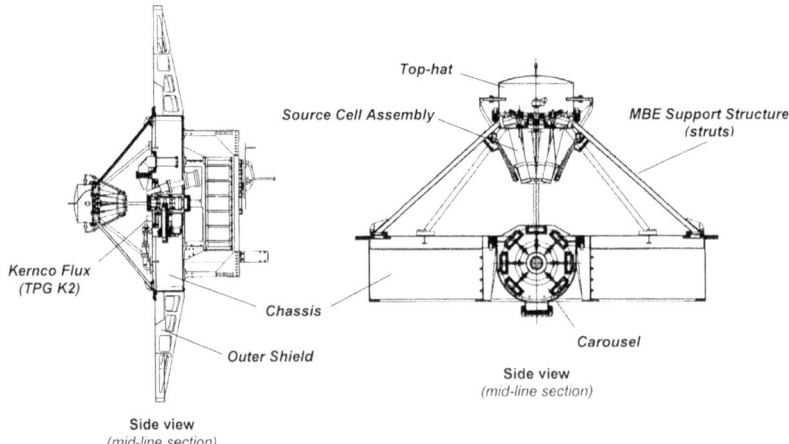

The main components of the Wake Shield Facility free-flying platform.

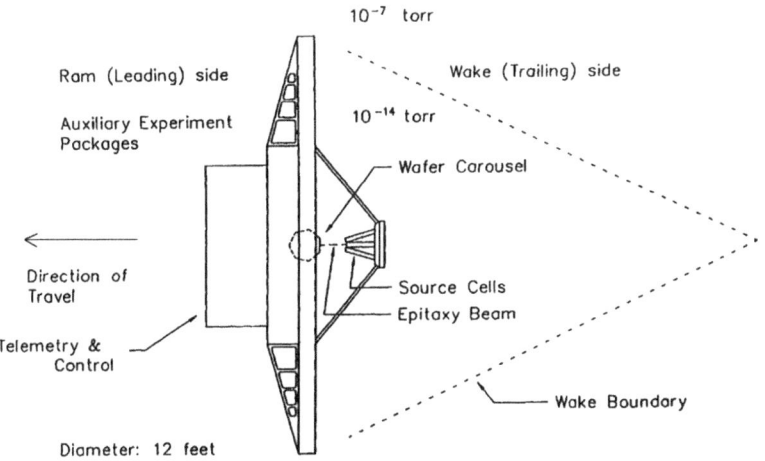

A demonstration of the MBE concept of the WSF.

The structure located between the carousel and the strut of the cell assembly was the chassis. The structure between the chassis and the outer edge of the platform was the outer shield. On the ram side, the outer shield provided some 65 square feet of high quality "real estate" with four attachment points, each capable of hosting up to 200 pounds of additional experiments. The chassis ram side housed the avionics and support equipment.

The WSF held the promise of orbiting factories to produce the next generation of semi-conductor materials, and the devices these would make possible. It would also mark the first time that the vacuum of space would be used to reproduce an industrial process in much improved form.

The inaugural flight of the WSF on STS-60 was bittersweet. *Discovery* lifted off on February 3, 1994, for an 8-day SPACEHAB mission. On the third flight day the WSF was unberthed from the rear of the payload bay, and that was when the troubles began.

As mission commander Charles F. Bolden clearly recalls, "In the 'faster, better, cheaper' mode, when they designed and built the Wake Shield Facility…one of the ways that they saved money was limiting the amount of pre-flight testing that they did. One of the crucial tests that they did not do was something called an EMI test, an electromagnetic interference test, whereby you put the satellite together in its flight configuration and turn it up and find whether there is interference between different components in it. What they satisfied themselves with pre-flight was that if we ran this test, if we powered it up when it was on the floor of the integration room down at the Cape and everything seemed okay, then we'd go with it. And it did. It worked superbly in the test site at the Kennedy Space Center. But when they bundled up all these feet of electrical cable, what we were reminded of post-flight in analysing the problem, was that you get a current generated around an electrical cable, and that generates a magnetic field which, if it comes in contact with another electrical cable, generates a current inside that cable. When we powered up the Wake

Shield Facility on-orbit and tried to turn on the attitude control system there was a 5 Hz signal that was generated by the power cables inside the electrical components of the device that shut down the attitude control system. So every time we tried to power it up, it would start spinning and cut itself off. We didn't have any clue what was going on, we just knew we could not reliably release it. We didn't know what would happen to it. If it got out of control, we may not have any way to control it. So the only option was to just put it out on the end of the arm and take as much data as we could, which we did. And we got some pretty good data."

Some of this concerned the Air Force-sponsored Charging Hazards and Wake Studies (CHAWS) experiment. With the robotic arm holding the WSF far above the cabin of the Orbiter, the experiment measured the plasma flow around the WSF. The intention was to characterize the build-up of electric fields around an orbiting vehicle in order to understand the interactions of the space environment with space systems, and the hazards these interactions create for satellite systems. Despite the let-down of not being able to release the free-flyer, it was clear that the technology developed for the WSF worked as advertised. However, with the platform on the arm, the vacuum in its wake was contaminated by the presence of *Discovery*. Upon returning the WSF to Earth, an advisory committee reviewed the anomalies and approved the necessary corrective actions.

On September 7, 1995, the WSF set off for space again, this time in the payload bay of *Endeavour* on STS-69. The deployment was scheduled for the fifth flight day. The process began by the arm hoisting the platform off its support structure and then positioning it over the port side of the payload bay, to let the atomic oxygen present in low orbit cleanse the side of the platform that would face the direction of travel. A checkout of the Attitude Determination and Control System (ADCS) that had caused so much grief on STS-60 showed that this time it was in good shape. Everything was set for release. All hearts aboard *Endeavour* and the ground must have stopped when it was discovered that the free-flyer and its carrier were not communicating properly. This was to be the sole radio link between the free-flyer and the Orbiter for data, telemetry, and television. After 2 hours of troubleshooting, ground control declared the WSF fit for release.

At 6:25 am CDT astronaut Jim Newman set the experimental facility free above Western Africa at an altitude of almost 250 miles. At last, the Wake Shield Facility was flying in space on its own. A few seconds later, it fired a small cold gas nitrogen thruster to maneuver away from *Endeavour*. This marked the first time in the history of the Shuttle program that a deployed satellite had maneuvered itself clear, instead of remaining in place while the Orbiter took the active role. At 3:33 pm the facility began its first thin film processing run. At about 7 am the following day, it was ready to start the fourth run when a sudden increase in temperature prompted an automatic shutdown to allow the excess heat to dissipate and the facility to cool down. At this point, the WSF pitched forward and put itself into safe mode.

On the ground, payload controllers quickly threw themselves into troubleshooting mode while assessing the impact of this event on further operations. When the issues were understood, the Shuttle managers granted a 24-hour extension to the schedule for the experiment because there was hope that it would be able to resume work and complete all of the intended thin films. The nominal cooling period of 8 to 10 hours between processing runs was also extended, in the hope of preventing a second safe mode. Although it was deemed that 12 hours would be sufficient time for the attitude control system to return to a nominal temperature, it was not before 20 hours that the MBE could be reactivated, just before 3 am on flight day seven. Unfortunately, the payload controllers were unable to initiate the flow of arsenic onto the substrate. This triggered a second shutdown and appropriate cooling off. Another full morning was spent in troubleshooting, then finally processing resumed at 3 pm. However, the fifth and final thin film growth was abandoned later the same day when a low reading was detected in one of the four batteries that powered the facility.

The next day, mission commander Dave Walker and pilot Ken Cockrell steered *Endeavour* alongside the free-flyer. Prior to capturing it, the two pilots fourteen times fired the jet thrusters of the Orbiter's reaction control system at distances of 290 and 200 feet as part of a test aimed at understanding the effects of thruster plumes against orbiting space structures. This test had been assigned to STS-60, but it was foiled by the inability to release the WSF. Charles Bolden explains what it was about, "As the Shuttle flies toward or away from anything, every time you fire one of its small jets, it sends out a wave from out of the jet. Every time it comes out, the thruster puts out stuff and this generates a force on anything that it hits. We had concern that the force, the thrusters from the Shuttle, if they were strong enough, might cause damage to the solar arrays or to some other component of the space station. We had built models to tell us whether it was going to be okay but we didn't have any in-flight data. So what we were going to do was take advantage of the Wake Shield. It had some very, very, very sensitive accelerometers on it, and as we approached the facility we were going to put the Shuttle in different attitudes and fire different rockets at it so as to measure the amount of force that the Shuttle imparted on the Wake Shield Facility, to help the designers of the International Space Station."

As expected, the attitude control system of the WSF performed well and collected data on the forces and pressures acting on it. Completion of the test cleared the way for mission specialist Jim Newman to use the robotic arm to capture the WSF and put it back on its support structure in the bay.

In fact, the WSF still had some work to do. At about 2 am on flight day nine it was unberthed again and the arm held it high above the payload bay for a rerun of the CHAWS experiment. Five hours later, it was mated with its carrier and latched into place, thereby concluding an eventful test.

The third test of the WSF was as one of the primary payloads of STS-80, flown by *Columbia*. On flight day four it was again show time. While several malfunctions had plagued the first two flights, the developers were confident the apparatus would operate flawlessly this time owing to a series of system improvements and a rigorous pre-flight test program.

Tom Jones unberthed the WSF at 2:56 pm CDT and held it over one edge of the payload bay with the business side facing in the direction of travel to cleanse it with atomic oxygen. Two hours later, it was maneuvered over one side of the bay to allow its ground controllers to verify and calibrate the attitude control system. At 7:38 pm it was released about 200 miles above the western Pacific, during the 51st orbit of the mission. Following a script rehearsed on the previous flights, shortly after release the ground controllers issued a 19-minute firing command for the small nitrogen thruster, which pushed the platform 20 to 30 miles away from the Orbiter.

The WSF free-flying platform being maneuvered by the robotic arm. Its Shuttle Cross Bay Carrier is visible underneath.

The ensuing 3 days saw the WSF performing flawlessly. By the end of the second day it had finished five of the planned seven MBE runs. The attitude control system solidly maintained the platform in the required orientation relative to the direction of travel. A newly installed scan filter removed the radio interference experienced by its predecessors. The upgraded communications system provided a reliable link between the WSF and its carrier, exceeding the anticipated performance. On flight day seven, it made further vacuum measurements. Then it was grappled by the robotic arm and stowed on its carrier. It was powered up again the next day and held by the arm at 45° to the direction of travel in order to collect 3.5 hours of data for an experiment to assess using the atomic oxygen in low Earth orbit to create films of aluminum oxide.

This proved to be the last flight of the WSF. Although a demonstrator for in-space manufacturing, the WSF was meant to be the first brick laid on the road of industrial and commercial applications in space. In fact, its proponents had in mind to conduct four test flights at annual intervals to demonstrate the manufacture of commercially significant quantities of products. Encouraged by the outcome of the test program, it was hoped that industry would exploit this technology.

In 1995 it was expected to fly an upgraded WSF-2 capable of processing a larger number and greater variety of thin films. It would have been directly commanded by a dedicated commercial payload command and control center. The following year, the even more capable WSF-3 would have had solar panels for longer use, additional central processing power, and even a robotic manipulator for the substrate samples. Finally, in 1997 WSF-4 was expected to grow up to 300 thin film wafers, which was considered to be representative of a production batch for a profitable industrial-scale operation. This demonstration would clear the way for the much awaited commercial phase. In this case, the free-flyer platform would be periodically visited by a Shuttle for retrieval of a finished production run (about 300 wafers) and replenishment of the raw materials for the next batch. Each Mark-II WSF was expected to have a useful life of 5 years.

Sadly, none of the above ever occurred. The WSF went into space three times and flew on its own twice, on one occasion suffering some problems and on the final test performing flawlessly. Just as had happened to the CFES/EOS, which also proved its viability, and the ISF, which was stifled while still on the drawing board, NASA and American industry in general lacked the vision that could have led, by now, to space not only serving as a place to explore but also as a place for commercial exploitation of its resources and environment.

9

Space Shuttle in Uniform

MILITARY SPACE SHUTTLE

Reporting the launch of *Discovery* for the STS-51C mission in the early afternoon of January 24, 1985, the *New York Times* said, "The sky was so clear that viewers could still see the speck of light five minutes after lift-off."[1] But the transparency of the mission was inversely proportional to that of the Earth's atmospheric shroud. In fact, almost nothing was known about the payload or the mission objectives, because for the first time NASA had broken its custom of issuing a press kit to the media and the general public to outline the flight. The agency did not even disclose the launch time, activating the large count-down display that sits on the grass of the KSC press site just 5 minutes prior to lift-off. Instead of thousands of viewers eager to be rattled, shaken and overwhelmed by the power of a Shuttle launch, the event was witnessed by only about two hundred people.

STS-51C was a mission like no other. In fact, it was the first classified manned mission flown on behalf of the Department of Defense. Although the objectives are still officially secret, informative leaks and reasonable speculations indicate that the payload was an electronic eavesdropping Magnum/Orion satellite developed for the CIA and operated by the National Reconnaissance Office.

The mission occurred at an awkward moment in the relationship between NASA and USAF. It was no secret that to secure Congressional backing for the Shuttle the agency had accepted the requirements which the military had imposed on the design. However, as the 1970s drew to a close and as NASA was shaping a USAF-compliant spaceplane, there were signs of discontent. It all started when an economic analysis indicated that to fulfill the promise of reducing launch costs by at least two-thirds as compared to expendable rockets of equal payload capacity, the Shuttle would have to be the sole means of launching into space. While this was music to NASA's ears, it worried the USAF. Launching rockets is a risky

---

[1] "Speck of light" was a reference to the ability of ground observers to stare at the hot-as-hell exhaust from *Discovery*'s three main engines even though it was already more than halfway through the ascent.

© Springer International Publishing AG 2017
D. Sivolella, *The Space Shuttle Program*, Springer Praxis Books,
DOI 10.1007/978-3-319-54946-0_9

business, and failures occur. What if the Shuttle fleet were grounded to rectify a generic fault? If the Shuttle were the sole means of launching satellites, then this would deny the country the ability to send up "national security" assets for intelligence and military communications. At the same time, the USAF was growing concerned about the increasing development costs and likely operating costs of the Shuttle. The frustrations were exacerbated by repeated slippage in the launch date for the first flight. The USAF was becoming increasingly skeptical of the agency's promotion of the Shuttle as a cheap, reliable, and available-on-short-notice means of accessing low orbit. In fact, to the USAF, the Shuttle was metamorphosing from an asset of strategic importance into a strategic vulnerability.

In response, by the end of the 1970s the USAF had initiated a vigorous campaign to protect national security interests by emphasizing the requirement for redundancy in terms of launch vehicles. The USAF set out to persuade NASA and Congress that at least until the Shuttle had proved its worth it would be wise to have another means of launching into space. After the release in September 1984 of a National Research Council study that expressed similar concerns about the Shuttle, the USAF solicited industry to make proposals for a new expendable launcher. Unable to keep the USAF at bay, the agency itself tendered two concepts which, unsurprisingly, would involve increasing the reliability of existing Shuttle hardware with the goal of raising USAF confidence in the Shuttle as the sole means of accessing space. In fact, in an effort to dissuade competition, NASA went so far as to threaten not to renew contracts with companies which worked directly with the USAF. However, both NASA concepts were dismissed as being nothing less than an insult to any rocket engineer worthy of their degree. NASA even tried to put pressure on Congress either to halt funding for the USAF expendable launcher program, or at the very least to postpone by one year the final decision on the winner of the USAF contest on the basis that more time was required in order to determine the "right" solution. But these machinations failed and the USAF selected Martin Marietta to develop the Titan 34D7 expendable launcher.

The fact that NASA could still cooperate with the USAF was demonstrated by the successful accomplishment of STS-51C. There was therefore room for some thawing of the ice-cold relationship between the two governmental entities. In fact, as early as February 14, 1985, the National Launch Strategy agreement was ratified by means of which the USAF renewed its commitment to dedicate one-third of the Shuttle flight manifest for the next 10 years to defense-related payloads. At the same time, NASA and the Department of Defense would work together to make the Shuttle a reliable service flying at an average rate of one launch every two weeks. This agreement also allowed the USAF to operate the Titan 34D7 to complement the Shuttle; as distinct from being in competition with it. It appeared that the decade-old controversy would have a happy ending. But the loss of *Challenger* in January 1986 exposed a wound in the NASA-USAF relationship that could never be mended.

"THE MISSION WAS PRETTY VANILLA"

If you like trivia about space missions, you will appreciate that Karol J. Bobko is the only astronaut to have participated on the maiden voyages of two different Orbiters. Having put *Challenger* through its paces as pilot of STS-6, Bobko, as commander of STS-51J, was now to help *Atlantis* to spread its wings.

Although a veil of secrecy fell over the national security mission, this time the payload was not so secret. In 1981 the Air Force had announced that it would launch a pair of DSCS-III communications satellites in mid-1985. Considering that the only launcher capable of hauling such a payload was the Shuttle, there was little doubt that these satellites were assigned to this flight. In fact, positive confirmation came from pictures the Air Force released in 1998 which showed *Atlantis*'s payload bay with an IUS upper stage carrying two of these satellites.

The secrecy of the mission was summed up by Bobko, "The mission was pretty vanilla…We went on time and we landed according to schedule."

THE LOST POLAR MISSION

As early as 1971, the decision was taken that the Space Transportation System would exploit both the Kennedy Space Center in Florida and Vandenberg Air Force Base in California as launching sites. The latter would be very suitable for weather satellites, or reconnaissance satellites such as the Key Hole and HEXAGON series, because by circling the Earth in an orbital path that flies over both poles they can pass over the entire surface of the planet several times per day, depending on the configuration of the orbit.

It would be possible to launch into polar orbit from Florida[2] but this is not done because the ascent would pass over densely populated areas and risk an exploding rocket raining down toxic fuel and hot debris. Even if the ascent went to plan, there would be the need to ensure the spent rocket stages fell far from populated areas. A requirement of the weather on areas to be overflown would be necessary, since high speed winds could easily push a falling spent stage far from the intended crash site. Hence a launch could occur only if weather conditions were optimal along all of the ascent ground track, which is an unlikely situation. This is why the highest orbital inclination achievable from KSC is 57°. In this way, an ascending rocket is able to fly over the Atlantic while paralleling the US East Coast.[3] In the case of the Shuttle, ascent abort scenarios had also to be considered. In fact, based on when the abort was called when flying north from Florida, the Orbiter might have to make an emergency landing on Soviet or Chinese territory. And clearly the DoD did not want an Orbiter carrying secretive spy hardware to land on the very enemy territory it was intended to spy on! Of course, polar missions could be achieved by flying southwards, but steps would have to be taken to safeguard the populations of Cuba and the South American continent.

---

[2] According to the laws of space mechanics, the lowest orbital inclination that a rocket can achieve is equal to the latitude of the launching site. In the case of the Kennedy Space Center, this means that a payload can be placed into any orbit ranging from 28.5° (the latitude of KSC) and 90° (polar orbit). If necessary, the payload can be subsequently maneuvered to a lower inclination but the high cost in terms of propellant makes it better to launch from a launching site at a lower latitude. For this reason, the closer to the equator the launching site is the wider is the range of orbital inclinations available for any given payload.

[3] This does not mean that launches across ground cannot be made. In fact, because Russia and China do not have launch facilities close to the ocean, they launch their rockets over large areas that are scarcely populated.

The only alternative was Vandenberg Air Force Base on the coast some 150 miles northwest of Los Angeles. It started out as a US Army armored division training site called Camp Cooke before World War II. In 1955 it was claimed by the Air Force as a secure launch site for testing long-range ballistic missile which were fired at targets in mid-Pacific. On October 4, 1958, it became Vandenberg Air Force Base (AFB), in honor of the late General Hoyt S. Vandenberg. Being on a flat plateau surrounded by hills and canyons, and fairly remote from populations, it was ideal for military space missions. A rocket could be sent into polar orbit by flying southward over the Pacific without overflying populated areas, on tracks that were clear of enemy or non-allied countries.

The Shuttle wasn't the first manned space program to be assigned to Vandenberg. In March 1966 construction began on Space Launch Complex 6 (SLC-6), a series of infrastructures to prepare hardware and crews for MOL missions.[4] Cost overruns, delays, and the emergence of new imaging technology caused the cancelation of this program and SLC-6 was abandoned. When the Shuttle came along, the USAF, with a close eye on the budget, proposed making SLC-6 Shuttle-worthy. This was approved in 1975 and requalification of the site started 4 years later. A mobile service tower, a flame trench,[5] and launch control center had already been built for the MOL but the Shuttle was more complex and additional facilities were built from scratch. In effect, Vandenberg was to replicate the facilities at KSC.

Richard W. Nygren, who in the early 1980s held the position of Assistant to the Director of Flight Crew Operations at the Johnson Space Center, says, "We realized we were going to have the same operations at Vandenberg as we were going to have at Kennedy, so we needed a facility for the crews to stay in pre-launch. We needed something equivalent to the KSC crew quarters out there. We needed to get a facility where the crew guys could be housed…We were going to be involved with a lot of their pre-launch testing with terminal countdown demonstrations. They were going to end up doing a flight readiness firing. We were going to get involved with a lot of their ground procedures. We were going to need a place for the Cape Crusaders[6] that were going to be there. We'd need office space. We were working the logistical build up for when the Orbiter got there and started doing testing, so we'd have all of the infrastructures in place to make things come together."

Although Vandenberg had to mimic KSC as far as operations were concerned, the local geography and air base layout would affect those operations. There were three different sites fairly separated from each other. North VAFB housed a 200 x 15,000-foot runway for the Shuttle to make a gliding approach at the end of its mission. This would also be used

---

[4] The Manned Orbiting Laboratory (MOL) was a 1960s Air Force program which had the ostensible goal of placing military personnel on-orbit to conduct scientific experiments to determine the "military usefulness" of flying men in space and, if the need ever arose, the techniques and procedures this would require.

[5] Flame trenches are placed beneath a rocket launch pad to discharge the exhaust away from the pad, so that it does not bounce back to the ascending rocket and flood its engines, potentially leading to destruction of the rocket in a matter of seconds.

[6] Behind every Shuttle crew there was a team of five to eight astronauts who served as the crew's point of contact between NASA-JSC and NASA-KSC. They were the eyes and ears to the Shuttle vehicle. Officially the Astronaut Support Personnel (ASP) they were nicknamed the Cape Crusaders.

by the modified Boeing B747 that carried an Orbiter on its back. Astronaut Richard Covey recalls the doubts his fellow astronaut pilots had about it. "We ran tests on making Shuttle approaches to Vandenberg. The runway was sloped, and everybody worried whether this would cause problems in flying that big glider down. The 1.5° slope is not severe, but the final approach of the Shuttle is only a 2.5° glide path, so 1.5° relative to that 2.5° can make a big difference…So we were trying to figure out if that was a problem. We ran a test out there. That was fun. We took the Shuttle Training Aircraft out and flew it around the runways." At wheel stop, the post-landing ground equipment would be positioned around the vehicle to prepare it for towing to the nearby Orbiter Maintenance Checkout Facility (OMCF) in which it would be processed for the next mission. Ancillary facilities such as the Hypergolic Maintenance and Checkout Facility (HMCF) and the Flight Crew System Facility (FCSF) completed the North VAFB site. Once declared ready for another flight, the Orbiter would be transported to SLC-6 at the South VAFB site.

At KSC, the Shuttle was moved by a huge crawler from the Vehicle Assembly Building to one of the two pads of LC-39, some 5 miles away. The local geography at Vandenberg demanded a rather meandering itinerary to bring the Shuttle from the North VAFB site to the South VAFB site. As Covey recalls, "The whole operations was going to be really weird out there. Getting from the runway to the launch pad, it was up and down hills and through the valleys because of the way California is built. They had made special provisions for being able to tow the Orbiter along roads and stuff to get it from the runway down to the launch pad. It was very interesting." After all, Vandenberg had not being built with the Shuttle in mind, and this was evidenced by the 16-mile journey through public and air base roads.

Shuttle Orbiter *Enterprise* being transported to the South VAFB site. The difficulty of moving the vehicle on the existing roads is readily apparent.

Operations at the South VAFB site would kick off with the delivery of the solid rocket motor segments via rail from the manufacturing facility in Utah. A modified harbor 2 miles away would receive the huge external tank, shipped by barge from the manufacturing facility in Louisiana, via the Panama Canal. These would have their own dedicated facility for storage and checkout prior to being moved to SLC-6. The former MOL mobile service tower was adapted to the Shuttle by reducing its height some 40 feet and moving it 150 feet back from its original location. In addition, its 40-ton crane was replaced by a 200-ton crane for the supersized and heavy elements of the Space Transportation System.

It is no secret that the Shuttle was significantly susceptible to weather conditions, and at Vandenberg the situation would have been even worse than in Florida. In fact, in winter, winds funneled by the surrounding hills can readily gust at up to 47 mph, which would have endangered the mosaic of thousands of tiles on the exterior of the Orbiter. To assure all-year-round operations, a large vertical shelter called the Shuttle Assembly Building, or the Windscreen, was added for assembly and protection of the Shuttle stack from the environment while awaiting launch. The site would resemble a giant clam, with the mobile service tower and the Shuttle Assembly Building tightly enclosing the Shuttle and its components.

At a cost of $80 million, the Shuttle Assembly Building was the largest single modification to the original design of SCL-6. It must be admitted this was a smart design, and it is shameful it was never adopted for the launch sites at KSC. In fact, on a number of occasions the temperamental Floridian weather inflicted severe damage to the Orbiter and external tank which required expensive repairs, either at the launch pad or upon being returned to the Vehicle Assembly Building. A good example is the 3-month delay which STS-117 incurred in 2007 due to a hail storm that caused some 2,000 damage sites on the protective foam of the external tank and minor damage to *Atlantis*. Given how vulnerable the Shuttle was to inclement weather, it would have made sense to build something like SLC-6, rather than repurpose the open air launch pads inherited from Apollo.[7] But the usual deficiency in funding obliged NASA to continue to use existing hardware, even though this would place a heavy toll on the operations and prompt criticism of the program.

At SLC-6, the Shuttle stack would be built directly on the launch pad, within the protection of the Windscreen and mobile service tower. This arrangement would not only provide a much-needed weather shelter, but would also offer a barrier to prying eyes as classified missions were being readied. The assembly would take place in the usual manner, with the solid rocket boosters being put together first, followed by the external tank being installed between the boosters. The stack would be completed by connecting the Orbiter to the external tank. In the meantime, the payload would be delivered to the adjacent payload preparation room for final checking prior to being transferred by crane to the payload change-out room, another mobile structure that would have rolled inside the Shuttle Assembly Building for payload installation into the Orbiter.

---

[7] The Shuttle launch pads were composed of a Fixed Service Structure and a Rotating Service Structure that closed around the stack on the pad. However, the rotating structure provided only limited weather shielding, as its main purpose was to allow loading of the payload into the vertically oriented payload bay.

Shortly prior to launch, the mobile service tower and Shuttle Assembly Building would be rolled to their "parked for launch" positions, respectively 375 and 285 feet away. Hydraulic jacks would raise the structures off their tie downs and the traveling drive system would transport each at speeds of up to 40 feet per minute. In total, it would have taken 40 to 50 minutes, including unlocking the tie downs, traveling, and relocking.

An aerial view of the South VAFB site with indications of the main infrastructures. Orbiter *Enterprise* was used to test handling procedures in preparation for the first Shuttle mission from Vandenberg.

Without a doubt, SLC-6 was the right type of launch pad for the Shuttle; complex but effective in guaranteeing protection from the external environment up to several hours before launch. It came however with a high price tag. From an estimated cost of $251.8 million in 1978 it had doubled by May 1986, chiefly owing to unforeseen setbacks.

For example, when STS-1 lifted off at KSC it was realized that more water than expected was needed for sound suppression purposes. At SLC-6 this translated into a $45 million facility to treat and safely dispose of 4,000 tons of water contaminated by the exhaust from the solid rocket boosters. There was also a blast protection issue for the adjacent launch control center. At KSC when the external tank filling procedure started, all but essential staff were evacuated from positions within 5 miles of the pad due to the hazard posed by a tank full of highly reactive hydrogen and oxygen. Also, the sound levels and vibrations at launch were so intense as to be fatal to anyone who was within 2 miles of the pad. As the launch control center at SLC-6 would be just 1,200 feet from the launch pad complex, it needed additional reinforcement. Another issue was the buildup of ice on the pad infrastructure owing to low temperatures in a highly humid environment. At Vandenberg, the local weather further exacerbated the problem, resulting in a \$12.8 million anti-ice system made of two jet engines to blast hot air onto the structure, but there were doubts whether this would actually work. By mid-1980s, therefore, the new facilities and improvements that were needed to make SLC-6 Shuttle-worthy were running considerably over budget.

However, the role of the facility was diminishing. In the early days of the Shuttle development, when the USAF was fully supportive, the plan envisaged 20 launches from Vandenberg annually. Two Orbiters would be exclusively dedicated to military missions, and there would be facilities for two pads. By 1981 these projections had halved, with now 10 flights annually and a single pad. Four years later, with support declining, it was expected that there would be just two launches annually, starting in 1988. But there was hope that after several years of operations the rate would ramp up to four or five flights annually.

After an initially planned launch date in October 1985 that had been postponed to January 1986, the first flight out of Vandenberg, designated STS-62A, was scheduled for March of that year.

A NASA press release issued on February 15, 1985, said, "Veteran Space Shuttle commander Robert L. Crippen will head the crew of mission 62A…Other crew members named include pilot Guy S. Gardner and mission specialists Dale A. Gardner, Jerry L. Ross and R. Michael Mullane." In addition, there were to be two payload specialists. As Jerry Ross recalls, "John Brett Watterson was going to be one of the payload specialists, and the other one would have been Randy T. Odle, another Air Force officer, but he'd been bumped by Edward Cleveland "Pete" Aldrige, who was undersecretary of the Air Force. So that would've been some pretty high-power folks flying with us on that flight."

The loss of *Challenger* in January 1986 put on hold for the foreseeable future any Shuttle flights from Vandenberg. By mid-1986 a report to the Senate explained that SLC-6 was not yet ready for such launches. One of the issues that remained unsolved was how to get rid of trapped liquid hydrogen underneath the launch pad basement. It was estimated that achieving a workable solution would cost at least \$8 million, and that no launches would be possible prior to July 1989. Over time, the first flight was postponed ever further into the future. Eventually, the Air Force decided to mothball SLC-6, essentially writing off the enormous investment made in it. Thus Vandenberg once again missed out on becoming a base for launching manned spacecraft.

How would STS-62A have looked had it not been canceled? Ross says, "It would have been a fascinating ride. We were going to go into an orbit at 72.5° inclination. Apogee was going to be something like 380 nautical miles with a perigee around 240 or so. It would have been a dual shift.…We had two main payloads. One was called Teal Ruby, and it was a prototype satellite…a staring mosaic infrared sensor satellite that was to try to detect low-flying air-breathing vehicles, things like cruise missiles, as a way to detect those approaching US territories. The other satellite…had a series of different types of ultraviolet and infrared telescopes on it and basically was trying to get background information about the environment of space so that they could use that information for designing interceptor missiles, seekers and things like that, in the future." From an astronaut's point of view STS-62A would have been a unique experience, as Mullane explains, "The idea of flying in polar orbit, oh, man, I was just looking forward to that so much. You're basically going to see the whole world. In a low-inclination orbit you don't get to see lot of the world. I was really looking forward to that."

The understandable disappointment of such a mission being canceled forever had, however, a bright side which cannot be underestimated. A few centuries ago, people could be burned to death for suggesting that the Earth rotates daily in its axis but this is now common knowledge. If we multiply the angular velocity of about 15° per hour by the radius of the planet,[8] then we see that an object on the surface at the equator has a tangential velocity in excess of 1,000 miles per hour! For a rocket engineer, this means that a rocket that lifts off at the equator will leave the pad with 1,000 miles of horizontal velocity in the direction of the planet's rotation. The practical result is that rocket performance is enhanced. This can be translated into a reduction in propellant, or into an increase in payload, or a higher orbital altitude. But the benefit to be gained from the Earth's rotation diminishes with increasing latitude, with the poles making no contribution at all. The poles would, therefore, be the preferred launching site for achieving polar orbit, as from there a rocket will not receive any push toward the east during its ascent. However, the impracticality of operating a polar launch pad means that a rocket for such an orbit launching from elsewhere must nullify the push toward the east received from the Earth's rotation. This translates either into an increase in propellant, a reduction in payload, or a lower orbit. Thus the farther a site is from the equator, the better it is for polar missions. At 51°N, Vandenberg was better than KSC at 28.5° because the eastward push would be substantially reduced.[9]

To recover some capability and allow the Shuttle to carry at least 32,000 pounds into polar orbit, as required by the DoD, it was decided to improve the solid rocket boosters. In fact, these were the only components that could be fully redesigned and made lighter. Even before *Columbia*'s maiden flight, from February through October 1981 preliminary studies were conducted to determine whether it would be possible to make the boosters out of the same composite materials (carbon fibers impregnated with epoxy resin) used for the large payload bay doors. Those studies confirmed the feasibility. Thus was born the Filament Wound Case (FWC) booster. The importance of building a database of accurate mechanical properties for composite materials was also stressed. In fact, while the behavior of carbon fiber/epoxy composite materials was well known, their mechanical properties had to be explored in even more detail for the FWC. In fact, this time the composite material would be used for a primary structure.[10] While modern aircraft such as the Boeing

---

[8] This is the radius as measured from the Earth's rotational axis, not from the planet's center.

[9] On the other hand the closer a launch site is to the equator, the better it is for launching into geostationary orbit. And the best latitude for launching deep space missions into the plane of the ecliptic is one at a latitude of 23.5° to cancel out the tilt of the Earth's axis relative to that plane.

[10] The Orbiter payload bay doors were not part of the primary structure designed to take up flight loads from the surrounding fuselage; they were only an aerodynamic fairing. The SRB casing was the primary structure of the booster itself.

B787 and Airbus A350 have their primary structure made almost entirely of composites, in the early 1980s there was no aeronautical application that made such an extensive use of composites on the primary structure of an aircraft, let alone on a spaceplane. FWC boosters were set to become the first such aerospace application.

Given the enormous stress that the high pressure burning propellant would induce on the structure of the booster, it was necessary to design a motor case that provided a structural strength comparable to the steel motor case of a conventional solid rocket booster. This required a campaign of additional tests to better predict the mechanical properties of composite materials prior to designing the booster cases. The beauty of composite materials is that they can be shaped according to the stress that they must withstand at each specific point of the structure. This is done by adding layer upon layer of carbon fibers in varying numbers and orientations. In other words, a piece of structure made of composite material can be highly customized for a particular stress distribution. While structures made of metal are designed based on stress, composite materials permit greater and more accurate tailoring, with the bonus of a considerable saving in weight.

An FWC booster would have had the composite shell manufactured with carbon fibers wetted by epoxy and wound in layers onto an aluminum mandrel. Each layer thickness and fiber direction would be selected to satisfy membrane stiffness[11] and joint integrity. Each segment end would be joined to steel rings by a pinned tang and clevis. The four filament-wound cylindrical segments making up one single booster would then be field-spliced by steel-to-steel joints substantially similar to those of the metal version. The composite boosters had the same interfaces with the Orbiter, the external tank, and the launch facilities.

At least a pair of FWC booster were manufactured and readied for STS-62A but the loss of *Challenger* promptly ended the program, to the relief of the astronauts. As Ross says, "They had the same joint design as the steel cases, and since the graphite ones would have been more flimsy, more flexible, we always were wondering what would have happened to us had we tried to launch with those, considering that in the *Challenger* accident the [hot] gas was seeping past the seals in those joints." Richard Covey agrees, "These were scary boosters, because they had a lot more flex in them. None of us were really sorry to see that go away after the *Challenger* accident. Then the DoD missions went away and we decided we were not going to launch out of California with the Space Shuttle."

The crew of STS-62A had lost their mission, but not the chance to fly a classified tasking. By the time STS-26 returned the Shuttle program to operational service, the DoD had already made clear that its services were no longer required. Every military payload would be delivered by an expendable launcher, most particularly a member of the Titan family. Only a handful of vital national security satellites which were too large and heavy for existing expendable rockets would be manifested on the Shuttle. The former STS-62A

---

[11] Membrane stiffness can be thought of as the difficulty of bending a laminate of composite material when a compressive force on the laminate plane is applied at two opposite sides. As an example, hold a sheet of paper with your two hands and bring the opposite sides close. Try this with a sheet of cardboard. While it will be easy to bend the paper, it will be necessary to apply a greater force for the cardboard. The membrane stiffness of the paper is almost nonexistent while that for the cardboard is considerably greater. The stiffness of a membrane can be radically altered by changing the number of layers and their thickness and orientation.

crew were reassigned to STS-27, with Robert "Hoot" Gibson replacing Crippen as commander and William M. Shepherd replacing Gardner as a mission specialist.

In the early afternoon of December 8, 1988, STS-27 began its long-awaited secret voyage into space. Even today, this flight by *Atlantis* is mysterious. What is known, is that Mullane used the robotic arm to deploy Lacrosse I, a side-looking all-weather radar surveillance satellite for the National Reconnaissance Office and the CIA. Such a satellite would provide ground profiles to program the terrain-following navigation systems of cruise missiles. It seems that a malfunction prompted *Atlantis* to return to the satellite but the details are secret. According to some sources, a spacewalk was conducted, possibly to release a jammed appendage.

"I WASN'T SUCH A HOT PILOT AS I THOUGHT"

On August 8, 1989, STS-28 launched with another classified payload. While early speculations had *Columbia* transporting an advanced imaging satellite, later reports and amateur satellite observations revealed that it was a Satellite Data System (SDS) relay similar in configuration to the Syncom/Leasats that were deployed earlier in the program. Crew member David C. Leestma is vague about what they did on the 5-day mission, "We had a payload for the Department of Defense, and then lots of in-cabin things that we did. Because it was a DoD mission, you did what you went up there to do and then you came back…They really only wanted us to be gone for 4 days but we talked them into one additional day simply because, if you are going to go to all that effort to get us up there, give us some more things to do and we'll do them."

Mission commander Brewster H. Shaw is of course tight-lipped, but says a funny episode occurred while landing at Edwards, "When we came down and I flared the Orbiter, I flared it, and I don't know how high we were…looking at the photographs, we weren't very high, but I basically leveled the vehicle off and then it floated and floated and floated and floated. So instead of landing at 195 knots the way we were supposed to, we landed at 155 knots. So here we are on the main gear on the runway at 155 knots decelerating fast and I've got to get the nose on the ground…and the nose goes 'bam' on the ground. I felt terrible about that, because I let this thing float for 40 knots' worth of deceleration." Jokingly, he admits "Here's where I learned I wasn't such a hot pilot as I thought." Despite his personal disappointment, a lot of data was gathered on Orbiter low-speed flying qualities that helped to increase safety on landing.

"WE WERE TREMENDOUS"

Like all astronauts who had flown DoD mission before him, STS-33's commander Frederick D. Gregory has sworn complete silence on what he and his four crewmates did aboard *Discovery* from November 23–28, 1989. However, he has said, "We had a crew that – I think it's the best crew that ever went through there. Every time it assembled, people would just come watch it, because it was like a ballet and we had so much fun… We were tremendous."

According to the experts, *Discovery* released a Magnum electronic intelligence (ELINT) gathering satellite, similar to the one deployed by STS-51C, which by now was running out of propellant for maintaining its station above the Indian Ocean. An interesting fact is that this was the only DoD flight to include two civilians, namely Story Musgrave and Kathy Thornton. The latter was also the only female astronaut to fly a mission for the DoD.

A MISTY MISSION

STS-36 was launched on February 28, 1990, after several scrubs due to the illness of the mission commander John O. Creighton[12] and poor weather. The only things we know for sure are that it had a crew of five, it lasted for a mere 4.5 days, and flew at an inclination of 62°. The latter is an important number because, as Creighton says, "Normally the highest inclination that you'll ever get is 57°, which keeps you just off the East Coast of the United States so that if anything bad happens, if you blow up or something, you're not going to rain debris on a major city in the United States." But for this flight the rules were relaxed to accommodate the national security payload. In another first of the Shuttle program, and indeed manned space exploration in general, *Atlantis* began the ascent as if heading for an inclination of 57° and then, out over the Atlantic and free of the solid rocket boosters, the three main engines on the Orbiter rotated the trajectory farther north to enter orbit at 62°. This "dog-leg" maneuver cost propellant and impaired performance,[13] but it restricted the risk to populated areas to a narrow portion of the north-eastern coast of the USA and Canada.

Even today this is all that anyone involved with this mission will say about it, but some captivating insights can be inferred from amateur observations and declassified documents. Days before launch, *Aviation Week* magazine speculated that the payload (code named USA-53) would be a large digital-imaging reconnaissance satellite. For amateur satellite sleuths, hunting a secret satellite and understanding its purpose is a challenge that is hard to resist.

One such sleuth was Canadian Ted Molczan, a technologist by education and top satellite tracker by hobby. As he says, "Members of an observation network which I organized, observed the satellite between the 2nd and 4th of March. It was deployed into a 62°, 254-km altitude orbit. Early on March 3rd it maneuvered to a 271-km altitude." What really caught his attention was, "Observers noted that the object was extremely bright, reaching a visual magnitude of minus 1 in favorable conditions." This hinted that the payload might not be what *Aviation Week* had suggested.

On March 7, three days after *Atlantis* had landed at Edwards Air Force Base, the Soviet press agency reported the satellite had exploded. The Pentagon backed up this assertion, adding that five or six pieces were being tracked where the spacecraft was supposed to be,

---

[12] This was the first time since Apollo 13 that a US manned space missions was affected by the illness of a crew member.

[13] For the Shuttle, every additional degree in inclination cost 625 pounds in useful payload.

and these would re-enter the atmosphere within 6 weeks. The career of the mysterious payload appeared to have come to a premature and inglorious end.

Then, as Molczan continues, "On October 19, 1990, I received a message from Russell Eberst stating that he, along with Pierre Neirinck and Daniel Karcher,[14] had found an object in a 65°, 811-km altitude orbit which did not match the orbit of any known payload, rocket body, or piece of debris. He suspected that the object could be a secret US payload and asked me to try and identify it." In accordance with a United Nations treaty, the US published the orbits of the objects which it inserted into space, and by a process of exclusion Molczan reached an amazing conclusion. "My analysis revealed that the orbital plane of the mystery object was almost exactly coplanar with USA-53 on March 7, 1990, the same date that the Soviets found debris from USA-53 on-orbit!"

Although the general belief was that USA-53 had failed and its debris fell back into the atmosphere, in reality the secret payload of STS-36 had maneuvered and was gathering intelligence. The mystery satellite was also spotted in November 1990, and amateurs found it again in 1996 and 1997 by then in a 66.2° orbit. Interestingly, in 2000 an observer studying orbital data from NORAD[15] found that in May 1995 the satellite was in an orbit ranging between 451 and 461 miles.

A further revelation came in a book published in 2001 which sought to expose the darker side of the US intelligence world. In *The Wizard of Langley. Inside the CIA's Directorate of Science and Technology*, Jeffrey Richelson[16] explains, "The payload was a stealth imaging satellite code named MISTY[17]…[which was] developed in exceptional secrecy subsequent to the 1983 decision by the Reagan Administration to establish a stealth satellite program…to reduce the threat to US satellites from the Soviet Union, whose antisatellite program was of significant concern during the early 1980s."

So STS-36's secret payload was nothing less than one of the many products of the Cold War, in this case taking stealth to new heights. In fact, the problem is that space is transparent and because spy satellites are relatively large and therefore reflect a lot of sunlight they are easy to spot optically. Furthermore, they can readily be tracked using radar and lasers. If the enemy knows when a spy satellite is going to fly over, they can hide military assets or stop suspicious activities until the unwanted observer is gone.

With an investment of $9.5 billion and a decade of research and development, MISTY was conceived. A second satellite was launched in 1997. A third one might have been launched also. This class of satellites is thought to have a mass of 37,000 pounds and its cloaking capability is based on the same "faceted surface" technology as the Lockheed F-117 "Nighthawk" and Northrop Grumman B-2 "Spirit" airplanes. An incoming radar

---

[14] Eberst, Neirinck and Karcher were other satellite spotters.

[15] The North American Aerospace Defense Command (NORAD) has its headquarters at Peterson Air Force Base near Colorado Springs, Colorado.

[16] At the time of this book's publication, Richelson was a senior fellow of the National Security Archive at George Washington University in Washington DC. Its goal is to gather declassified US documents obtained by the Freedom of Information Act to help to shed light on some of the most tantalizing and mysterious espionage programs of the United States.

[17] Despite being written with upper case letters, this is not an acronyms. It is a code name. It might not be the satellite's real name at all.

beam or laser beam would be deflected in different directions, significantly reducing the signal returned to tracking systems on Earth. Interestingly, several patents have been filed since the 1960s designed to reduce the detectability of satellites. In September 1994, US patent 5,345,238 "Satellite Signature Suppression Shield" offers a reasonable explanation of how stealth technology might have been applied to MISTY. It could be a large inflatable cone coated with radiation-reflective material deployed on a rotating arm. The cone would be rotated below the satellite to avoid ground detection while circling the planet and then moved out of the way when over a target to allow the sensors to scrutinize the panorama below. As of today, it is not known what kind of intelligence MISTY satellites supplied and how important it was for some of the most significant conflicts the US has fought during the last two decades.

U.S. Patent   Sep. 6, 1994   Sheet 5 of 6   **5,345,238**

F I G. 4

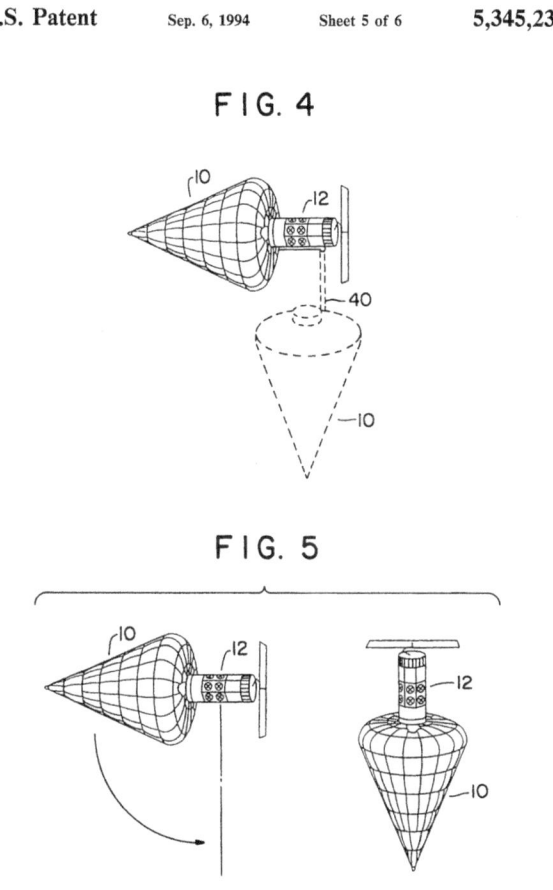

F I G. 5

A possible way in which MISTY might work.

ESPIONAGE DECEPTION

STS-38 was a textbook mission of espionage deception. Lift-off occurred at night on November 15, 1990, with Richard O. Covey as the commander of another highly classified mission for the Department of Defense.

As Covey recalls, "Going to the launch pad in the dark…is always an interesting experience because you see things that you don't necessarily see in the daytime, like the hydrogen burning off and away, and the lights, the way things are lit up is very, very interesting and surrealistic out at the launch pad with the big xenon lights and all of the burn-off and the hissing and the gases and stuff that are moving around out there. It's pretty neat." However, about the mission itself Covey is rather vague and generic, so it becomes an exercise in reading between the lines, "We didn't get very high, and you can read a lot into that. We didn't go very high because we couldn't go very high, which says we probably had a heavy payload…We had to do our most serious and significant work the first day in deploying a payload."

It is likely that the payload was a secret SDS-2 military communications satellite, similar to the one released by STS-28. In fact, although *Aviation Week* stated prior to the mission that the payload was a Magnum satellite like those launched by STS-51C and STS-33, images in the public domain shows the aft portion of the payload bay of *Atlantis* to be devoid of the platform that a Magnum satellite would have required. Its absence supports the speculation that SDS-2 was the satellite whose release Covey mentioned. But this was not *Atlantis*'s only guest. Leaks suggest there was a second satellite, named Prowler. Designed as a stealth satellite, its mission would have been to maneuver close to other nations' communications satellites in order to inspect their design and perhaps hinder their operation.

Such speculations are supported by the observations made of *Atlantis* on-orbit by Ted Molczan and his network of amateur sleuths. In fact, this may be one of the best stories of deception in space. In accordance with well-established procedure, SDS-2 was deployed 7 hours into the flight by rolling it out of *Atlantis*'s payload bay. When deploying commercial satellites, the next step was for the Orbiter to move to a higher orbit, in order to slow down and withdraw to a safe distance before the satellite fired its perigee kick motor. In this case, *Atlantis* raised its orbit only very slightly, with a delta-V of less than one-tenth of that of a nominal mission. Furthermore, the satellite hunters noticed that SDS-2 did not fire its engine, it remained loitering in low orbit. Some 22 hours into the mission, *Atlantis* lowered its altitude and gradually caught up with SDS-2. Seen from the ground, it seemed as if *Atlantis* was carrying out station-keeping maneuvering with the satellite. What Molczan and his colleagues believe, is that at some point between SDS-2 being deployed and *Atlantis* lowering its orbit, the Prowler must have been deployed. The fake station-keeping was a ruse to confuse the Russians, just as an illusionist fools an audience by distracting them from seeing how the trick is executed. By giving the Russian ground stations something interesting to watch, it was possible to sneakily deploy the Prowler and fire its rocket engine some 12 hours later. This scenario is reinforced by the fact that the maneuvers by *Atlantis* in proximity to SDS-2 were observable to the Russian listening station in Cuba. It is very likely that this "sleight of hand," of suggesting that *Atlantis*'s was standing by to fix a problem with its newly deployed communications satellite, successfully masked the deployment of an even more highly classified payload.

The plan was for *Atlantis* to land at Edwards in California but continuing adverse weather ruled this out. As Covey recalls, "So everybody started looking at landing in Florida. It was really weird to think that Florida was going to be the alternate weather site, as opposed to a primary site."

In the afternoon of November 20, *Atlantis* gently touched down on Runway 33, marking not only the first time since the *Challenger* accident that an Orbiter landed at KSC but also the first landing there by *Atlantis*. Landing a Shuttle is never an easy task, and even more so when half of the runway is covered in smoke. As Covey says, "One of the things they do in Florida during the fall, is they burn the underbrush in the pine forests, a very controlled type of burn, just to get everything down." By the time *Atlantis* had executed the de-orbit maneuver and was committed to returning to Florida, winds near KSC had shifted and were blowing a thick layer of smoke across the southern end of the runway. Covey recalls the approach to the runway, "The Sun starts getting down, the smoke starts getting a little thicker, and pretty soon, because of the refraction of the light off the smoke, you can't see through it. So we're flying around, and when we rolled out on final, I couldn't see the runway. I couldn't see the visual aim points. All I could see through the smoke was what we call PAPI lights." This was good, because even though he could not see the runway, the PAPI lights[18] helped Covey to maintain the Orbiter on the correct landing path until visual contact was made with the runway. "We fly all the way down. We go into the smoke and we get down at our pull-out altitude, and when we go in the smoke, I can still see those lights but I can't see anything down towards the runway. We come out and we pull out below the smoke, and there it was. There was the runway." Moments later, he performed a perfect touchdown in one-of-a-kind conditions never before experienced by a Shuttle commander. Jokingly, Covey notes, "I think, technically, I get to log an instrument approach on that."

A MILKSHAKE IN SPACE

STS-39 was the first unclassified Department of Defense Shuttle mission. Some of the primary objectives included around-the-clock observations of the atmosphere, gas releases, Shuttle engine firings, subsatellite gas releases, and the orbital environment of the Orbiter in wavelengths ranging from infrared to the far ultraviolet. The results would help the Strategic Defense Initiative Organization (SDIO)[19] to develop space-borne platforms to detect ballistic missiles traveling towards American territory. The payload bay of *Discovery* was filled with hardware for the two principal experiments, known as the Air Force Program (AFP) 675 and the Infrared Background Signature Survey (IBSS).

---

[18] PAPI stands for Precision Approach Path Indicator. It is a system of lights on the left of the runway which enables a pilot to ascertain whether he is following the correct approach path.

[19] The SDIO was created by the Department of Defense in 1984 to oversee the development of the Strategic Defense Initiative (SDI), a missile defense system intended to protect the USA from a nuclear attack by ballistic missiles. Part of the system would include orbital platforms to detect and shoot missiles aimed at American soil. The SDI was publicly announced by President Ronald W. Reagan on March 23, 1983, and soon nicknamed the "Star Wars" program.

As mission specialist Guion S. Bluford explains, "The AFP-675 was a collection of experiments designed to measure background infrared and ultraviolet emissions, identify contamination in the Orbiter environment, and demonstrate X-ray imaging." While these five experiments were to remain in the payload bay for the full mission, the IBSS would enjoy a ride in space by itself. "IBSS was mounted on a deployable Shuttle Pallet Satellite (SPAS-II)[20] platform," Bluford notes. "It was an experiment designed to collect infrared, ultraviolet and visible data for use in the development of ballistic missile defense sensor systems. Phenomena to be observed included OMS and RCS engine firing exhaust plumes, the Orbiter environment, the Earth and its background, chemical and gas releases, and celestial calibration sources." Although the main component of IBSS would fly on SPAS-II, "Two elements of IBSS, the Critical Ionization Velocity (CIV) experiment and the Chemical Release Observation (CRO) sub experiments, were mounted in the cargo bay… The CIV consisted of four canisters containing different gases. These gases would be released into the payload bay and would be observed by the deployed IBSS. The CRO was three subsatellites, containing different chemicals. After CRO deployment, these chemicals would be released by ground command and observed by the IBSS."

The diversity of payloads and nature of the research made it a fairly demanding flight, as Bluford says, "We had to do rendezvous, multiple translational maneuvers, extended station-keeping, and deployment and retrieval of the SPAS with the RMS. This involved precision Orbiter maneuvering, IBSS/SPAS commanding, observation sequences and multi-body management in a very intensive timeline. A great deal of coordination was required on the flight deck, synchronizing Orbiter and SPAS-II maneuvers and documenting key events. There were approximately 36 hours planned for rendezvous and proximity operations." It is easy to agree with Bluford when he says, "It was quite a challenging flight plan and training was intensive." And it is no surprise that STS-39 was one of the most complex deployment and retrieval missions of the Shuttle program. To accomplish all of the mission objectives and conduct the demanding flight plan, six crew members were split into two teams on 12-hour shifts. The commander, Michael L. Coats, was free to adjust his working hours as required and join either shift.

On April 28, 1991, *Discovery* was launched and its crew set about the aggressive flight plan. As Bluford recalls, "We had an uneventful early morning launch and the Red Team started the mission by initializing and checking out the AFP-675 and the IBSS…On flight day two, we did AFP-675 operations and unberthed the IBSS/SPAS payload by keeping it attached to the RMS. The next two days we deployed the IBSS and made numerous observations of OMS and RCS plumes, and CRO and CIV gas releases."

By the time of STS-39, the Shuttle program had become proficient in deploying and retrieving free-flying platforms, as well as stranded satellites. Those missions not only accomplished their primary objectives but also helped the Shuttle to prepare for the greatest and most complex deployment and retrieval mission ever. After releasing the SPAS-II/IBSS, *Discovery* withdrew to a position some 5.4 miles behind the free-flyer, on the same velocity vector. This was called the far field point. Later, the crew maneuvered the Orbiter to aim its nose and the payload bay in the direction of travel, and therefore toward SPAS-II/IBSS. On receiving confirmation that SPAS-II/IBSS had its imaging system trained on the

---

[20] The SPAS-II was an improved version of the SPAS-I that flew on STS-7 and STS-41B.

Orbiter, the crew performed a 20-second firing using one of its OMS engines to allow the SPAS-II/IBSS's imaging system to obtain data on rocket exhausts seen against the background of the Earth's atmosphere. This was vital information for the future development of space-borne sensors for detecting intercontinental ballistic missiles. Apart from pushing *Discovery* north of its orbital track in an out-of-plane maneuver, this marked the first time that an OMS burn was performed using only one of the two engines. To set up the next observation whilst remaining aligned with the leading free-flying platform, *Discovery* made a fast-flip yaw maneuver by commanding the jet thrusters of the RCS to turn the nose through an angle of 180°, this time to face south. Owing to its rapid execution, this maneuver was known as the "Malarkey Milkshake," in honor of John Malarkey, the rendezvous guidance team leader who worked on the planning of STS-39, including this fast-flip milkshake-type maneuver. The first OMS burn made the Orbiter slide northward, but after a displacement of only a mile the second such burn pushed it back southward. Another milkshake maneuver was executed to face *Discovery*'s nose north again for the third OMS burn, which was to prevent the Orbiter from overshooting the orbit of the free-flyer. These maneuvers left *Discovery* station-keeping behind SPAS-II/IBSS. Two further out-of-plane translation sequences were performed, and then a series of firings by the primary thrusters of the RCS, with the free-flyer carefully monitoring the plumes.

Later, *Discovery* released the first of three CRO subsatellites to obtain infrared, visible light, and ultraviolet data of chemicals that could be released by spacecraft for obscuration purposes. The data would also be useful for characterizing signatures of propellants escaping from damaged boosters. Each subsatellite of this type was to be released to permit simultaneous observations by the SPAS-II/IBSS payload and from Vandenberg AFB, which commanded the operations of the tiny satellites. In this way the data harvesting would be maximized.

With the first CRO experiment underway, *Discovery* transitioned to the so-called near point in order to repeat the burn plume observations but this time just 1.2 miles from the free-flyer. Two full out-of-plane translation sequences interspersed with a few Malarkey milkshakes concluded the OMS/RCS plume observation objectives of the mission.

As *Discovery* gave chase to SPAS-II/IBSS, the Orbiter released the second CRO. The third and final CRO was deployed after the robotic arm had grappled the free-flyer.

The maneuvering on this mission set a record tally for OMS and RCS burns that would never be broken. In addition to the orbit insertion and de-orbiting OMS burns and several RCS firings for normal attitude control activities, the 2 days of milkshake operations added 14 OMS and 41 RCS burn sets.[21] The final days were devoted to further work with the AFP-675 experiments. In the evening of May 6, the exhausted but satisfied crew emerged from *Discovery* at KSC to conclude an intense but fruitful 8-day mission.

STS-39 is one of the finest examples of the great maneuverability of the Shuttle, and of NASA's ingenuity in devising and executing a challenging flight plan.

It was also the operational debut of a "smarter" Orbiter. From 30 seconds prior to lift-off through to wheel stop on the runway, an Orbiter was managed by a set of five computers. These were called general purpose computers because they could execute different

---

[21] For a detailed description of the Orbiter's RCS and OMS, refer to Chapter 6 of my previous book *To Orbit and Back Again: How the Space Shuttle Flew in Space.*

software as necessary for a given phase of the mission. For instance, during ascent and re-entry four computers would run the guidance, navigation and control software, checking on one another some 400 times per second to verify they were all at the same point in the software script and providing the same results. On-orbit, two computers would run the guidance, navigation and control software while a third ran software to operate the onboard systems and, if necessary, the various payloads. As a precautionary measure, a fourth would hold the re-entry software lest an emergency require a quick end to the flight. Regardless of the mission phase, one computer was always loaded with the so-called backup flight software, a simpler and alternative software to safely return the Orbiter to Earth. All of the computers were identical and connected to one another to provide two layers of redundancy at the hardware level, but the primary avionics software and the backup software provided protection at the software level, lest some undiscovered bug the primary software occur, in particular during ascent and re-entry.

Each computer consisted of a Central Processing Unit (CPU), an Input/Output Processor (IOP), a megabyte of memory and various other components housed inside a casing which was "hardened" against electromagnetic interference. While the CPU performed the instructions to control onboard systems and manipulate data, the IOP formatted and transmitted commands to the systems, received and validated response data from the systems, and maintained the status of the interfaces between the CPU and the other computers. Thus the CPU was the "number cruncher" and the IOP did all the interfacing with the rest of the computers and vehicle systems. The computers were able to perform their functions by control logic embedded in a combination of software and microprogrammed hardware.

Within a few years of initiating the IBM AP-101B design for the GPC in January 1972, it became evident that an improved GPC would be required. Studies to upgrade the existing AP-101B started in January 1984 and culminated in the mid-1990s with the introduction into service of the AP-101S on STS-39. From a configuration point of view, the big difference was that the new computers incorporated the CPU and the IOP in a single avionics box, thus halving the size and weight, and also reducing the power requirements. From a performance point of view, the upgrade provided 2.5 times the memory capacity and up to three times the processor speed with minimum effect on the flight software other than to increase the 400,000 operations per second of the original computers to 1,000,000 operations per second in the new ones.[22]

MILITARY MAN IN SPACE

Although *Atlantis*'s tenth mission was flown for the Department of Defense, the fact that the National Reconnaissance Office was not involved meant it did not need to be classified, and the mission objectives and payload were fully disclosed well ahead of launch.

The STS-44 mission began in the early evening of November 24, 1991. The first flight day saw the flawless deployment of the primary payload, which was a satellite for the Defense Support Program (DSP). Once an IUS had delivered the satellite to geostationary

---

[22] For a detailed description of the Orbiter's avionics system, refer to Chapter 1 of my previous book *To Orbit and Back Again: How the Space Shuttle Flew in Space*

orbit, it would undertake real-time detection and reporting of missile launches, space launches, and nuclear detonations by employing infrared detectors to sense the heat emitted from the rocket plume against the background of the Earth. It was a new addition to an existing fleet of warning satellites. The data was used by the Tactical Warning and Attack Assessment System at NORAD[23] in their mandate to provide aerospace warning and control for North America.

With their payload bay relieved of 37,000 pounds, *Atlantis*'s crew performed an equipment power down (referred to as a Group B power down) in order to preserve sufficient onboard cryogens for the remainder of the mission, which was planned to last 10 days, excluding the usual 2-day extension option required by the flight rules. As part of this power down, four CRT monitors on the flight deck were deactivated, three general-purpose computers and an inertial measurement unit were switched to standby, and three multiplexers were turned off.

The crew focused on a battery of nine experiments for the Space Test Program, a project that the DoD had started in 1965. The objectives of the STP were to provide space flight opportunities for advanced DoD research and development experiments that were not authorized to fund their own flights, and to plan flights for payloads on either STP-provided flights or those of NASA or other DoD programs. It was really a pathfinder for defining man's military role in space.

For example, the Military Man in Space (M88-1) experiment was to evaluate how an observer could use special optics and communications equipment to enhance air, naval, and ground force operations by making visual observations of fixed or mobile military sites and facilities from space and then communicating those observations to ground personnel. To fully evaluate the benefits of a space-borne observer, the trial was coordinated with ongoing DoD exercises.

M88-1 was split into three investigations. The Maritime Observation Experiments in Space (MOSES) involved observing maritime targets such as ships, ports, and the wakes of ships using small-aperture long-focal-length optics. The astronaut would be in communication with the location to be observed for pre- and post-pass briefings. The Battleview experiment involved making observations of US armored mechanical formations, military ground sites, and flying aircraft. Also in this case, the astronauts were to relay their observations in real time. Finally, the Night Mist experiment used an encrypted UHF communications system to link the crew in space with people on the ground.

Work on these and several other operations began with one of best ever wake-up calls. It was traditional to awaken Shuttle crews at the start of each flight day with a song dedicated either to one of the astronauts or to the entire team. On flight day two it was the theme tune of the TV series *Star Trek – The Next Generation*, with a new voice over by Captain Jean-Luc Picard (actor Patrick Stewart) announcing, "Space, the final frontier. This is the voyage of the Space Shuttle *Atlantis*. Its ten-day mission to explore new methods of remote sensing and observation of the planet Earth. To seek out new data on

---

[23] In collaboration with other agencies, NORAD safeguards the sovereign airspaces of the USA and Canada by responding to unknown, unwanted, and unauthorized air activity approaching and operating within these airspaces.

radiation in space and a new understanding of the effects of microgravity on the human body. To boldly go where 254 men and women have gone before."

Along with five career astronauts, a sixth seat had been reserved for a guest. As mission commander Frederick D. Gregory remembers, "We also had Tom J. Hennen on board, an Army Warrant Officer." This was a first, since, as Gregory points out, "We had never flown a military non-officer before, so this was pretty unique and it was actually pretty exciting, because he was from the photo interpretation field." In addition to being knowledgeable about terrain and aerial observations, Hennen had some formal training in geology. This rather conspicuous background proved to be his ticket to space as STS-44's payload specialist for the Terra Scout experiment. The objective was to evaluate the ability of a specially trained person to detect objects on the ground at predefined targets using Earth observation tools. The results would then be compared with those of an observer at the same sites using conventional methods. The experiment was to assess whether having an observer in space would provide an advantage on the battlefield.

The planned 10-day mission proved impossible, because a malfunction by one of the three inertial measurement units midway through STS-44 mandated that the flight be curtailed at just 7 days.

HONORABLE DISCHARGE

On December 9, 1992, *Discovery* touched down at Edwards AFB to conclude the 10-day STS-53 mission. Though the primary payload was classified, it is believed it was another satellite for military communications, essentially the same as those released by STS-28 and STS-38. The rest of the flight, however, was taken up by a battery of unclassified experiments, conducted both in the cabin and in the payload bay, which embraced disciplines such as life sciences, meteorology, communications, and fluid mechanics. This marked the end of the Shuttle's services to the nation's defenses, as it was the final mission whose primary payload was carried for the DoD.

In fact, following the loss of *Challenger* in January 1986 and the almost 3-year hiatus in operations, the USAF lost all confidence that the Shuttle could adequately satisfy its needs in a timely and cost-effective manner. As several payloads required a Shuttle ride, DoD missions continued until STS-53 closed the backlog. By then, the USAF had completed its return to expendable rockets and no longer needed NASA.

As a civilian agency known for its openness, NASA had never been comfortable with the cloak of secrecy required for national security missions. As T. K. Mattingly, commander of STS-51C, has reflected of the onset of secrecy, "JSC and the whole NASA team has worked very hard at building a system that insists upon clear, timely communication. The business is so complex that we can't afford to have secrets…So [then we switch over] to a classified mode where we have a limited number of people and we don't talk about all these things. I had some apprehension about could we keep the exchange of information timely and clear in this small community, when everybody around us is telling anything they want, and we're kind of keeping these secrets." Mattingly's pilot, Loren J. Shriver, was thinking along much the same lines, "One of the strong technical parts of NASA programs like the Shuttle, is that it is so open, everybody just keeps data and information on everybody else so that when the process is all through, everybody is pretty well assured of having the information that they know they need. We were concerned that just the opposite was going to happen, that because of the classification surrounding the mission, that people were going to start keeping secrets from each other, and there was a potential that some important product or piece of information might not get circulated as it should." Nevertheless, the agency dealt with this situation in the best way possible.

Each mission on behalf of the Department of Defense had its own small, tight-lipped closed community of astronauts, trainers, and support personnel. They would plan and manage the mission with the same dedication and efficiency as for a regular civilian flight.

As John O. Creighton, commander of STS-36, recalls, "There were probably less than 25 people in all of NASA that knew what we did. The Administrator of NASA and the Flight Directors [knew]. Most of the people in Mission Control didn't know specifically what we did. Obviously the crew and a couple of people on the training team knew." Richard M. Mullane, a mission specialist on STS-27, observes, "All the software was classified. The people that were working on it, the MCC [in Houston], everybody supporting it, had to have clearances. But that was pretty transparent to us. We would go to a simulator. We would see our software. We would do our thing and launch the mission or launch the payload. It was pretty transparent to us, the security aspects of it." STS-38 commander Richard O. Covey, adds another insight, "Outside of the core group in the [Astronaut] Office that was either flying the missions or had leadership positions and had the appropriate clearances, very few people knew what was going on those DoD missions. There were further levels of knowledge of what was going on and what the payloads were designed to do…even within the crew. So there were some things that I, as the commander of the mission, got read into that no one else on the crew did, and there were some things that the crew knew about and that some of the people working the payloads knew about, that other people working the payloads did not know about. So even in the MCC they had a limited knowledge of all the things that were actually involved in the payload operations. So that was a different environment."

In an effort to ensure that important pieces of information would be circulated, the role of coordinator for DoD flights was introduced and assigned to Covey. As he points out, "You know, all of the missions were highly classified, and each crew, as they were selected, was read into the particular program that they were in, the DoD program that they were supporting. But there needed to be someone who was aware of what all of those missions were going to be doing and working that interface with the appropriate agencies within the DoD to make sure that the crew issues that may cross all of those were being taken care of. So we had a very small staff which was primarily focused on managing the classified materials in the Astronaut Office about these missions." These people required Department of Defense security clearances above the regular Top Secret level. As Covey continues, "They helped to manage all of that activity…They'd sit in on all of the meetings, the standard types of meetings for any payload that might fly on the Shuttle in the course of mission preparation, but because they were of a classified nature they would be held in special environments or different places. We had people that could go and attend all those, other than just the crew for a specific mission."

As only astronauts who were part of the military or had served in the forces were assigned to a DoD mission, they were already familiar with the way in which secret missions were managed. Jerry L. Ross, a mission specialist on STS-27, recalls. "The problem with working classified programs is that you have to be very careful of who you share information with, and how you do that. You have to work within secured facilities which are swept so that you don't have any inadvertent electronic signals or voice going outside. That constrains you significantly on how and where you can do your business. When we traveled, it was on basically classified orders. We could not tell people where we were going, or why." As STS-28 mission specialist David C. Leestma adds, "Sometimes you had to disguise where you were going. You'd file a flight plan in a T-38 for one place and you'd go somewhere else, just to try to not leave a trail for where you were going, or what you were doing, or who was the sponsor of this payload, or what its capabilities were, or what it was going to do. You just had to be careful all the time of what you were saying. Everything was done in a classified mode. Our flight data file was classified. You couldn't bring it home."

This additional layer of secrecy increased the burden on the astronauts at both a personal and a private level. As Shriver frustratingly remembers, "I couldn't go home and tell my wife what we were doing, anything about the mission. For [the ordinary] missions, everyone in the world knew exactly what was going on. NASA's system is so wide open, they could tell their wives about it, their family knew, everybody else in the world knew what was on those missions. We couldn't talk about anything. We could not say what we were doing, what we had, what we were not doing, anything that would imply the launch date, the launch time, the trajectory, the inclination, the altitude, anything about what we were doing in training. All that was classified. We couldn't talk about anything. The Air Force didn't even want…the names of the crew released. We weren't going to be able to invite guests for the launch in the beginning. This is your lifetime dream and ambition. You're finally an astronaut. You're going to go fly the Space Shuttle. And you can't invite anybody to come watch…It was an interesting process. We finally talked them into letting us invite – I think each one of us could invite thirty people, and then maybe some other car-pass guests who could drive out on the causeway. But trying to decide who among all of your relatives and your wife's relatives are going to be among the thirty who get to come and see the launch, well, it's a career-limiting kind of decision if you make the wrong decision!"

Nonetheless the DoD classified missions had their own advantages. As Mattingly enthusiastically notes, "To keep the launch time classified, they wanted us to make all our training as much in the daytime as at night, so someone observing us wouldn't be able to figure this out…They did convince me that if you watch these signatures, you could figure it out and it is secret because we said it was…So they had us flying equal day and night. I didn't mind, because it meant I got to fly more. We didn't split the time, we just doubled it!"

Apart from the veil of secrecy where everything was done and said behind closed doors, the training remained mostly unchanged. As flight control instructor Lewis J. Swain recalls, "From a standpoint of vehicle and training, it was done the same way. The only thing the DoD complicated was the environment. I worked the first DoD flight [STS-51C] as a control instructor. Actually I was a control instructor for other missions at the same time. From the standpoint of what you did and what you trained the astronauts to do, from a control propulsion instructor point of view, it was pretty much the same apart from the payload being different. The vehicle and those sorts of thing were pretty much standard.

The systems worked the same way…The vehicle still operated the same way…As an instructor, it was not that much of a difference from one [normal] flight to the DoD flight. You still instructed the same way, and the systems still basically functioned the same way."

Payload planning was significantly affected by the classification status. Larry D. Davis performed a large number of roles in mission planning, and recalls, "What we had to do, we had a special facility behind two locked doors that we had to go into to have meetings or run computer simulations. We had files in there. We could not take any information out of there. It had its own storage. If you wanted to do any work, you had to go sit at office areas within that facility. The communications had to be on STU [Secure Telephone Unit] phones, which were scrambled phones. You had a lot more face-to-face meetings with the customers than you might with a payload which wasn't secret. There was a lot more paper-work. If you wanted to send a product to the DoD customer, it was not a matter of putting something in the mail and sending it to them. It was a lengthy process. That was why there were a lot more face-to-face meetings. Either you would go to their facility or they would come to yours, and you would use secret carriers to bring the data in. They would take care of it, and get it to the meeting, or they would bring theirs to the meeting, and you would meet, and then everybody would go home with no paperwork. That was very unusual."

The Mission Control Center itself had to adapt. James Brandenburg, who worked there during the Shuttle program, explains, "It did impact the facility, because we had to make modifications to protect data security. The third floor of the control center was used for the support of the DoD flights, and we had to keep all the information that was classified sepa-rate from the regular unclassified information. For instance, we had cable runs under the floor. The ones that supported the classified flights were separate from the ones that sup-ported the non-classified flights. That was pretty much across the board; everything, even the voice loops on the key sets, the ones that could have classified information on them were separate from the ones that didn't. So there was a lot of impact to prepare the facility for the DoD flights."

Yet the most difficult obstacle that NASA had to face was not the classification and the security, it was managing the expectations and requirements of the USAF. The USAF had been forced to use the Shuttle, and they made NASA feel their dislike whenever they could. As Glynn S. Lunney recalls, "The Air Force community came to us with these requirements. One, they didn't want to be here in the first place. And two, they wanted to show that we civilians could not handle the security which they needed. So they just used to kind of drive us in circles with one requirement after another, after another, after another, about how we should run missions out of here. They would give us requirements, and we'd try to figure out how to do them, then we'd tell them what it cost. That would drive them crazy and they wanted ten dollars' worth of security for a buck's worth of price. I'm exag-gerating, but it was that kind of argument."

Security costs at NASA was an ever present issue which reached its apex when it was apparent how cheeky the USAF was becoming towards NASA. Lunney recalls a conversa-tion that he had with General Dick Henry, who at the time was running the USAF Space Division. He had told Henry that the price tag to implement all of the security require-ments would be $35 million. "That turned out to be a bombshell, because the colonel who had been interfacing with us had been telling him we were saying it would cost $100 mil-lion; three times that." What happened was that instead of honestly representing NASA's cost valuation, the USAF colonel had inflated those numbers to include, as Lunney put it, "all of the things that the Air Force wanted to do for this and other programs, IUS, and a bunch of other stuff" and added in "the kitchen sink and the bathtub and the commodes upstairs," thus inflating the budget for security by JSC. After that meeting with the general, Lunney wryly points out, the colonel never showed up again and actually resigned from the Air Force.

For Lunney and NASA, that was a minor victory on a very treacherous path. "I don't know what all the motivations could be for that, but there was a tendency for [military liaisons] to paint what we were doing in the worst possible light, and there was a tendency for people, when they were unhappy with the integration process or anything else, like security, to call Washington headquarters." The irony was that the space agency needed USAF payloads to generate traffic to keep the Shuttle flying, and the USAF needed the agency to transport its payloads. As Lunney observes, this relationship was something the DoD had difficulty with. "They are very sensitive to command and control, and chain of command. I mean, we deal with it in NASA for executing programs, but they are more sensitive to it in terms of controlling things that they think should be under their command. I think they just had a longstanding difficulty with trying to have something that was very important to them executed by this other government agency called NASA and they didn't really have the control. The DoD doesn't control NASA. So the loss of control, or the degradation of control, just was against their grain. It was against the way they've been taught to think and operate all their careers. That gave them a great deal of difficulty, and it manifested itself in this resentment, this constant positioning to make us look bad. But it derived from their loss of control…lack of control, lack maybe more than loss. It wasn't like they lost it, but they didn't have the control that they used to have in the past, and it just upset them a great deal."

Despite the skirmishes between NASA and the USAF, with STS-53 the Shuttle was being discharged from military service with honor. We can trust Mattingly when he says, "What those programs did was spectacular. They were worth classifying, but when the books are written and somebody finally comes out and tells that chapter, everybody is going to be proud…The missions were worth doing, they really were. Just to know that you had a chance to participate in something that magnificent is really kind of interesting."

For the remainder of the program, the Shuttle would, from time to time, carry out some secondary experiments that were classified but the era of the Shuttle in uniform was now definitely over. NASA and DoD could finally go separate ways ending their troubled relationship without regrets.

10

Something That Nobody Had Ever Done Before

ORIGINS OF THE SATELLITE ON A TETHER

On May 25, 1961, before a joint session of the US Congress, President John F. Kennedy made what become known as the "Moon Shot Speech." In it, the youthful president challenged the mightiness of the American aerospace industry to put a man on the Moon and return him safely before the decade was out.

The boldness of this assignment was amply illustrated by the fact that NASA had logged a mere 15 minutes of space flight experience to date. In fact, on his Mercury mission on May 5 aboard a capsule named *Freedom 7*, Alan B. Shepard Jr had spent only about 5 minutes in weightlessness at the peak of a ballistic arc. For the rest of the flight, America's first astronaut endured the high accelerations of ascent and re-entry. Shepard's flight was less impressive than the full orbit of the Earth by Yuri A. Gagarin on April 12 on behalf of the Soviet Union, but NASA was seeking to match this achievement.

Although in those days space was known as "the high frontier," the Moon was a much more remote target than NASA had been considering. To gain vital experience in working in space and exploiting celestial mechanics to reach the Moon and return safely, the agency initiated the Gemini program. With the exclusion of the first two test flights, the remaining ten missions were all flown by two-man crews who tested their ability to survive in space for a fortnight, this being the length of time that a full lunar landing mission would take; rendezvous and docking procedures, which would be essential for a lunar mission; and spacewalking techniques which, amongst other things, would provide a rescue option in the event of a docking failure. It was a fast-paced program, because many capabilities had to be proven in time for the agency to meet the deadline for a lunar landing within the decade.

Despite the hectic pace of events, NASA was able to give thought to the future of human space exploration. It envisaged a human expansion into the solar system, with interplanetary spaceships and large outposts for scientific research, the processing of resources mined from all around the solar system, and the manufacturing of various goods for consumption back on Earth. While weightlessness might favor the analysis and

© Springer International Publishing AG 2017
D. Sivolella, *The Space Shuttle Program*, Springer Praxis Books,
DOI 10.1007/978-3-319-54946-0_10

exploitation of physical and chemical phenomena that are masked on Earth by the predominance of gravity, it was recognized that, in the long term, weightlessness would make living in space awkward from a human physiology perspective. This is why early space station concepts envisioned that at least a part of the structure would spin so that the resulting centripetal force would emulate gravity and provide relief to the astronauts.

One way to simplify the design of a spinning space station is to link together two sections using a long cable or tether, then rotate them about their common center of mass. The advantage of a tether is that it removes the need to build a large truss-like structure, thereby reducing the mass and simplifying the overall design. Recognizing that such a configuration required further investigation in the real space environment, NASA assigned an experiment to the Gemini XI mission.

Launched on September 12, 1966, mission commander Charles "Pete" Conrad Jr and pilot Richard F. Gordon Jr rendezvoused and docked with an Agena upper stage at the end of their first orbit. On the next day, Gordon linked the vehicles together by a 118-foot-long tether made of Dacron. The Gemini then undocked and pulled back from the Agena, unreeling the tether and drawing it taut for the subsequent execution of two tasks.

The first assignment was to adopt the so-called gravity gradient attitude in which the differential gravitational attractions and centrifugal forces acting on separated but connected masses caused them to line up in the local vertical direction whilst circling the Earth. It is a convenient attitude because is stable against perturbations and thus requires only a minimum, if any, expenditure of propellant to maintain. It was being considered for station-keeping, since it would greatly reduce the fuel consumption of active maneuvering. The second assignment, once a stable gravity gradient had been achieved, was for the Gemini to fire its thrusters to start a cartwheeling motion about the center of mass of the two-body system and thereby create a small but noticeable degree of artificial gravity.

The first task proved to be a challenge because the pull of the cable inhibited the establishment of a stable gravity gradient attitude. The effort was abandoned in order to attempt to spin up the system. This was also more difficult than expected, but with some effort Conrad achieved a stable rotation rate of 55° per minute. The astronauts were able to detect the artificial gravity when it caused loose objects to slowly drift in a straight line towards the rear of the cockpit.

Intrigued by the behavior of a tether, NASA decided to repeat the gravity gradient task on the next Gemini mission, which would be the last of the program. Once again a spacewalker linked a tether to an Agena and, as before, straightening the tether was difficult. With persistence, the crew of James A. Lovell and Edwin E. "Buzz" Aldrin Jr achieved the desired attitude, confirming its properties even in the face of induced perturbations. Gemini XI and XII proved that tethers systems can work in space, but also highlighted the need for a deeper understanding of their behavior.

No tether experiments were conducted during the Apollo program, whose focus was Kennedy's challenge. This was accomplished on July 20, 1969, when Apollo 11 touched down on the Sea of Tranquility and mission commander Neil A. Armstrong became the first human to leave a boot imprint upon the dusty lunar surface. Several more landings followed, but a succession of budget cuts and rapidly decaying public interest sealed the fate of the program – and indeed the dreams of space stations and interplanetary missions.

Nevertheless, studies of tethered systems continued on a low budget. For instance, they might facilitate the rescue of astronauts stranded on-orbit. And there was a plan to link together the main body of Skylab, known as the Orbital Workshop, with the Apollo Telescope Mount, a suite of telescopes for observing the Sun which were installed on a frame that was derived from the descent stage of the Apollo lunar lander, but this idea was soon rejected in favor of physically connecting the two elements.

Tethers were relegated to the back burner of human space exploration plans until interest resumed in the mid-1970s with such a vigor that it led to the planning and execution of two of the most brilliant and complex Shuttle missions ever flown.

It all started in September 1974, when the Smithsonian Institution Astrophysical Observatory published an article entitled "Shuttle-borne 'Skyhook': A New Tool for Low-Orbital Research," penned by Giuseppe "Bepi" Colombo, a professor of applied mechanics at the Padua University in Italy who was hired by NASA in 1961 for his talents as a mathematician, a physicist, and an engineer. As the Shuttle program was taking shape, Colombo saw an opportunity to develop a package of instruments that could be unreeled on a tether from an Orbiter for about 100 km and trailed through the upper atmosphere at altitudes that were unattainable by satellites.

Above an altitude of 30 km, molecular diffusion supersedes wind and turbulence as the means of mixing gases. The constituents separate according to their molecular or atomic weight, with some such as oxygen, hydrogen and nitrogen taking part in a complex and rich photochemistry driven by sunlight. Large high-altitude balloons are routinely used to study the stratosphere, but they rarely exceed 50 km. Satellites are also extensively used for atmospheric research, but they cannot orbit below 200–180 km because the tenuous atmospheric drag is sufficient to cause re-entry in a matter of days. Below this altitude, atmospheric properties and behavior can be investigated by sounding rockets, but the observation window is geographically localized and limited to a couple of minutes.

But a tethered package unreeled deep into the atmosphere from the payload bay of a Shuttle orbiting at a safe altitude, could carry instruments to directly observe this otherwise inaccessible region. The inertia of the Shuttle would tend to overcome the drag on the small tethered subsatellite, enabling it to make observations across a wide geographical area for several days. This would yield insights into questions about the chemical composition of the atmosphere, the coupling mechanisms between small and large scale motions, the global wind field of the lower atmosphere, patterns of electric current circulation, energy fluxes, and so on.

Although the Skyhook concept would be an unorthodox use of the Shuttle, it was very attractive to the scientific community. Colombo's reputation spoke volumes for the soundness of the proposal, and NASA decided it was time to resume studies of space applications for tethered systems. In the ensuing years, papers were published and workshops were held; the former to apply mathematics and physics in order to understand the complex dynamics of such systems and the latter to explore a range of innovative applications.

For instance, the configuration for atmospheric studies could also be used to chart at higher resolution the distribution of the Earth's magnetic and gravitational fields to complement the measurements by satellites. Or by replacing the subsatellite with an instrumented aerodynamic model, engineers would gain the largest open, continuous, wind tunnel for prolonged studies of hypersonic aerodynamics, spacecraft re-entry optimization, aerobraking techniques, and characterization of materials for spacecraft and hypersonic vehicles. Furthermore, a tether capable of conducting electric current would permit the simulation and stimulation of plasma electrodynamic phenomena in the ionosphere at a scale impossible for a ground-based laboratory. Because plasma permeates the universe, electroconductive tethers could be used to mimic important processes that occur in the solar system and beyond; for example, plasma interactions with the surface of a celestial body, its atmosphere, and its intrinsic magnetic field, all of which give rise to shockwaves, sheaths, and currents.

An intensive lobbying activity by Colombo and his collaborators in Italy at the *Centro Nazionale di Ricerca* (National Research Science) and at the *Gruppo Sistemi Spaziali* (Space System Group) of Aeritalia, a major aerospace firm based in Turin, led President Ronald W. Reagan and Italian President Amintore Fanfani to sign in 1983 a bilateral agreement to develop, construct, and execute a tethered mission on the Shuttle as early as 1987. The loss of *Challenger* in January 1986 put this project on hold, but finally, on July 31, 1992, *Atlantis* lifted off with a tethered system in its payload bay. The seven crew included Dr. Franco Malerba, the first Italian to fly in space, as the payload specialist to oversee all the operations regarding the innovative experiment. Regretfully, Colombo did not live to see this. On February 20, 1984, he lost his battle with cancer. He was just 63 years of age.

TETHERED SATELLITE SYSTEM 101

In NASA parlance, the hardware for the first Shuttle flight of a tethered system was the Tethered Satellite System-1, or simply TSS-1. Stowed in the forward half of the payload bay of *Atlantis*, it consisted of three essential elements housed between a U-shaped Spacelab type pallet and an MPESS platform: the tether deployer system, the satellite itself, and the science experiments complement.

Developed by Martin Marietta for the NASA Marshall Space Flight Center, the deployer system was the body and soul of TSS-1 because it provided the structures and mechanisms, electrical power distribution, communications and data handling, and thermal control subsystem for the checkout, deployment, and operational control of the tethered satellite. The tether was wound onto a 4.44-inch-diameter, 3.7-foot-wide shaft which had 3.17-foot-diameter flanges at each end. While TSS-1 would use a 13.8-mile tether, the reel was sized to accommodate 68.4 miles of cable with a maximum diameter of 1 inch. The reel motor and brake assemblies were coupled to the starboard flange of the shaft. The motor would serve as a generator for braking torque during deployment and then as a motor to haul the satellite in during retrieval operations.[1] The brake would halt activities

---

[1] This is analogous to an elevator motor where the elevator is respectively descending and ascending. In the first case, the motor acts as a brake to control the rate of descent and counteracts the downward pull of gravity. When ascending, the motor acts a generator that hauls the elevator up against gravity.

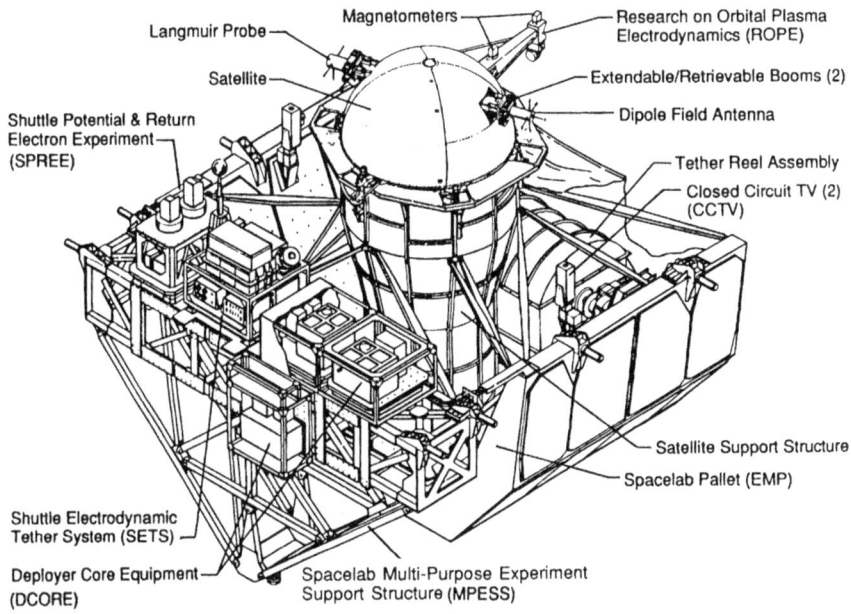

The main hardware components of the TSS, along with science experiments on the MPESS platform.

either upon command from the crew or automatically upon exceeding preset velocity limits or in the event of a loss of power in the brake assembly or the rate sensors. The brake assembly had a lock to preclude rotation of the reel during the ascent to orbit and also to maintain the tether at a preset tension just prior to commencing deployment operations. Without this provision, the tether might go slack and become entangled within the deployer mechanisms, which would either prevent deployment or cause damage to the tether and deployer. Upon completion of the retrieval operation it was not necessary to maintain the tether at a preset tension, so the lock would not require to be engaged.

At one end, the tether was wound around the reel assembly shaft and at the other it was connected to the satellite, which sat above a canister on top of a 40-foot-long deployable boom that was folded into the canister boom within the Satellite Support Structure (SSS), a cylindrical structure whose top end expanded to accommodate the satellite canister and held it in place during non-deployed operations by means of six restraint latches and three shear wedges.[2]

---

[2] These latches and wedges prevented the satellite from moving radially or along the vertical direction of the SSS. Their design allowed for the radial dimensional changes which the satellite would be subjected to while on-orbit, preventing unacceptable high stress when restrained in place.

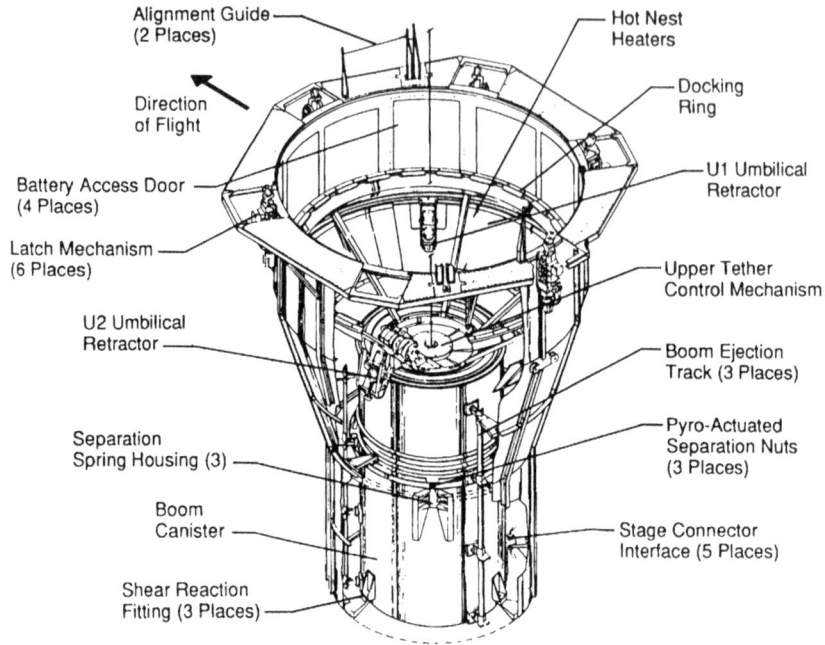

Alignment Guide
(2 Places)

Direction
of Flight

Battery Access Door
(4 Places)

Latch Mechanism
(6 Places)

U2 Umbilical
Retractor

Separation
Spring Housing (3)

Boom
Canister

Shear Reaction
Fitting (3 Places)

Hot Nest
Heaters

Docking
Ring

U1 Umbilical
Retractor

Upper Tether
Control Mechanism

Boom Ejection
Track (3 Places)

Pyro-Actuated
Separation Nuts
(3 Places)

Stage Connector
Interface (5 Places)

A detailed view of the Satellite Support Structure.

The boom was an open lattice structure with a square cross-section that consisted of 12 individual bays. It was extended and retracted using an elaborate and ingenious mechanism. Each bay was a truss structure of four vertical foldable longerons and rigid horizontal battens on each side to create a top and bottom rigid bay frame end. Between the two bases, four flexible battens would hold the bay open when it was extended and the tension in diagonal cables transferred the loads from the flexible battens to the rigid bases, thereby completing the cube of a deployed bay. During the extension of the boom, rollers located on each corner of the rigid horizontal batten end frames would engage with the internal thread of a large rotating deployment nut inside the boom canister. As the nut rotated, it would push the top rigid frame of the deploying bay upwards. At the same time, blocks at each corner of the flexible batten frame were being captured by S-tracks within the canister boom and forced to unfold inward while also stretching out and thrusting the longerons upwards. As the top end frame rose up the nut, the upper end frame of the next bay would be engaged by the nut. This deployment would repeat until the boom was fully extended. The retraction process worked in the opposite manner, but with a spring-loaded frame at the bottom of the rotating nut being engaged at the end of each folding process in order to hold each collapsed bay in a compact state.

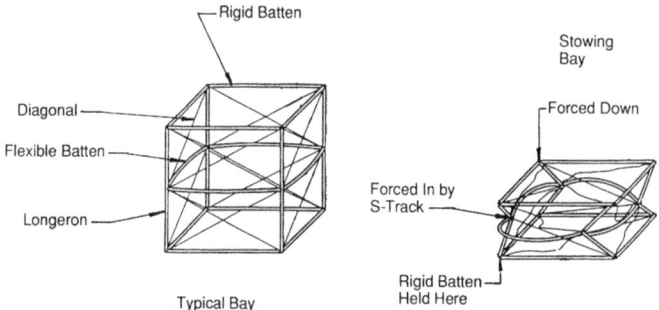

The structural components of a bay on the deployable boom.

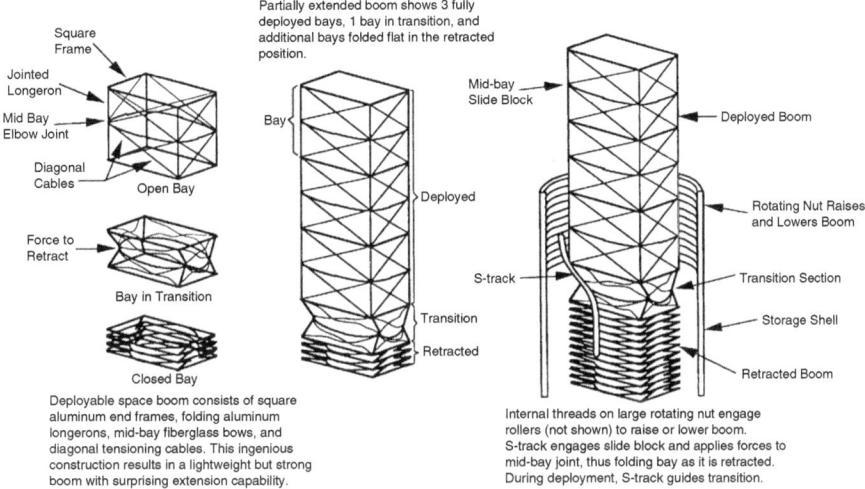

The boom deployment sequence.

As the tether unwound from the reel shaft, it would be routed through the Lower Tether Control Mechanism (LTCM) which had a tensiometer to measure the tether tension, and a tether measurement wheel to keep track of the lengths of deployed and retrieved tether and the rate at which the tether was passing through the unit. It then continued through the extended boom and passed into the Upper Tether Control Mechanism (UTCM), which was on the boom canister just beneath the satellite. Its function was to guide the tether from the boom to the satellite and provide a constant boom-to-reel tension for reeling operations. Both control mechanisms included cutter assemblies to sever the tether in a contingency situation that necessitated such action. Likewise, the boom and canister each had a mechanism that would have ejected them from the payload bay, with or without the satellite docked.

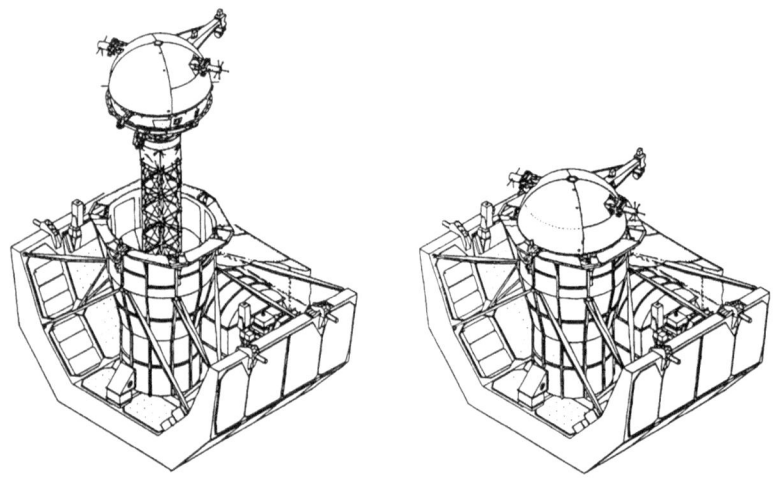

The configuration of the boom when extended (left) and retracted (right).

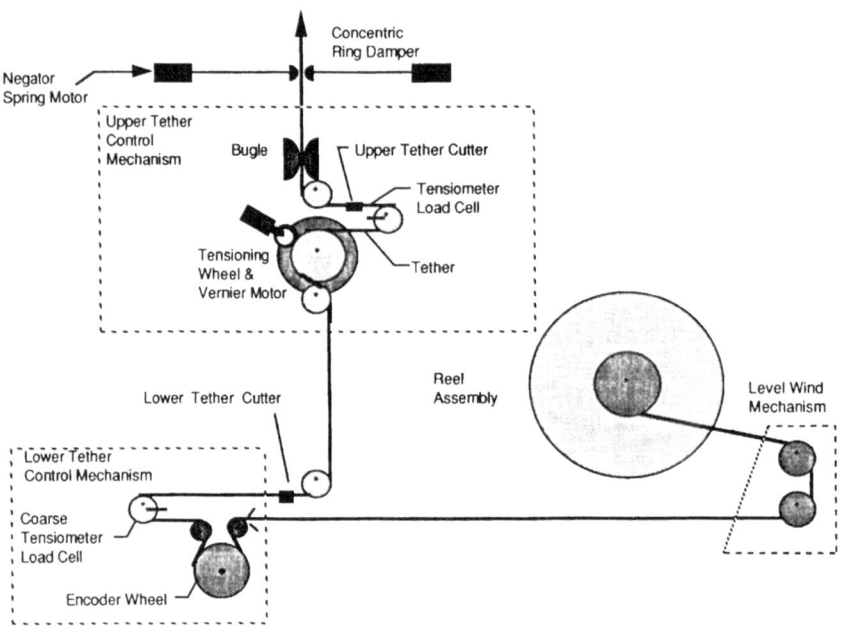

A schematic of the two Tether Control Mechanisms.

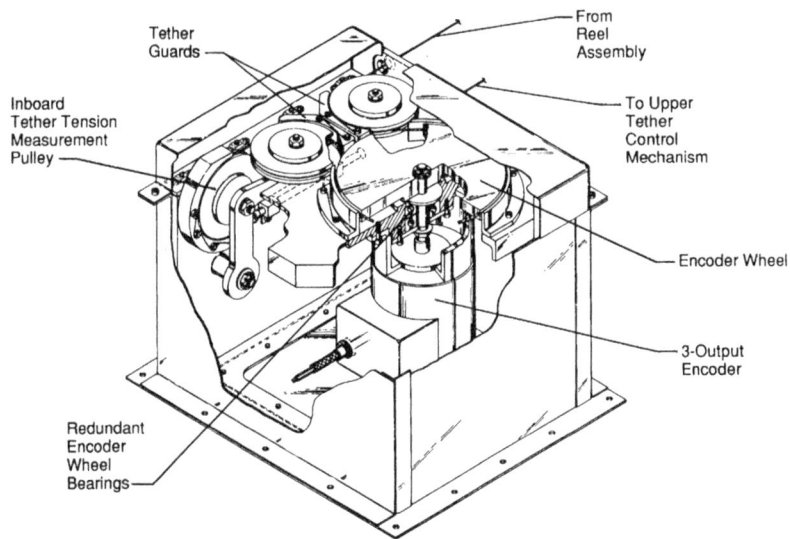

An internal detailed view of the Lower Tether Control Mechanism.

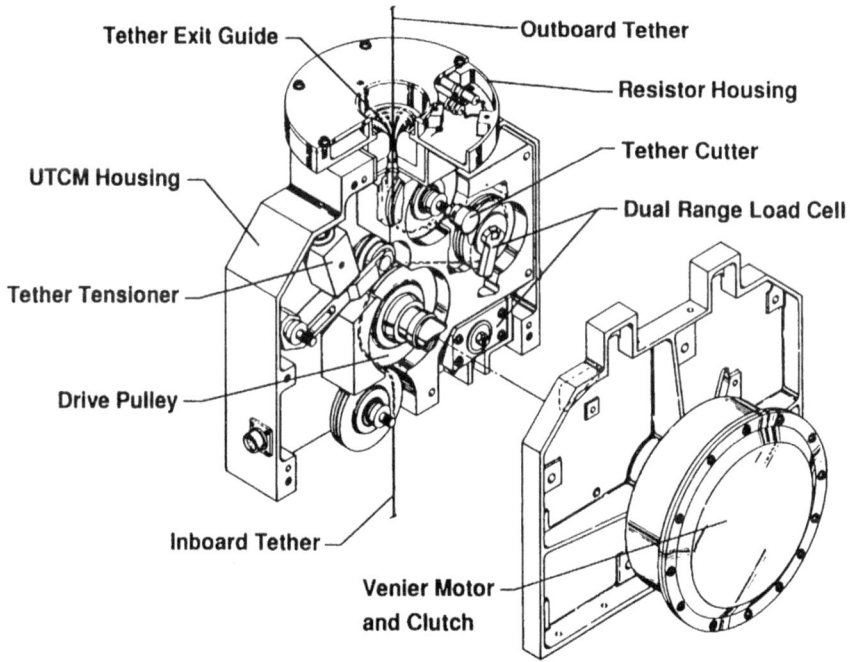

An internal detailed view of the Upper Tether Control Mechanism.

The complex deployment and retrieval operations were under the supervision and control of the Data Acquisition and Control Assembly (DACA), which was the brain of the deployer. Control was by closed-loop control laws that compared a preselected mission profile with data on tether tension, tether length, rates of tether deployment and retrieval, and the supply voltage to the motor. An optical shaft encoder within the LTCM provided digital readouts that were converted by the DACA into actual tether length and tether velocity parameters. Differences between the actual values and the intended profile were corrected by sending commands to the reel motor.

The most visible part of the TSS was the 5-foot-diameter white spherical satellite built by Aeritalia in Turin, Italy, under the management and supervision of the Italian Space Agency. It comprised two hemispheres divided by an equatorial plane, and its surface was covered by aluminum alloy panels coated with an electrically conductive paint.

The upper hemisphere (opposite to the tether attaching point) was known as the Payload Module. It provided an easy-to-access volume in which to install a mission-specific array of experiments. It also contained two 7.67-foot extendable booms and one 39-inch fixed boom to accommodate additional experiment and sensor packages. Opposite the fixed boom was a shorter mast for the S-band antenna to communicate with the Shuttle.

The lower hemisphere, called the Service Module, housed support systems for power distribution, data handling, telemetry, navigation, and attitude control. It also contained the Auxiliary Propulsion Module that was composed of a pressurized tank of gaseous helium and the associated plumbing (valves, pressure regulators, filters, heater, etc.) to feed three sets of cold-gas thrusters for attitude control and to maintain the appropriate tether tension during deployment and retrieval of the satellite in close proximity to the Orbiter.

The hardware on both hemispheres was held in place at various attaching points on the interior walls of the satellite's outer skin and on internal divider walls, and was arranged for easy access during installation and testing. Thermal control was mainly passive, by multi-layer insulation (MLI) blankets on the external and internal walls of both modules except for the bottom half of the Service Module which was a radiator section to reject onboard heat into space. Electric heaters placed at key points within or surrounding the internal hardware provided active thermal control.

A pair of umbilical mechanisms provided power transmission (umbilical U1) and data and command telemetry (umbilical U2) prior to deployment. After separation from the extended boom, a set of four silver-zinc batteries within the Service Module would feed the satellite's electric system while an S-band antenna mounted on the tip of the fixed boom provided communications between the satellite and the Orbiter.

TSS-1 EXPERIMENTS AND SCIENCE OBJECTIVES

Colombo envisaged his Skyhook reeling down a tethered satellite to explore the most inaccessible portion of the lower atmosphere, but the TSS-1 subsatellite was intended to be spun away from the Earth to perform large-scale plasma electrodynamic studies within the ionosphere, and to reproduce related phenomena that occur throughout the solar system.

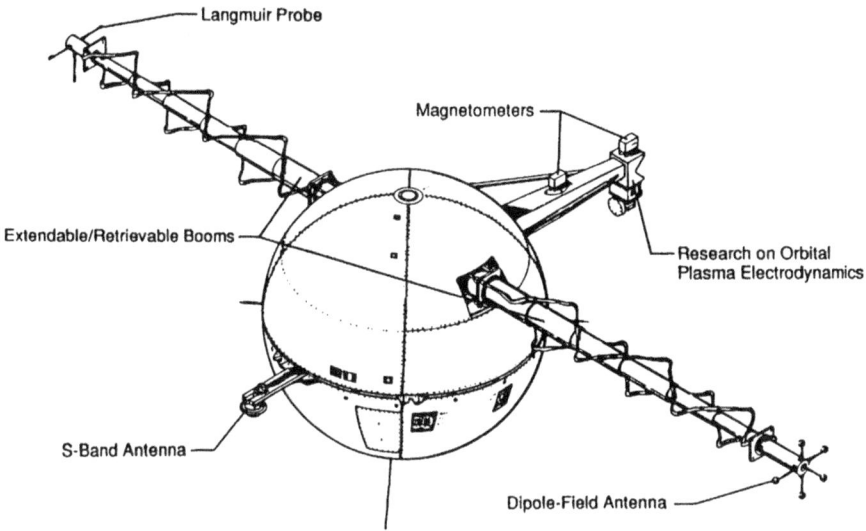

An overview of the TSS.

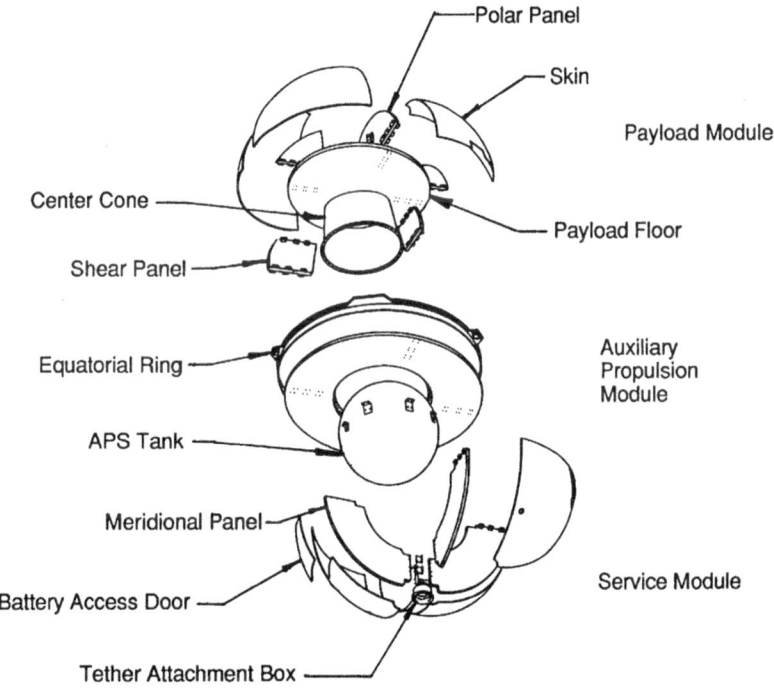

An exploded view of the TSS.

Plasma is known as the fourth state of matter, a condition in which the atoms of a gas are so intensely heated that electrons, which circulate around the nucleus, gain sufficient energy to escape, leaving behind a positive ion.[3] This process is called ionization. In other words, in a plasma most of the atoms have been ionized and are surrounded by a cloud of negatively charged electrons whose kinetic energy allows them to roam free. It is worth noting that plasma can include a proportion of neutral atoms that retain their full complement of electrons. Despite the presence of positive (the ionized atoms) and negative (the electrons) charges, macroscopically speaking a plasma is electrically neutral. This does not mean that a plasma is of little interest. On the contrary, within the plasma environment charges are affected by external electric and magnetic fields and these in turn "feel" the electric currents and magnetic fields of the plasma. Complex and difficult-to-model interactions arise that make plasma a fascinating and complex state of matter which, since the 1920s, has found numerous applications. Neon lights and television plasma screens are among the most popular examples at a consumer level.

Although plasma does not occur naturally on Earth, it is the predominant state of matter in the universe. In fact, thanks to plasma observations of the universe, we can observe invisible structures and investigate processes ranging from star formation to the evolution of galaxies.

In the case of Earth, the ionizing effects of solar ultraviolet radiation on the gases of the atmosphere mean that plasma predominates at altitudes in the range 53 miles to several tens of thousands of miles. This region is aptly called the ionosphere. It primarily consists of electrons and atomic oxygen ions. It is thanks to the ionosphere that over-the-horizon radio communications are possible, because the waves of radio energy bounce off the ionosphere and are reflected back to the ground far away.

The ionosphere continues well into the magnetopause, that region of space where the magnetic field of the Earth yields to that of interplanetary space. On the night side of the planet, this area is distorted by the solar radiation to form a tail akin to that of a comet, extending far beyond the orbit of the Moon. The area of overlap between the ionosphere and the magnetopause is called the magnetosphere (even though it is not spherical). Mutual interactions between the electric and magnetic fields external and internal to the plasma produce a magneto-hydrodynamic (MHD) generator fueled by the strong electric currents that are caused by the ionosphere interacting with that part of the solar wind that leaks into the magnetosphere.[4] These fields, in turn, produce a general circulation of the plasma (equivalent to a current system) and this is how Mother Nature paints the polar night sky with jaw-dropping auroras. However, when the currents generated by these fields exceed the capacity of the plasma to carry the charge, a double layer plasma is formed, a structure in which two layers of excessive electric charge, one positive and the other negative, exist

---

[3] If a neutral atom loses a single electron, it becomes an ion with a single positive charge. If the ion still has electrons and is further heated, it may continue to shed electrons and become progressively more ionized until, in the extreme case, it is fully ionized. If hydrogen, the simplest of atoms, is ionized, the loss of the single electron leaves behind a proton.

[4] The solar wind is mostly diverted away from Earth by the planet's magnetic field. However at the poles, the field allows some of the plasma to interact with the lower atmosphere.

in close proximity. Between the two, an electric field is generated that can accelerate some electrons and ions to even greater energies.

By making the TSS-1 tether capable of conducting an electric current, one of the primary objectives was to reproduce the MHD phenomenon on a large enough scale to study electrodynamic interactions occurring within the ionosphere. Measurements were to be taken across a wide range of conditions, depending on the density of the ionosphere, which varies along an orbit from day to night and with the inclination of the Earth's magnetic field. The current and voltage across the tether were expected to change at least two-fold during a full orbit. As plasma double layer structures are thin and weak, they are hard to observe directly. However, because it is possible to follow their particle distribution, it was hoped the subsatellite would detect their presence by passing through and disturbing them, with a tell-tale being a variation of the electrons accumulating on the satellite as the double layers collapsed near it.

To understand how a current could flow through the tether we need only visualize a direct current generator whose components, in the most basic configuration, are a magnet and a conductive wire. As relative motion between the wire and the magnetic field occurs, a so-called electromotive force is applied to the electric charges within the wire, thereby inducing a current to flow.[5] A space-borne tether system works on this principle. The tether is the wire moving through the stationary magnetic field of the Earth, generating, therefore, a potential that drives the current from the positive to the negative pole. By virtue of its electrically conductive paint, the subsatellite would become the positive pole collecting free electrons from the surrounding plasma and routing them through the tether to the negative pole located in the Shuttle's payload bay. Because a circuit must be closed to produce an electric current, there were two electron cannons to fire the collected electrons back into space. These were expected to travel along the Earth's magnetic field lines back to the satellite and complete the circuit. Clearly, this is a good way for an orbiting spacecraft such as a space station to generate electric power in a very efficient manner. However, as we will see, there is a downside.

Features such as sheaths and wakes were expected to take shape in the vicinity of the subsatellite as it perturbed the density, temperature and electrical properties of the surrounding plasma. A plasma sheath resembles a cloud of electric fields enveloping the subsatellite, and a plasma wake has the same shape as the wake that trails a boat in water. Both were expected to alter the distribution of electric fields, electrons, and ions at and near the surface of the subsatellite, as well as affect the current or voltage through the tether. It was also predicted that sheaths and wakes would influence and interact with one another, producing further changes, and that by actively varying the electric charge of the subsatellite new phenomena would be triggered.

Another phenomenon that TSS-1 was to investigate concerned the generation and propagation of plasma waves within the ionosphere. Generally speaking, a wave is a coordinated disturbance within a medium. While the medium in which a wave travels does not necessarily move as a whole in the direction of the wave, the wave creates a disturbance of

---

[5] The wire and the magnetic field move with respect to one another. Regardless of whether the wire moves (usually by rotation) through a stationary magnetic field or the magnetic field moves (again usually by rotation) about a fixed wire, the result is the same with the generation of a current.

the constituents of the medium in a direction which depends upon the medium and the origin of the wave itself. Within the atmosphere, plasma waves are one of the causes of an exchange of energy and particles between the ionosphere and other regions of the atmosphere.

Due to its electrical nature, plasma readily interacts with electromagnetic waves, such as radio waves. Therefore injecting a known amount of radio energy into the space plasma and measuring how its propagation is affected provides information on the composition, distribution, and motion of the plasma. Once it was fully deployed the TSS-1 tether would be the longest antenna ever placed on-orbit. By modulating the current that flowed through it, low and ultra-low frequency radio waves would be sent into the ionosphere that could be received by stations on the ground to undertake additional investigations into plasma physics.

Several interdependent experiments spread between the subsatellite and *Atlantis*'s payload bay comprised the toolbox to accomplish the scientific goals set for TSS-1. Installed on the MPESS platform in the payload bay, there were the Deployer Core Equipment (DCORE), the Shuttle Electrodynamic Tether System (SETS), and the Shuttle Potential and Return Electron Experiment (SPREE). DCORE contained the two aforementioned electron cannons to allow the electrical potential of the satellite to be varied by controlling the current that flowed through the tether in response to the electrons that were projected by the cannons. SETS was to determine the ability of the tether to collect electrons, and how it was affected by electrons emitted within the Orbiter generated plasma. SPREE was to look at the full ion and electron energy distribution to investigate how the current was generated, and how it was affected by the return currents to the Orbiter. Both SETS and SPREE were also to characterize Shuttle-induced environmental effects. In fact, the Shuttle has often been compared to a comet nucleus enclosed within a halo of water vapor and other ions emitted by water dumps, thruster firings and outgassing of vacuum exposed hardware. This halo inevitably interacts with the surrounding plasma. Analysis of the Shuttle's influence prior to, during, and after deploying the subsatellite would enable the investigators to adjust the data that was gathered and distinguish natural variations from disturbances induced by the Shuttle.

The TSS-1 Payload Module had four experiments. The Satellite Core Equipment (SCORE) would work in tandem with the payload bay-based DCORE to control the electrical current flowing between the subsatellite and the Shuttle via the tether. The Research on Orbital Plasma Electrodynamics (ROPE) was to study the interactions that occur when a large conducting body moves through collision-less space plasma at supersonic speed. It would also examine the behavior of ambient charged particles in the ionosphere and ionized neutral particles around the satellite. The Research on Electrodynamic Tether Effects (RETE) would investigate the physical processes in the charged region of space surrounding the subsatellite. Finally, the Magnetic Field Experiment for TSS Missions (TEMAG) was to study the magnetic signature of the tether current and its closure through the structure of the sheath that surrounded the subsatellite. In addition, there were two theoretical science investigations that did not have hardware in space and instead relied upon observations made from the ground to detect radio waves emitted by the tether for further studies of ionospheric plasma.

Along with characterizing the electrodynamics of a space-borne tethered system, TSS-1 was also to simulate and experiment with a variety of plasma phenomena that commonly occur around celestial objects and are impossible to replicate in ground laboratories whose limited size prevents the modeling of such large-scale processes. For example, it would mimic a scaled-down version of the interactions between the planet Jupiter and its moon Io. Unlike Earth's Moon, Io orbits Jupiter well within its magnetosphere, traveling at hypersonic speed through the magnetic field and plasma environment. The surface of Io is covered by volcanos that erupt tonnes of gases that form a tenuous ionized atmosphere. As Io sweeps through the Jovian magnetic field, this generates a potential of some 400,000 volts across its conducting atmosphere, inducing massive currents of the order of 5 million amperes flowing between the two bodies and penetrating deep into the lower Jovian ionosphere. Since the process that generates this current is the same as that which would occur with the subsatellite, the experiment would provide insight into the dynamics of a planetary-scale process.

FUNDAMENTALS OF SPACE TETHERED SYSTEM DYNAMICS

Unreeling a tethered subsatellite from the payload bay of a Shuttle on-orbit might at first sight appear to resemble a kite being drawn into the wind by a running child, but the two systems could not be more different. While the kite flies under the influence of lift and drag created by the airflow lapping its surface, a tethered subsatellite relies exclusively on the so-called gravity gradient attitude. Let us imagine two spacecraft that are orbiting the Earth at different altitudes. The lower spacecraft will experience a larger gravitational attraction than its centrifugal force. The situation is reversed for the upper vehicle.[6] If the two spacecraft are linked using either a flexible tether or a rigid structure then they will be forced to travel together as one, and the difference in gravitational and centrifugal forces at the two endpoints will orient them radially to the center of mass of the Earth. In fact, the lower craft will be pulled closer to Earth owing to the greater gravitational force and the upper one will be pulled away by the greater centrifugal force. They have adopted the gravity gradient attitude.[7] At the same time, this tug of war between forces develops a tension which acts between the center of masses of the two spacecraft along the axis which connects them. Another way to visualize this tension is by remembering that if it were not for the connecting structure or tether, each spacecraft would follow its own independent orbit, with the lower one traveling faster than the upper one. By linking them together, the

---

[6]The gravitational force follows the inverse square law, so the lower is the orbit the greater is the attraction felt by a spacecraft, but because the centrifugal force follows a linear proportionality, the farther away the orbit is from the Earth's center the greater is the centrifugal force acting on the spacecraft.

[7]Note that despite being known as the gravity gradient, centrifugal force is an equally vital player in this tug of war between contrasting forces. In fact, up to one-third of the total force acting on the system is due to the centrifugal force.

vehicles are constrained to move at the same orbital speed of the system's center of mass.[8] The result is that the lower spacecraft, which is below the overall center of mass, is forced to move at a slower speed than it would otherwise have if it were free. For the upper craft, the situation is reversed. In fact, being above the system's center of mass it is compelled to move faster than it would if it were free. Unlike objects hung under one gravity condition, in which their weight is supported by the tension on the cable, a space-borne tether creates tension from the difference between the free trajectories and the actual trajectory of the two vehicles while they are connected to each other.

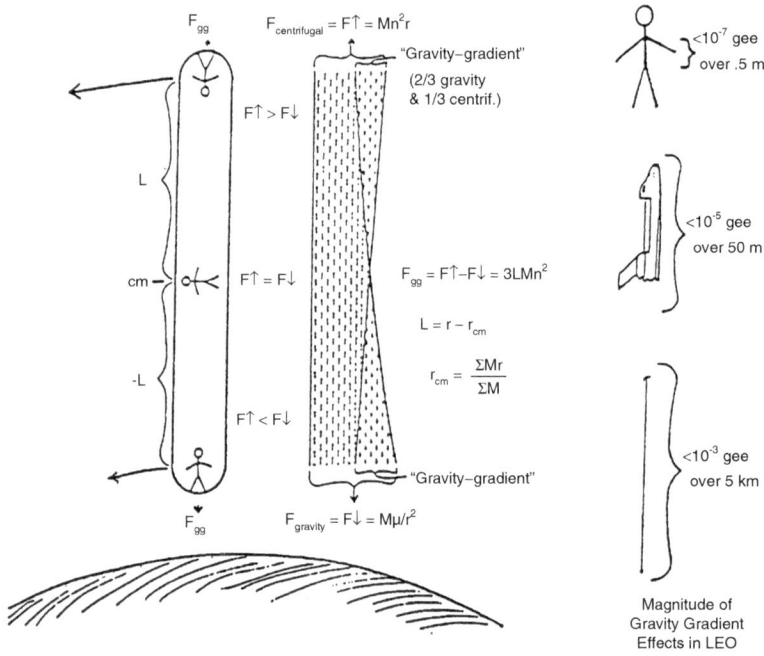

The gravity gradient concept.

The gravity gradient is also responsible for the deployment of the tether. Once the sub-satellite is given an initial impulse away from the Orbiter,[9] at a certain distance the combination of gravitational and centrifugal forces overcomes the friction in the deployer and

---

[8] This is the same as the point where the gravitational and centrifugal force balance. However, it must be noted that for very long tethers, much longer that those used by the TSS mission, this does not hold true anymore because of the large mass of the tether.

[9] For this reason, the satellite was equipped with four in-line thrusters.

the tether unreels without the need for thrusters.[10] An interesting effect on the Orbiter from the tension on the tether is to induce the so-called hang angle. As the tether applies a tension force on the endpoints, unless the Orbiter's center of mass and the tether's attachment point are aligned with the tension vector, the tension will apply a torque that rotates the Orbiter until the attachment point, the Orbiter's center of mass, and the tension vector are lined up. As the planned attachment point for the tether boom was forward of the Orbiter's center of mass, the Orbiter would adopt a stable nose-forward, positive pitch attitude. The angle between the local vertical axis and the Orbiter X-body axis is called the hang angle. With a stable 12.4-mile tether, the Orbiter was expected to adopt a +25° pitch attitude. Different attachment points and tether lengths would change this angle.

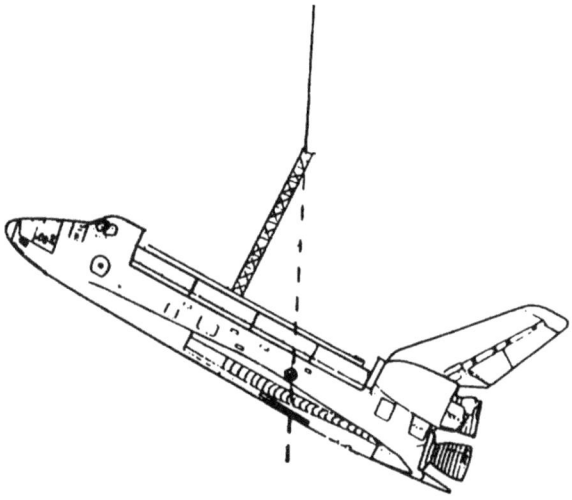

The hang angle of induced Orbiter attitude stabilization.

If the system is displaced from the local vertical but remains in the plane of the orbit (in-plane displacement), a downtrack component of the tether tension develops which reduces the velocity of the leading spacecraft and increases that of the trailing one. However, with the gravitational and centrifugal forces acting predominantly on the lower and upper spacecraft respectively, the system rapidly returns to its original radial attitude. The same is true if the system is displaced on a plane perpendicular to that of the orbit

---

[10] Note that gravity, a downwards pulling force, is responsible for pulling the satellite outward. How is this possible? Remember that as the Orbiter and satellite change their positions, the system's center of mass remains on the same orbit. As the Orbiter is pulled downwards, in order to keep the system's center of mass on the same orbit the satellite has to go upwards. Hence, with the help of the centrifugal force, the satellite is pulled upwards as a consequence of gravity acting on the Orbiter.

(out-of-plane displacement) but the system is less stable.[11] Here, the system will oscillate about its center of mass until it resumes the vertical orientation. At low amplitudes the frequency of this swinging motion, called libration, happens to be independent of the tether length. The net result is that the tether does not tend to swing faster than the space-craft at the endpoints (like the chain on a child's swing), instead they all move rigidly together (like a dumbbell). Libration is therefore aptly defined as a rigid pendulous motion of the system about its center of mass. It was to play a significant role during the deployment and retrieval of the tethered subsatellite.

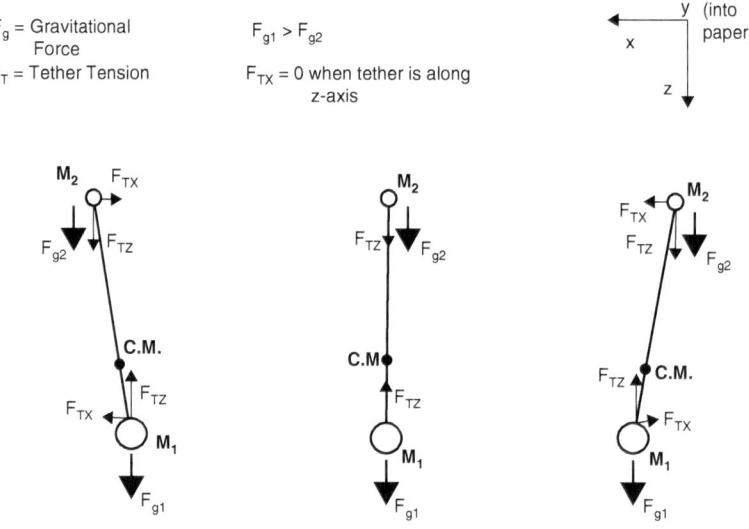

Gravity gradient stabilization by a tether.

Let us now imagine the TSS and Shuttle traveling together on the same orbit. The center of mass of the system is contained somewhere within the Orbiter, so we can safely refer the trajectory of the system to this point. If now we begin to deploy the subsatellite upwards, the Orbiter will move downwards because the center of mass of the system must remain where it was prior to initiating the deployment.[12] Since the Orbiter is now in a

---

[11] The gravity gradient is not limited to tethered systems. In fact, any sizable spacecraft will be subjected to it if an uneven distribution of masses of its internal components reproduces the same condition as two masses on two slightly different orbits. If a non-Earth oriented attitude is required, the spacecraft's attitude control system must fight against the gravity gradient.

[12] As we have not applied any external force to the Shuttle-satellite system prior to deployment, the center of mass cannot change its orbital altitude. The satellite deployment brings about a shift in mass distribution but not in center of mass position. That is why the Shuttle has to move downwards. Obviously, given the different order of magnitude in mass between the Orbiter and the satellite, the displacement of the former is minimum compared to that of the latter. In fact, upon full deployment of the 12.4-mile tether, the Orbiter was expected to have fallen just 328 feet below the system's center of mass.

lower orbit, it travels faster and starts to pull on the subsatellite. In turn, the subsatellite is now in a higher orbit and seeks to slow down. The tension in the tether causes the gravity gradient to align the system with the local vertical with a libration motion.

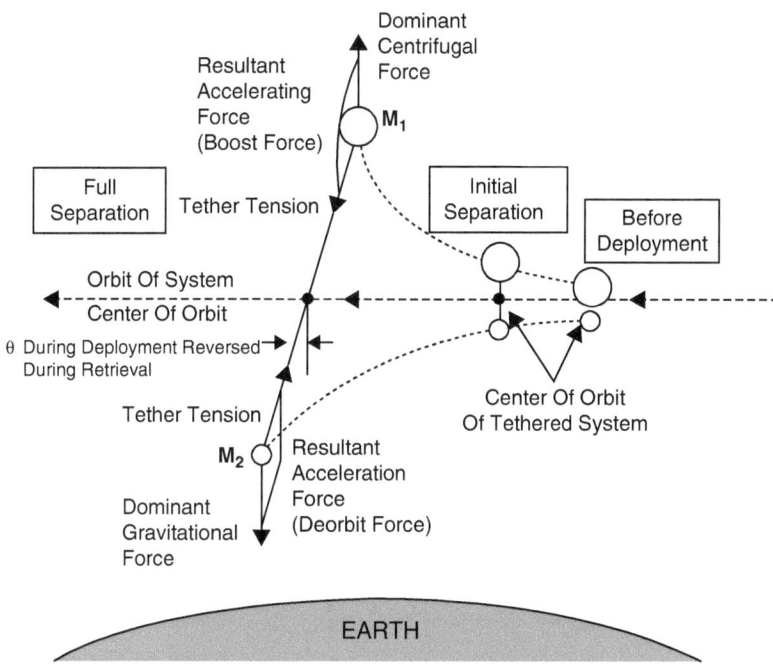

The dynamics during tether deployment. Although here the subsatellite (M2) is assumed to be unreeled towards Earth rather than deep space, the concept remains the same.

As the deployment continues, the two components draw away from the center of mass of the system, feeding on the libration oscillation. The reverse is the case during retrieval, when the subsatellite drags on the lagging Orbiter. However, the situation is now made more complicated by the transfer of momentum which occurs between the two components. In deployment, the gravity gradient maintains the system aligned with the local vertical. As the local vertical changes along the orbital path, the gravity gradient causes the system to rotate around its center of mass with a period identical to that of the orbit. This rotation bestows angular momentum on the system.[13] As the tether is reeled in during retrieval, the conservation of angular momentum causes the two components to increase

---

[13] The angular momentum of a body is the product its moment of inertia (how the mass is distributed about specific axes) and its angular velocity. Unless a rotating system is perturbed, its angular momentum remains constant.

their rate of rotation about the center of mass,[14] thereby increasing the libration. As the system will have a very small inherent damping, once such an oscillation is initiated it will continue and the amplitude will increase as the tether shortens. The retrieval therefore, must be very carefully controlled. This can be achieved by slowly decreasing the reel rate so that the gravity gradient can return the system to vertical slowly. If the motion is slowed just right, the tether will come to a halt in a vertical position with no residual libration.[15] In any case, the Orbiter will be able to control libration by performing timed translations.

There were other oscillation modes for a tether that had to be considered for the TSS experiment. For instance, the tether could stretch and compress as a result of its inherent elasticity, making the two connected masses bounce to and fro on the axis of the tether. This is the so-called "bobbing" mode. The Orbiter and subsatellite could also rotate about their respective centers of mass, creating a pendulous oscillation[16] with each mass swinging back and forth around its attachment point. There was also the possibility of wave-like oscillations typical of a string under tension, that would either travel along the axis of the tether or cause it to emulate a skipping rope. And then there was the local yaw attitude and spin mode of the system. The length of the tether and its tension define the frequencies at which such modes can develop. If, for a given length, two or more frequencies resonate, then the tether could potentially become uncontrollable. A lot of effort was put into modelling the TSS throughout its operational envelope, to determine the best way to control oscillations.

Of particular concern was the skipping rope motion, expected as a consequence of the electromagnetic drag generated by the interaction of the tether with the Earth's magnetic field. As explained above, by crossing the Earth's magnetic field lines the tether causes an electric current to flow from the subsatellite to the Orbiter. Since the laws of thermodynamics don't allow you to get something for nothing, the downside is that to create electric power, kinetic energy must be extracted from the tether. This induces a decelerating electromotive force (drag) along the tether, which slows down the system.[17]

---

[14] This is similar to a skater pivoting on its axis with both arms stretched out. As soon as the arms are pulled in, the moment of inertia of the skater is reduced. In order to maintain a constant angular momentum, the angular velocity must increase. Thus the skater spins faster.

[15] It is worth noting that orbit eccentricity is also a source of libration. In fact, as the system's center of mass travels along an elliptical orbit, the tethered spacecraft will be subjected to continuous changes in both its orbital velocity and altitude. Hence a continuous imbalance in the gravitational and centrifugal forces acts as a constant perturbation to the gravity gradient equilibrium, thereby generating libration.

[16] Do not confuse this mode with the libration rigid pendulous motion where the two masses rotate about the center of gravity of the system formed by the tether and the masses. In the pendulous mode, each mass rotates about its own center of mass with the tension in the tether being the restorative force. In contrast, in the libration rigid pendulous motion the restoring force is the balance between gravity and centrifugal force.

[17] If the current were pumped from the Orbiter to the satellite instead, the electromotive force would actually increase the velocity of the system. It is evident that this could be a good way for a spacecraft to change its orbit without using propellant. Tethers have been analyzed extensively as a means of de-orbiting satellites when their usefulness has expired.

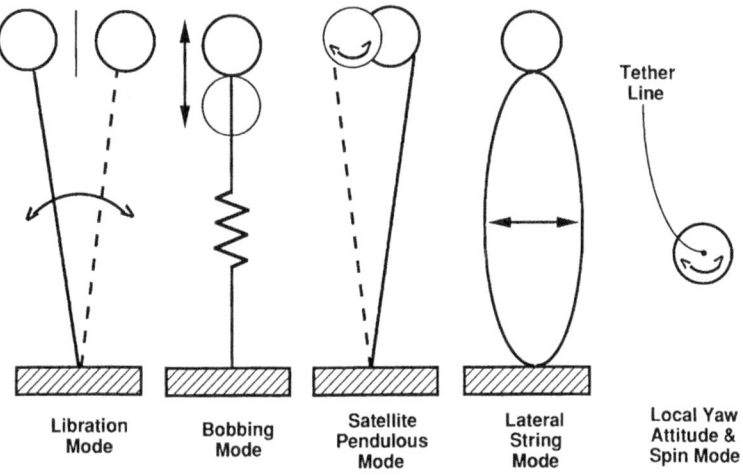

A schematic overview of the main oscillating modes typical of a tethered system.

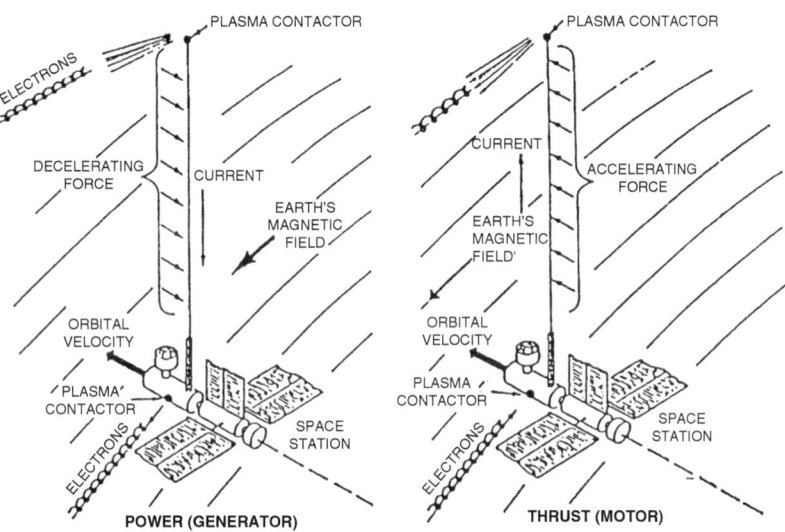

The principle of electromotive force generation on a conductive tether.

This decelerating drag is normal to the Earth's magnetic field lines, but because the magnetic field is an inclined dipole, the field lines will rarely be perpendicular to the velocity vector of the Orbiter and a significant portion of the electrodynamic drag will act out-of-plane and this, in combination with aerodynamic drag, will trigger a skipping rope motion. It was also recognized that a skipping rope motion would be initiated by interactions between the Earth's magnetic field and the electric field that surrounds the tether. Just as in the case of the libration motion, the small intrinsic damping properties of the tether would maintain the skipping rope oscillation, and its amplitude would grow ever greater during the retrieval operation due to conservation of angular momentum.

Analysis showed that this would cause two potential problems. Firstly, a skipping rope could inhibit docking the TSS by introducing large subsatellite angles due to the coupling between the subsatellite pendulous mode and the first lateral skipping rope mode at a 0.25-mile tether length. Secondly, the large skipping rope amplitude could become entangled with the Orbiter, particularly if the subsatellite was unable to dock and needed to be jettisoned. A means of controlling the subsatellite and damping the skipping rope oscillation had to be designed and baselined to ensure mission success. The multifaceted solution called for initiating an Orbiter yaw maneuver as the tether reached the critical length of 0.25 miles, slowing the tether retrieval, controlling the attitude of the subsatellite during the period of resonance, and using a passive damper to suppress the skipping rope motion as the subsatellite drew closer. The Orbiter yaw maneuver was to be employed at a distance of 1.5 miles to reduce the amplitude of the skipping rope to less than 65.6 feet; a motion that would be manageable during final retrieval and docking operation. The time of the maneuver, the Orbiter yaw rate, and the number of rotations would be determined by the ground on the basis of the received telemetry and crew observations.[18]

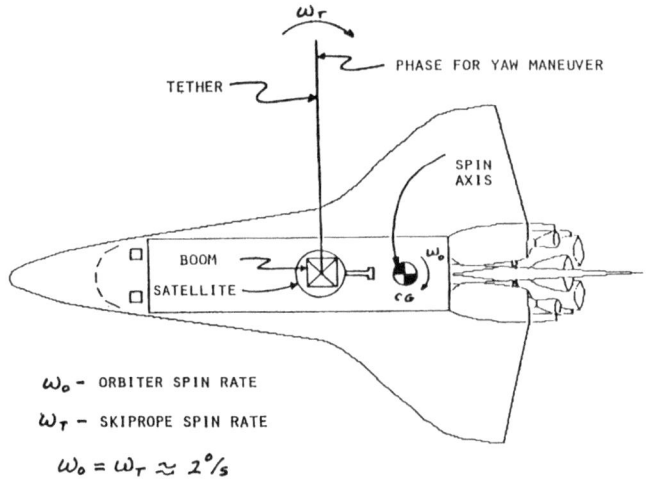

An Orbiter yaw maneuver.

---

[18] This is akin to playing with a jump rope and shaking it out of phase in order to damp out its vibrations.

The attitude control system of the subsatellite had pitch (in-plane), roll (out-of-plane), and yaw thrusters, in addition to in-line thrusters to augment the tether tension during deployment and retrieval near the Orbiter. Originally the thrusters were to be for pure libration (in-plane and out-of-plane) and yaw control, but the attitude control requirements developed when it was realized that at about 0.27 miles a coalescence of the skipping rope and the subsatellite attitude frequency (pendulous) would occur with a significant transfer of energy. During this phase, the attitude of the subsatellite would be excited by about 6° for each 3.3 feet of amplitude in the skipping rope, and without attitude control a skipping rope motion amplitude which exceeded 23 feet would drive the attitude of the subsatellite beyond recovery limits. To achieve this control without completely redesigning the propulsion system, the out-of-plane and in-plane thrusters were canted at angles of 20° and 30° respectively. The unwanted translational motion that this produced was eliminated by firing the thrusters in pairs, one on each side of the subsatellite.

During tethered operations, the software aboard the Orbiter which controlled the payload would take telemetry from the subsatellite and issue thruster commands. As a backup solution, the crew would command the firings based on telemetry that was shown on a dedicated display. Simulations showed that both approaches would work well for a skipping rope amplitude of up to 65.6 feet.[19]

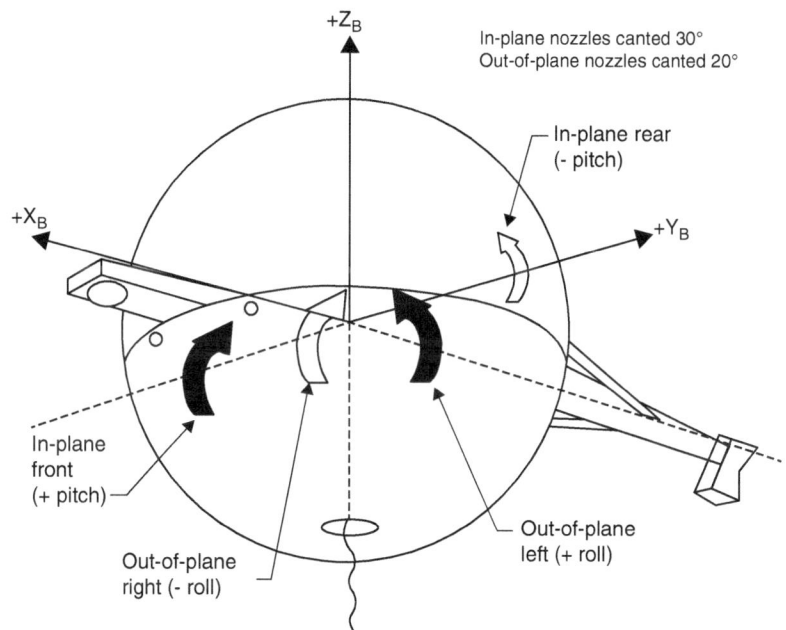

The in-plane and out-of-plane thrusters of TSS.

---

[19] This is why the yaw maneuver of the Orbiter was designed to damp out the skipping rope to a maximum of 65.5 feet by the time the satellite had reached a length of 0.25 miles.

The requirement for a passive skipping rope damper stemmed from the fact that even if the attitude of the subsatellite could be controlled, the conservation of angular momentum meant the skipping rope motion would persist and indeed increase as the subsatellite approached the docking ring on the extended boom.

It was soon realized that passively damping the tether would not be easy, since if too much damping was introduced this would create a new end point (fixed point or node point) that would undermine any beneficial effects of damping. Designing and verifying a tether damper for short lengths was a major challenge. The solution was a triangular yoke and ferrule which was connected to three individual negator motors attached to the docking ring. The negators were a constant spring-loaded system that rolled in and out as the tether moved the yoke. Because the negators had to act as a damper system in a vacuum at low temperatures it was necessary to fully understand their characteristics so that an analytical prediction could be made of how effectively they would damp out the skipping rope motion as the subsatellite was reeled in. As a result, extensive engineering tests were conducted to prove the concept and establish pertinent data characteristics. The test results were correlated with simulations, and models reflecting this correlation were developed for skipping rope simulations. In addition, the hardware was subjected to the standard qualification/acceptance tests in thermal vacuum conditions. The last thing anyone wanted, was to be surprised by the behavior of the system on-orbit.

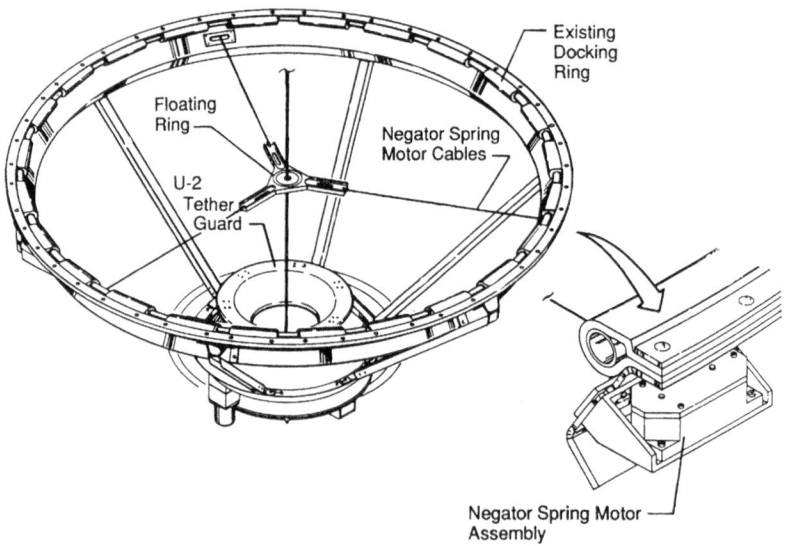

The skipping rope docking ring damper.

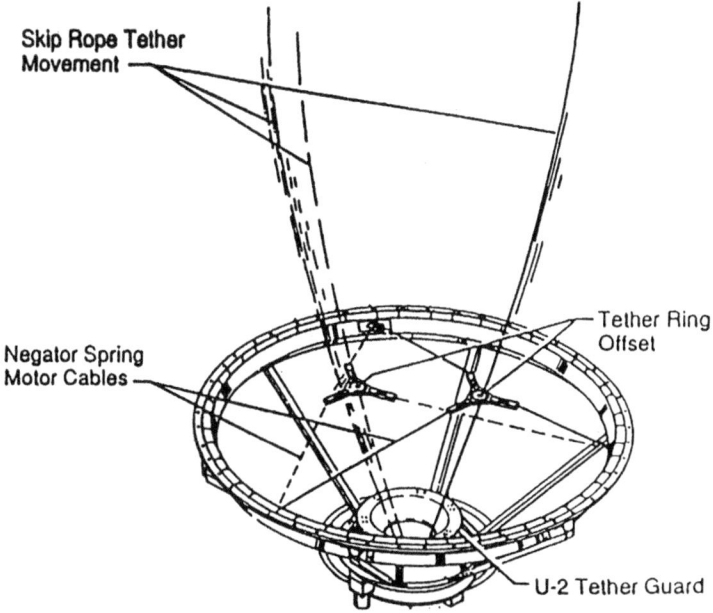

Skip Rope Tether
Movement

Tether Ring
Offset

Negator Spring
Motor Cables

U-2 Tether Guard

A visualization of the skipping rope docking ring damper in action.

THE EVENTFUL FLIGHT OF STS-46/TSS-1

After deploying the EURECA free-flyer, an ESA platform of experiments that was to be retrieved by a subsequent mission, *Atlantis*'s crew got ready for TSS operations. Once the deployer had been checked out, the subsatellite was powered up and the U1 umbilical was disconnected. In order not to waste time on the satellite's batteries the boom was promptly extended to its full length, an operation that required 11 minutes. The U2 umbilical was commanded to disengage from the satellite but the retraction mechanism did not function. This would prove to be the first of many anomalies that the crew had to overcome in the ensuing hours. Speculating that it might be a thermal issue, *Atlantis* was maneuvered to expose the connector to the Sun to warm it up. A second attempt at disconnection was made a few hours later, this time in conjunction with a translational pulse by the Orbiter, and to everyone's delight the umbilical was finally released.

The deployment came to a halt after unreeling just 5.1 inches of tether due to (as was revealed by post-flight analysis) a jam in the Upper Tether Control Mechanism. The tether was then retrieved and a second attempt was initiated by using a different sequence of events and two sets of thrusters for a total thrust of 4 newtons. This time the satellite emerged from the payload bay. As the tether continued to unreel, mission specialist Jeffrey A. Hoffman was awed, "It was a spectacular sight, the satellite and then this tether linking it [to the Shuttle]. When the Sun set and everything turned red, it was just glorious."

TSS-1 on its way.

When the tether reached 558 feet there was a decrease in the deployment rate and in the deployer reel's motor current. It was a clear indication that an anomaly was developing somewhere, and 29 feet later the deployment came to a grinding halt. As Hoffman recalls, "All of a sudden it started to get all these wiggles in it. Wiggles mean that there's no longer tension in the tether, that it has gone slack for some reason…The tether wasn't broken, we could see that. The tether had jammed, in fact. The satellite had a jet of nitrogen gas to pull it away, so that was still on. But it had bounced back. Now it was starting to tilt over. The jet that was coming out, instead of pushing it away from us, was now pushing it over to the side."

But Hoffman and his colleague Claude Nicollier managed to halt the motion and return the satellite to a stable attitude. It was a typical contingency scenario they had been trained for, but in the back of everyone's mind was concern that the slack tether might become entangled with the Orbiter (called the Spaghetti effect), damaging its structure and compromising the safety of the mission.

Although a more complex attitude control system had been added to the satellite as the designers gained a better understanding of the potential tether dynamics, there had not been time prior to launch to develop and implement an autopilot that could govern it. As Hoffman says, "We basically had to control the attitude just by looking up at the satellite. We practiced this a lot in the simulator. The satellite would be pitching and rolling, and you'd look up and try to time exactly when it would get to the end of its pitch. Then you'd say, 'Right roll, now.' Somebody else would be on the computer and would have to enter the command. We got pretty good at it. It was a ludicrously primitive way of trying to control a satellite. We were basically in what should've been an automatic attitude control loop, but they didn't have time to build it, so we were doing it all manually."

In the meantime, on the forward flight deck, mission commander Loren J. Shriver was maneuvering *Atlantis* to remain beneath the satellite and avoid, at all costs, the tether reaching the 45° angle limit relative to the Orbiter's vertical. As he says, "That was our red line. We would have had to cut the tether, which we didn't want to do, so I was madly trying to fly the Shuttle to get back underneath it…trying to control it. It was certainly the wildest time that I've ever had in space. We were really up against the wall. We got very close to the red line…but we managed to get it back under control and finally brought everything to a halt."

With the satellite once again in a stable attitude, the decision was taken to retrieve 33 feet of the tether in order to evaluate the deployment system prior to resuming the deployment. Since both the Lower Tether Control Mechanism and the Upper Tether Control Mechanism were reporting nominal parameters, whatever had impaired the deployment must have been in the reeling mechanisms. The momentum to be gained from that 33 feet might be needed to overcome the obstacle. The deployment was resumed, this time in manual mode and at a faster rate. As Hoffman says, "Of course, starting up the tether fast means the whole thing goes unstable again, so again we're fighting to control the attitude of the satellite and to get it back." In the meantime, the tether deployment rate continuously decreased from the initial 7.8 inches per second until the system stopped with 840 feet of tether out. Throughout the deployment, a higher-than-expected friction was observed. This, together with the satisfactory data from the upper and lower control mechanisms, strengthened the suspicion that there was a problem with the reel assembly. Since the deployment had been under manual control, an effort was made to restart it in the automatic mode but this failed. Under the watchful eyes of at least one astronaut seeking any anomaly in the behavior of the satellite and tether, the decision was made to leave the system for about 10 hours to enable the engineers to figure out how to proceed.

At least at the beginning, it was an unsettling situation because, as Shriver recalls, with 840 feet of tether out, "That was right in the middle of this so-called unstable zone. So we are sitting there worrying about, 'Well, this thing, we're going to have to cut it loose. There's no way we can hang on to it.'" To everybody's satisfaction, the unstable zone didn't live up to the expectations. "Well, we noticed that as soon as we stopped jerking it around and doing things to try to unjam the tether, it would just go to one place and stay there. It was totally stable." In fact, only an apparent twist in the tether required periodic firing of the satellite's yaw thrusters to retain a fixed attitude. This, however, did not degrade the overall stability of the Orbiter-satellite tethered system.

Once again it was proposed to retrieve the satellite in order to be able to gain the momentum to break through this new anomaly. However, retrieval halted abruptly at 735 feet, and all attempts either to continue with the recovery or to proceed with the deployment failed. The satellite was stuck! The astronauts and engineers now faced either having to cut the tether or to have Hoffman and Franklin Chang-Diaz attempt a contingency EVA. This would have seen Hoffman climb the boom to haul the tether in hand over hand while Franklin wrapped it up. In preparation, they began the usual pre-breathe protocol. As air was expelled outside to lower the cabin pressure to 10.2 psi, an unexpected phenomenon developed. As Hoffman notes, "It hadn't occurred to anybody that the air emerging from the cabin was all going out in one direction, so it was actually propulsive. Under most circumstances you'd never know it, but we had a tether sitting up there. After we'd been letting air out of the cabin for 20 minutes we looked out, and the tether was way over to the side because the Shuttle had been pushed over to the side. It was only by 10 or 20 yards. Without the tether, it wouldn't have made any difference. But again, that was an instability. We knew how to correct it and when the tether swung back we took out the velocity, but it was interesting that nobody had thought of it when they told us to start the cabin depress."

Before Hoffman and Chang-Diaz could complete their preparations for an EVA, Mission Control proposed a new strategy, which consisted of retracting the boom by one bay and then extending it again with the reel brake active. This was deemed safe by the deployer engineers because the boom had a significant structural safety factor and its normal mode of operation during extension included pulling the tether off the reel with the brake applied. Since the motor extending the boom was more powerful than the reel brake, this action was able to free the tether and allow a final retrieval of the satellite.

No more attempts were made, because there was clearly a fault in the system that could not be solved while on-orbit. It was the presence of a manual backup mode for tether operations that had enabled the system to be recovered, instead of having to be jettisoned. As Hoffman recalls of the original design, "It was going to be completely automatic. All you would do is push a button and it would go up. Then you'd push a button and it would come back. There wasn't any sort of manual control. It had an attitude control system [so] that it could yaw back and forth, but it had no control over pitch and roll, and this is just typical of the design philosophy when it was first done. They just never designed for any sorts of contingencies. So we said, 'Suppose something is going wrong, and we want to stop the deployment in the middle.' They said, 'The only thing you can do is put the brakes on.' The problem is when the thing really gets going, it's coming out at several meters per second. That's pretty fast. If you just slam the brakes on, it is going to go wildly unstable. They basically hadn't designed for all these contingencies. As we did more and more simulations, and we learned more and more about the system, we came up with more and more scenarios where you need these manual capabilities. In the end, as things turned out, we ended up using every single manual capability which we finally had them build in."

Four days after *Atlantis* returned home, NASA formed the TSS-1 Contingency Investigation Board to review the anomalies, to determine the probable cause, and to recommend measures to prevent a recurrence. The investigation was swift. Indeed, by the beginning of October the Board presented its final findings. It had uncovered a series of gross mistakes in design and engineering.

The initial failure to deploy the satellite after only 5.1 inches and the subsequent inability to move in either direction at 735 feet were both traced to tether jamming within the Upper Tether Control Mechanism (UTCM). As explained earlier, this was to guide the tether from the boom to the satellite while maintaining a constant boom-to-reel tension for reeling operations and supply tension measurements. One of the primary components was a vernier motor used to overcome inboard system friction and to drive a gripper pulley through a clutch, which in turn pulled the tether off the reel during deployment. In the initial "flyaway," the tether became slack within the UTCM because the vernier accelerated the tether faster than it could be pushed out, causing the tether to overlap and wrap itself around the vernier pulley, jamming it. A ground test carried out by the investigators using both a UTCM mockup and actual flight hardware showed that the inertia of the accelerating tether was enough to cause a wrapping involving up to 5 feet of tether.

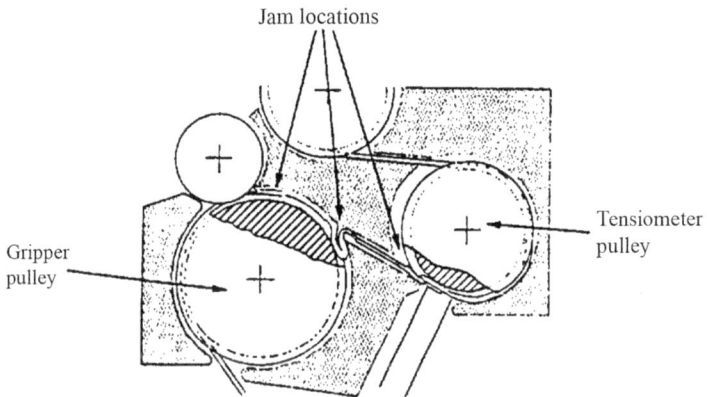

The tether jam in the UTCM.

A second cause concerned the activation of the tether control laws prior to turning on the vernier motor. With the system operating under control laws, the length of the tether being reeled out (or in) was being compared to a preset value of tether length at that time. During the first flyaway attempt, the tether length being deployed was less than was predicted by the control laws, both because of the inertia of the tether and the inability of the satellite's thrusters to induce sufficient tension. That is, the tether was not being pulled out fast enough. When this condition was sensed by the control laws, there was a rapid reduction of inboard tension because the system sensed there was no change in deployed tether length. This condition greatly increased the chance of the tether going slack within the UTCM.

A third reason was the higher-than-expected stiffness of the eyesplice that secured the tether to the satellite interface. This stiff section of tether extended down into the UTCM, where it acted as a "column" during the initial flyaway attempt and increased the potential for jamming.

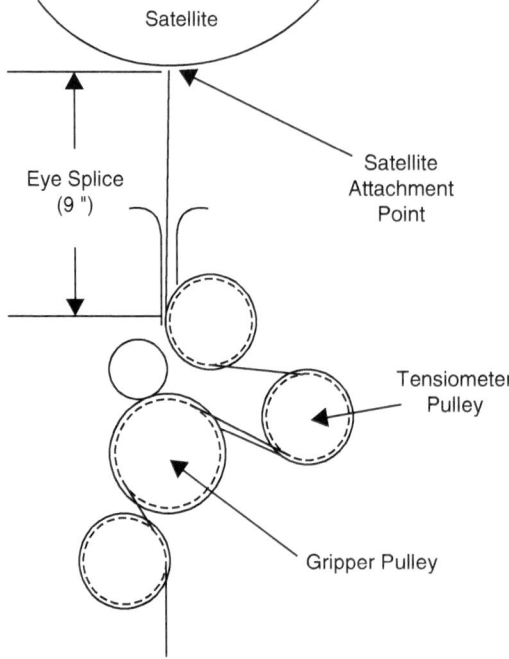

The tether eyesplice configuration.

Finally, the investigators found that the pre-flight ground tests had not accurately simulated the acceleration of the satellite in a zero-g environment. In fact, because only actual flight-worthy hardware was available for testing, the engineers had been understandably reluctant to explore off-nominal scenarios. As the satellite was being reeled in, this cleared the tether jam from the UTCM. A revised command sequence was then issued for the initial deployment. First, two sets of thrusters were activated to produce 4 newtons of thrust, sufficient to impart a strong tension on the tether, and then the vernier motor was powered up ahead of activating the control laws, further increasing the tension.

The second jamming occurred at a deployed length of 735 feet due to having left the satellite in this position for a prolonged time. Thermal expansion and contraction of the boom had induced another slackness in the UTCM, and it jammed when the vernier motor was activated. Collapsing one bay of the boom and then reinstating it freed the jam and allowed a safe recovery of the tether and the satellite.

The stops at 587 and 840 feet were attributed to a last-minute modification to the support structure of the reel assembly. As the flight hardware was being put together and integrated into *Atlantis*'s payload bay, a new coupled load analysis was run to identify the stresses and strains acting on the payload with this change in the revised configuration. The analysis revealed that one bolt that was common to the deployer reel mechanism and support structure would incur a negative margin of safety[20] at touchdown, and the hardware could not be certified for flight until this was rectified. This kind of issue is typically resolved by adding some form of reinforcement, but to do so would have necessitated a major disassembly of the deployer structure that had already been integrated with the rest of the payload and fitted in the Orbiter. Another option was to use a same-size bolt with higher strength, but the unavailability of such an item in NASA inventory meant a long procurement time. Both options implied a significant postponement of the mission which, considering the normal pressure to launch on time, was unacceptable. A third and simpler alternative that would save the schedule was found in replacing the current fastener with a shear wedge. Acting a C-shaped clamp, the shear wedge would relieve the bolt of some of the loads, ensuring a positive margin of safety for any condition. The hole that the fastener was inserted into was a through hole, providing some room to accommodate a longer bolt than the original.

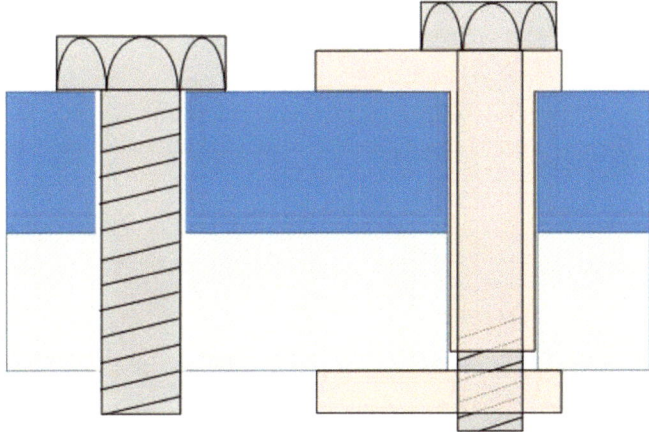

The fastener configuration for TSS-1: (left) nominal fastener in joint assembly, (right) modified joint assembly.

As said earlier, the 12.4-mile tether was wound around a shaft in a configuration which Hoffman likens to a fishing reel, "It was essentially like a big spinning reel, a fishing reel. Like a spinning reel, as the tether is wound on you need a mechanism to make it go back and forth and back and forth evenly. It's called a level wind device. On a fishing reel they have a little catch that holds the line; it moves first to the left, then it gets all the way to the left and moves back to the right and then to the left and so on. It is geared together with the reel, so that as the reel turns the little level wind mechanism moves back and forth. There was a similar sort of mechanism on this big reel of the tethered satellite. The level wind was geared together with the big drum, and it moved back and forth."

---

[20] A negative margin means that in a particular condition the structure will be subjected to a higher loading than that for which it was designed, thus increasing the likelihood of failure. Owing to conservatism in the design of any structure, a negative margin will not necessarily endanger safety, but certification rules out such negative margins.

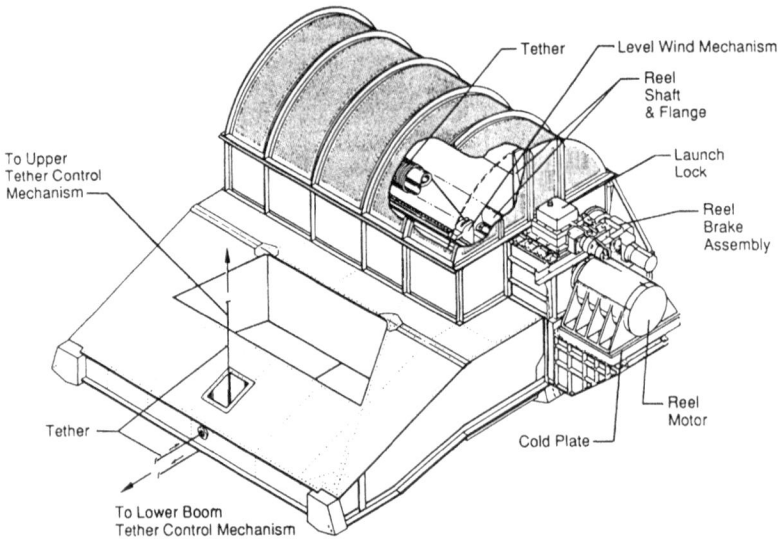

The level wind mechanism location within the reel support structure.

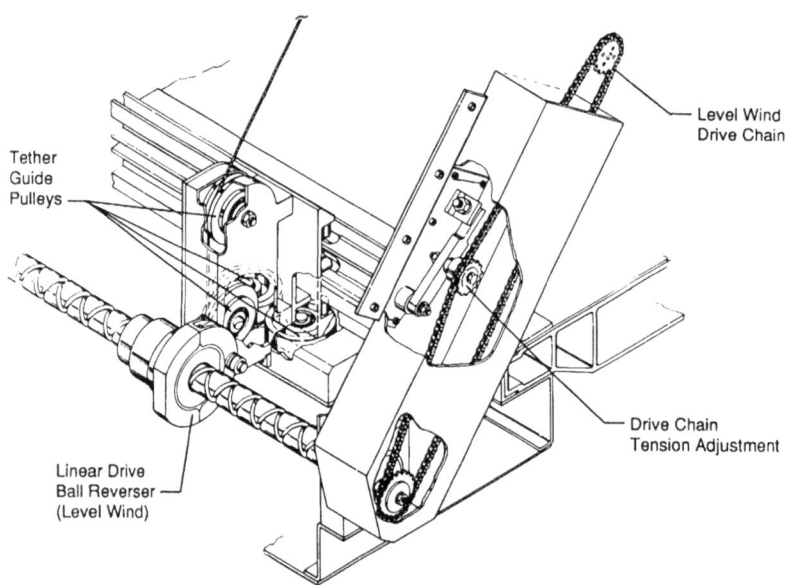

A detailed view of the level wind mechanism.

On-orbit, as the level wind device moved from one side to another, it struck the longer-then-intended bolt, causing the first halt at 587 feet of deployed tether. At the time of installing the shear wedge, the possibility that it might obstruct the level wind mechanism had not been appreciated. Old drawings which did not reflect the actual hardware configuration had been used in designing the modification, and these did not suggest encroachment on the dynamic envelope of the level wind mechanism. In fact, the existing assembly drawings, which may have revealed the interference, were not up to date pictorially (project requirements and contractor policy did not require updating top assembly drawings until after three modifications) and did not contain a view that would show the interference. Following installation, a new payload system integration verification test should have been conducted in order to confirm that the modification would not have any impact on the proper working of the deployer, but the amendment was deemed minor and the verification test was waived off because it would have postponed the launch. Another contributing factor was the late discovery of the negative safety margin. Had it been identified at an earlier time, the preferable option of using a stronger fastener would have been implemented or, in the event of the shear wedged being adopted, the interference would have been spotted by system testing prior to shipment for integration on *Atlantis*.

Eventually, the system was able to move past the 587-foot mark. Hoffman says, "Because the people on the ground thought that it was a kink in the tether, we wound it back and then we said, 'Now let's take a running start.' We came smashing in and actually bent the bolt…When they took the whole thing apart, this bolt was bent way over." Slippage in the system allowed the deployment to continue out to 840 feet, at which time it reached its final hard stop and further deployment became impossible.

In their report, the investigators acknowledged the complexity of the TSS and the various design problems, and accepted that the system was unable to be fully verified prior to the flight because of the considerable length of the tether, vacuum, thermal, gravity gradient, zero-g, and other environmental factors that could not be simulated on Earth on that large scale. Aptly, one of the most important recommendations was, "Always verify even the smallest change made to the hardware late in the program and particularly at the launch site." In fact, we can safely say that for just one bolt the mission failed! They also emphasized the need for ground testing to fully explore the dimensions of the expected flight environment, particularly in off-nominal situations.

The unfortunate blunder of the bolt highlights the need to involve all disciplines whenever a change is made, and confirms the adage that it is better to do the job right first time in order to preclude having to make last-minute changes.

TSS-1 was not a complete failure. It proved that the tether could indeed conduct current and thus generate an electromotive force. While its intensity was insufficient to trigger significant tether dynamics, skipping rope amplitudes of less than 3.3 feet due to other excitation sources were observed. Libration modes, satellite pendulous oscillations and longitudinal slack/taut tether dynamics were also all experienced. It was proved that the manual procedures put in place for contingency conditions were sufficient for the crew to control the system and keep it stable and safe, in particular near the Orbiter both during deployment and retrieval.

STS-75/TSS-1R "THIS CAN'T BE HAPPENING AGAIN"

On February 25, 1996, looking outside the windows of *Columbia*'s aft flight deck, Jeff Hoffman had a déjà vu. Before his eyes the small white spherical TSS satellite was perched on top of the extended boom. Flashing in the darkness, it signaled that it was alive and ready to redeem itself from the shameful flight of 4 years earlier and so prove its worth. "The project office at Marshall started making plans for a re-flight. It was a question whether NASA really wanted to do that, but I think partly because of the Italian involvement they felt we had a responsibility…We had the non-advocate review, and a lot of people talked about the interesting science and engineering that could be done with this, so NASA decided to go ahead and have the re-flight. There was a strong scientific case to fly the tether again. It was a fascinating mission, both because of the technology of how to control these long tethers in space and also the interesting ionospheric physics that would be done using all the experiments on the satellite and on the Shuttle. I think there was a good case for the value of the mission, and the engineers who redesigned the equipment made a good case to show that they really did understand the sources of the failures and had properly addressed them."

Before the TSS could be flown again, some hardware revisions were made on the basis of the lessons learned from the first flight. Naturally, the deployer mechanisms received the bulk of the attention. Since the TSS Investigation Board was unable to clearly identify the reason for the initial failure of the U2 umbilical to disconnect, the decision was taken to delete it and reassign its functions to the U1 umbilical. And to prevent a recurrence of tether jams within the UTCM, the eyesplice at the satellite interface was shortened from 9 inches to 3 inches to reduce its stiffness and hence its interference with the UTCM. The vernier motor speed controller was modified for a gradual ramp up force at the start of tether deployment, to more closely match the movement of the tether outside the UTCM. The nominal initial deployment sequence was changed to that used successfully on the second flyaway attempt; namely having four satellite thrusters fire prior to flyaway in order to increase tension on the tether, followed by turning on of the vernier motor prior to activating the control law. The mechanical interference between the level wind mechanisms and the protruding bolt was eliminated by a combination of shortening the bolt and revising the mechanism for greater clearance. All level wind components that suffered damage or stress were replaced. Additional ground testing equipment was also employed to undertake more realistic simulations of the system, particularly to better emulate satellite inertia and acceleration on-orbit.

As Hoffman notes, "Amazingly we turned everything on, the tether started on its way, and everything was beautiful. It started moving away from us. When it got a few hundred meters away we turned off the nitrogen jets, which are what provided the initial tension. Once the tether gets long enough, gravity provides tension and that pulls the rest of the tether out." Mindful of the exhausting events of 4 years earlier, the crew ran two shifts so that there would always be someone watching the tether as it was deploying, ready to intervene. "We just watched it gradually deploy…Actually as the tether got longer and longer it was going out faster and faster. It was moving pretty fast, a few meters per second, by the time it got all the way out. What was fascinating to me, was that the tether developed a very large bend in it. It was like a huge arc through the sky. It wasn't going out straight at all." In fact, simulations had shown that a combination of electrodynamic and aerodynamic forces and the Coriolis effect[21] would induce a pronounced rearward deflection of the tether as it was being deployed, and despite comments from the crew stating that this was "huge," it was in fact within estimated nominal amplitude.

The deployment continued for the next 5 hours with the tether behavior matching predictions. The pendulous motions and bobbing were within expected frequencies. An unexpected out-of-plane libration at the beginning of deployment was promptly taken out by Orbiter maneuvering. By the time the tether was close to being fully extended, a current of 1 ampere was already flowing through its conductive copper wire, which was more than the scientists were hoping for.

Then another déjà vu struck Hoffman, "It was within 1 km of its final length, at which point we were going to put on the brakes and just let it sit there, then we would start all the experiments that we were to do. I was recording this huge arc in the tether through the camera when I started to see little ripples in the tether. It was this horrible feeling of, 'No, this can't be happening again.'"

Looking out of the aft windows of the flight deck, Hoffman realized the tether had broken within the extended boom. He immediately radioed, "Houston, the tether is broken. It's broken at the bottom. We're in no danger."

Despite the deep feeling of disappointment for the loss of the satellite, the priority was to guarantee the safety of the Orbiter. Had the tether broken at any point but the bottom, it would have been immediately severed from the deployer and the Orbiter would have executed an evasive maneuver to prevent the tether from snapping back, but because the tether had broken in the boom there was no such danger.

Hoffman was able to inspect the tether which remained in the mechanism, "I was able to hook up a very powerful train of optics, telephoto lenses, and take a close look at the broken end of the tether. I could see that it was brown and charred, so we knew before we ever came home that it almost certainly had been a short circuit that had melted the tether."

The TSS-1R Mission Failure Investigation Board, established just 2 days after the failure, reached the same conclusion as soon as they received the remaining section of the tether and the available telemetry. The report issued at the end of May said the "tether failed as a result of arcing and burning of the tether, leading to a tensile failure after a significant portion of the tether had burned away." The Board determined that the arc occurred in the Lower Tether Control Mechanism and discharged a current of 1 amp. This event occurred during a passive mode of science operations, with minus 3,500 volts of direct current on the tether conductor. The arc continued intermittently for 9 seconds, as the breached portion of the tether ran at 3.3 feet per second through the remaining deployer mechanisms and into the 39.4-foot-tall deployer boom, where the space plasma provided the current return path. This arcing produced significant burning of most of the tether material in the vicinity of the arc, compromising the structural strength provided by the

---

[21] The Coriolis effect is a fictitious force that is used to simplify calculations involving rotating systems, in particular in relation to the Earth's surface whose tangential velocity depends on latitude.

The frayed end of the TSS tether is seen at the end of the supportive boom.

layer of Kevlar. At that point, the nominal load on the tether was sufficient to separate the tether at the location of the burn, while it was within the deployer boom.

What caused the arcing? The 13.7-mile-long, 1-inch-thick tether might not appear to be a big deal to the casual observer, but it was a unique piece of hardware whose manufacturing complexity had been greatly underestimated. To conduct an electrical current and at the same time withstand the tension induced by the gravity gradient, the tether was a composite structure with an inner Nomex core enclosed by a bundle of ten strands of thin copper wire. Since the maximum length of an individual strand was about 2.24 miles, end-to-end joints connected the strands to produce the total required length of 12.4 miles. A butt welding procedure was devised to join the wire strands without increasing the overall diameter of the conductor. A layer of extruded Teflon insulated the copper and supported a layer of Kevlar which gave the tether the required structural strength. Similar to the copper, ten Kevlar sections were spliced together to create a single layer. A second layer of Teflon completed the tether cross-section for protection from abrasion and the aggressive atomic oxygen of the orbital environment.

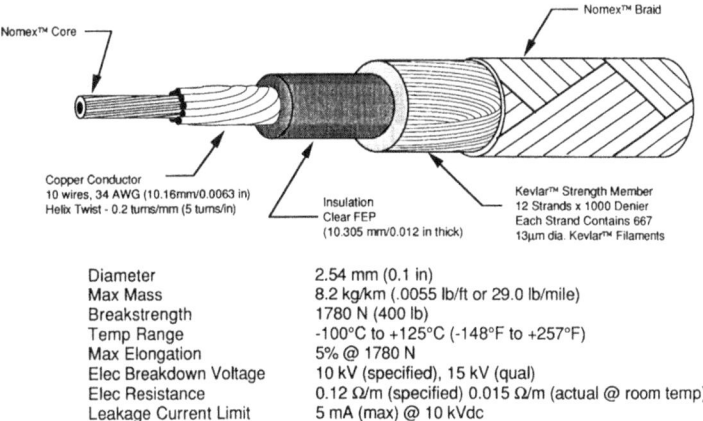

Nomex™ Core

Nomex™ Braid

Copper Conductor
10 wires, 34 AWG (10.16mm/0.0063 in)
Helix Twist - 0.2 turns/mm (5 turns/in)

Insulation
Clear FEP
(10.305 mm/0.012 in thick)

Kevlar™ Strength Member
12 Strands x 1000 Denier
Each Strand Contains 667
13μm dia. Kevlar™ Filaments

Diameter	2.54 mm (0.1 in)
Max Mass	8.2 kg/km (.0055 lb/ft or 29.0 lb/mile)
Breakstrength	1780 N (400 lb)
Temp Range	-100°C to +125°C (-148°F to +257°F)
Max Elongation	5% @ 1780 N
Elec Breakdown Voltage	10 kV (specified), 15 kV (qual)
Elec Resistance	0.12 Ω/m (specified) 0.015 Ω/m (actual @ room temp)
Leakage Current Limit	5 mA (max) @ 10 kVdc

The cross-section of the tether.

Although the damaged portion of the insulation was destroyed by the burning, the Board found sufficient evidence from tests and analyses to realize that either foreign object penetration or damage to the Teflon insulation layer occurred during either manufacturing or handling, and they cited this as the probable cause of the breach of the insulation that exposed the conducting copper layer and prompted the arcing. The production and inspection records documented the difficulties that were experienced in fabricating the tether. Numerous problems occurred in the extrusion and braiding process. For example, because the tether manufacturing was carried out in a normal environment rather than in a clean room,[22] it was easy for metallic and non-metallic contamination to find its way into the insulation layers. In fact, aluminum shavings of sufficient size to breach the insulation were found in several locations in the deployer mechanism. Damage to the copper conductor was likewise also found on the returned section of the flight tether. Deep penetration of the contamination into the tether was attributed to the process that wrapped the tether onto the reel assembly. As layer after layer of tether was added, relatively high loads were induced on the already wrapped ones, particularly the deepest layers. These forces were orders of magnitude greater than that required to push debris into the tether, resulting therefore in the Teflon and Kevlar layers being pierced and creating fatal breaches in the insulation of the copper wires.

A similar condition must have occurred in the Lower Tether Control Mechanism. Indeed, in weightlessness, floating metallic debris under the influence of electrostatic forces must have found its way into the pulleys where, despite the weak forces acting on the tether, an unfortunately positioned foreign object could be pushed through the insulation. Although the crew had been trained to cope with arcing by connecting the tether to a

---

[22] It is common practice to manufacture, assemble, and store space-worthy hardware in clean rooms which maintain strict control of temperature, humidity, pressure, and particulates. Being akin to the operating theater of a hospital, it is not surprising that people who work in clean rooms must wear over garments similar to a scrub, inclusive of gloves, shoe covers, hat, and face mask.

shunt resistor in order to reduce the voltage potential driving the arc, they would not have been able to prevent the failure. In fact, with a delay of 6 seconds in sending telemetry to the ground and the sampling rate of the onboard data, there was simply insufficient time for anyone to see the data, evaluate it, and take action before the tether failed.

On the plus side, the tether break demonstrated the validity of using a tether for an exchange of momentum. As previously explained, in the gravity gradient the Orbiter and the satellite were traveling at the orbital velocity of their common center of mass. Prior to the break the satellite, being above the system's center of mass, was traveling at a higher speed than that required to remain at that altitude. With the link broken, the satellite was free to use that excess of energy to climb into a higher orbit whose perigee was the altitude at the time of the break. In other words, the satellite changed its orbital altitude without any fuel expenditure. The opposite happened to *Columbia*. It was traveling at a lower speed than required by orbital mechanics, with less energy than was required to remain at that altitude. Upon severance of the tether, the Orbiter entered a lower orbit whose apogee was the altitude at the time of the break. Though not an objective of the mission, the concept of using tethers for propellant-less orbital changes was thereby demonstrated.

Following the failure, the crew and Mission Control evaluated the opportunity of recovering the satellite by getting near to it and having two spacewalkers manually mating it with the docking ring but concern about the amount of propellant that such an attempt would require, as well as how the almost 12.4 miles of tether still attached to the satellite should be stowed in the payload bay, promptly led to a rescue being rejected; not least due to the risk of the Orbiter becoming entangled in tether.

The satellite remained in visual range of *Columbia*, as Hoffman recalls, "We did get a couple of amazing sightings of the tether. Just because of the orbital mechanics of tethers, it went to an orbit where its apogee was about 140 km above the apogee of the Shuttle. Gravity had pulled it straight. Even though it was all coiled and twisted as it broke, gravity straightened it out. Mission Control would call and tell us when the tether was going to fly over us, so we actually have a few photographs of it." In fact, even from the ground it was possible to see the satellite and its attached tether. "We live in the satellite age, we've all seen satellites fly over. They are tiny points of light. With the tether flying over, you saw a line, this luminous line moving through the sky. It was eerie! It gave you goosebumps to see something with actual physical dimension moving through the sky."

With *Columbia*'s crew moving on to the rest of the mission activities, a new free-flying satellite science mission was planned to complement the data which had been gathered during the 5 hours prior to the tether breaking. It significantly contributed to meeting a part of the primary mission objectives. Of course, the loss of the satellite was a bad blow for everyone involved in the project. Thinking back to those events, Hoffman says, "There was a lot of good science that we didn't get to do. I think what I was most disappointed with was that there are all sorts of instabilities that you have to deal with when you are reeling the tether back in. We had spent literally hundreds of hours in flight techniques meetings and in the simulator figuring out how to damp out all these instabilities and do a controlled retrieval of the tether, and we never got a chance to see how well it would work."

Sadly, STS-75 was the final tethered mission of the program. Despite the many applications that tethers could bring to space science and exploration, the project was curtailed by the loss of the satellite, the evident complexity that such a system entails, and the ever-present challenge of winning the necessary funding. The close coupling of basic orbital mechanics, dynamics, control, mechanisms, orbital environment, and the potential for each part of the system to interact with each other and create various dynamic situations that were not fully understood, all conspired to make STS-46 and STS-75 some of the most complex missions ever flown by the Shuttle.

As Hoffman would reflect years later, "That was an absolutely fascinating project because it was something that nobody had ever done before. It was like learning how to go to the Moon…Just how do you do it? Nobody knew how to control a tethered satellite… We were using the Shuttle to do things that it had never been designed to do."

11

More Power and Time Needed

INTRODUCTION

Time and power are amongst the most valuable resources to manage on a spacecraft, so much so that mission planners rarely leave much of either to spare. Even before *Columbia*'s maiden flight, and considering the enormous costs involved in launching a single mission, it became evident that the Shuttle performance would soon require to be extended in order to pack as many activities as possible into each mission.

In response, in 1979 the NASA Johnson Space Center in Houston, Texas, started an 8-month Orbital Service Module Systems Analysis Study to investigate concepts that would add power, thermal control, and attitude control to the baseline Orbiter for extended stays on-orbit and a greater variety of mission objectives. The starting point was the NASA STS Mission Model dated October 1977, a document which outlined traffic projections for 1981–1984, providing estimates of orbit, weight, payload, and schedule for the early missions. The research missions, principally Spacelab, were of particular interest because they displayed an increasing requirement for time on-orbit and power for the payloads.

The three fuel cells that powered the baseline Orbiter were able to continuously supply 21 kW, fully two-thirds of which was needed by the Orbiter itself; only one-third would be available to the payloads. As the European industry had little (or no) experience in making fuel cells for space application, it was specified that Spacelab's power would be supplied by the Orbiter. This meant that Spacelab and its subsystems had to function on 4.5 kW, regardless of whether it was a pressurized module, an all-pallet configuration, or a mixture of the two, with just 3 kW for the experiments. The NASA STS Mission Model, however, foresaw that the electrical energy requirements of the Spacelab experiments would be in the range 17–33 kW, with an average of 29 kW for at least 80% of the 29 Spacelab missions projected up to 1984. Of course, it was expected that as NASA gained flight experience its understanding of the Orbiter power management would improve to a point where it was realistic to assume 29 kW would be sufficient for all Spacelab flights. The greater uncertainty in forecasting a manifest beyond 1984 made it harder to predict the power requirements in the longer term, but the ambition to undertake a large number of power

© Springer International Publishing AG 2017
D. Sivolella, *The Space Shuttle Program*, Springer Praxis Books,
DOI 10.1007/978-3-319-54946-0_11

hungry investigations in microgravity materials processing suggested the gross energy requirements would steadily rise to at least 100 kW by the mid-1990s, when space processing facilities were expected to be open for business.

A similar trend was perceived for mission duration, with flights being projected to last 45 days by the start of 1984 and even longer thereafter. Obviously the baseline Orbiter would not be able to satisfy these requirements. In their study, JSC engineers settled on a flexible evolutionary growth path to increasingly augment the power and duration capabilities of the baseline Orbiter.

The initial thought was to increase the number of cryogen tanks for the fuel cells, but this was soon discarded because it would drastically reduce the useful payload of a mission. For example, adding ten sets of cryogen tanks would boost the power by 7 kW and the mission duration to 23 days, but reduce the payload from 20,000 pounds to less than 9,000 pounds.

THE POWER EXTENSION PACKAGE

A more satisfying concept was to exploit the free energy coming from the Sun on the day side of the orbit by adding a solar power facility to the Orbiter itself. Called the Power Extension Package (PEP), this was the Array Deployment Assembly (ADA) and the Power Regulation and Control Assembly (PRCA), plus all of the necessary interface and display control equipment. Normally, both the major units would be at the forward end of the payload bay, above the Spacelab tunnel to provide good visual clearance from the aft flight deck window for crew operations using the robotic arm. This installation would have reduced the length of the power cabling and minimized interference with the apparatus in the aft part of the payload bay. However, based on the payload envelope, the ADA could be fitted anywhere in the bay and the system would be compatible with an all-pallet Spacelab configuration.

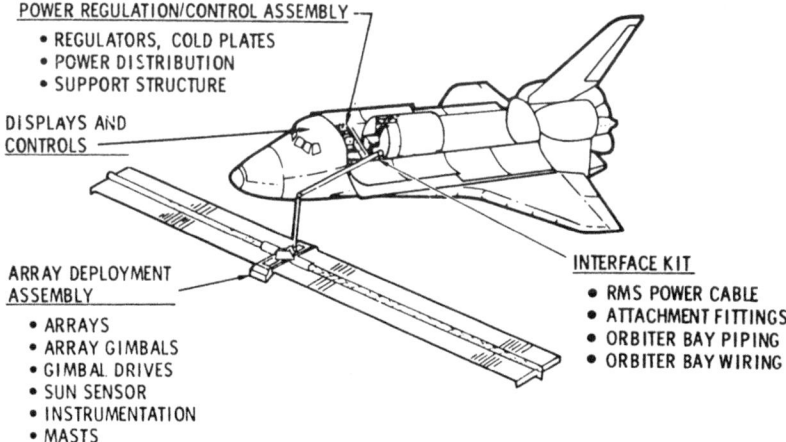

The main components of the Power Extension Package.

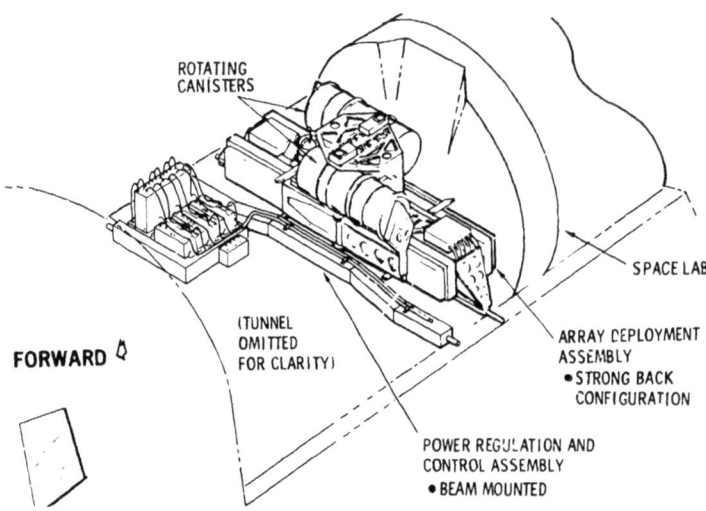

ROTATING
CANISTERS

FORWARD ⟨

(TUNNEL
OMITTED
FOR CLARITY)

SPACE LAB

ARRAY DEPLOYMENT
ASSEMBLY
• STRONG BACK
CONFIGURATION

POWER REGULATION AND
CONTROL ASSEMBLY
• BEAM MOUNTED

A detailed view of the ADA and PRCA installations within the payload bay.

The core structure of the ADA was a box beam that spanned the full width of the payload bay and fitted onto the sidewalls by way of standard bridge fixtures. The box housed two solar array wing assemblies, bolted to opposite sides of the beam, each of which had its own deployment mast canister and diode assembly package. On-orbit, the robotic arm would grasp the ADA at the so-called slip-ring/RMS grapple fixture assembly and hoist it vertically out of the payload bay. It would then be translated to the deployment position. The crew would visually monitor the unfolding of the solar arrays. Then it would be moved to its operational position, chosen to satisfy mission and payload orientation requirements while maximizing exposure to the Sun. Once in place, the arm's joint brakes would be kept locked to rigidize the whole structure and minimize the dynamic loading on the flimsy solar arrays.

Deployment of the solar arrays would start by rotating the two canisters through 90°, perpendicular to the long axis of the core structure. In fact, packaging issues had required the canister to be stored along the length of the core structure, as opposed to perpendicular to the arrays. A support assembly was added to permit each canister to pivot around one end and orientate itself along the deployment direction of the solar array. Each canister contained the deployable solar array mast, which consisted of a composite triangular truss stored helically within the canister itself. Mast extension and retraction would be regulated by a redundant motor that drove a two-speed gear box. Guide wires would control the unfolding of the array, and when the 11.4 x 118-foot solar array was fully deployed it would be kept under tension by the mast itself using springs to assure the necessary flatness. With the arrays deployed, the canister support assembly would control the dynamic response to Orbiter-induced loads in the array plane and perpendicular to it. Each array would have 50 hinged segments made of 0.8 x 1.6-inch solar cells attached to a flexible substrate.

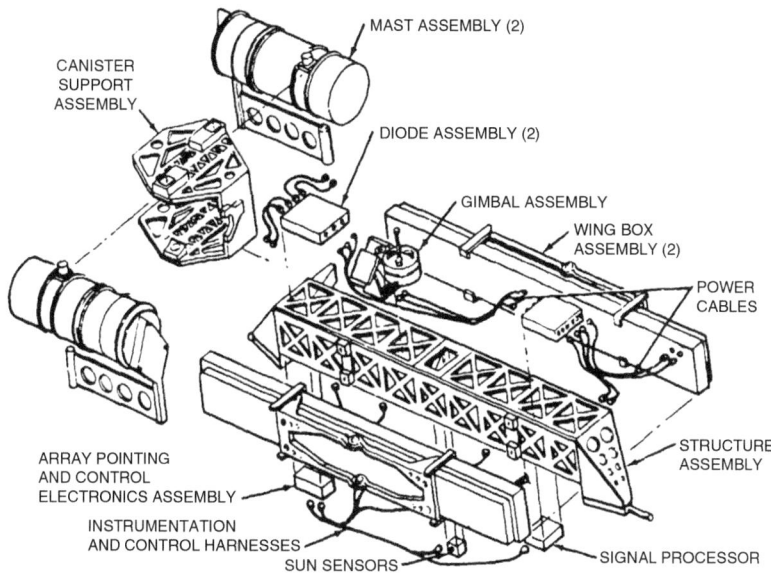

A detailed view of the Array Deployment Assembly.

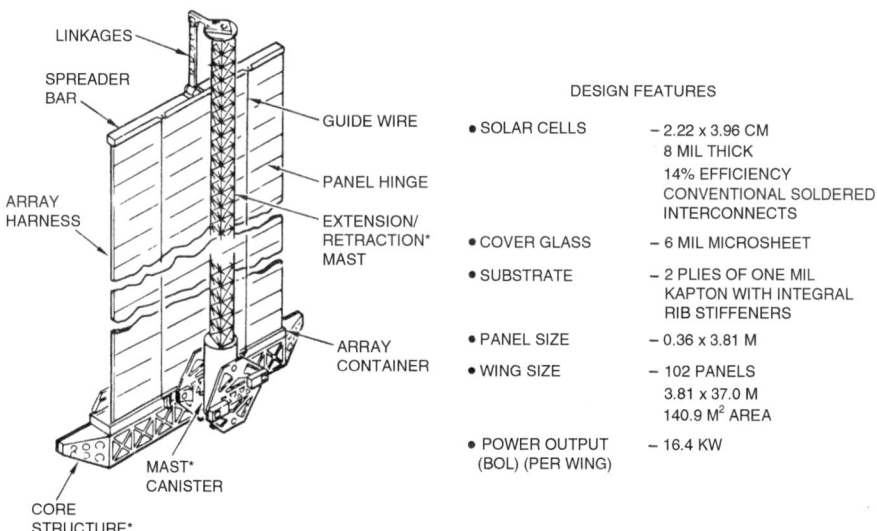

The PEP solar array wing characteristics.

In sunlight, power generated by the solar arrays would be supplied to the Orbiter across the slip-ring/RMS grapple fixture[1] to an umbilical connector on the robotic arm end effector, physically mated to the assembly. It would then flow along cables attached to the arm and terminate at the shoulder mounting on the sill of the payload bay, where it would be transferred by cable to the PRCA to feed power to the Orbiter main buses.

The PRCA consisted of six voltage regulators mounted on three cold plates which would reject waste heat, three shunt regulators, data bus couplers, and power cables installed on a beam support structure. It would be installed at the front of the payload bay to reduce the length of the cables originating from the ADA/RMS assembly and minimize interference with other payloads.

In flight, the orientation of the arrays would be controlled by the ADA avionics, which included Sun sensors, signal processor, and pointing and control electronics assemblies. To maximize exposure to sunlight, the slip-ring assembly was capable of a full 360° alpha angle rotation; i.e. a full revolution about the axis perpendicular to the plane of the arrays. A motor on each canister offered a second degree of freedom for a beta angle rotation of ±90° about the array deployment axis. To optimize both operational flexibility and power generation, the ADA could be maneuvered around the Orbiter by the robotic arm into any convenient location.

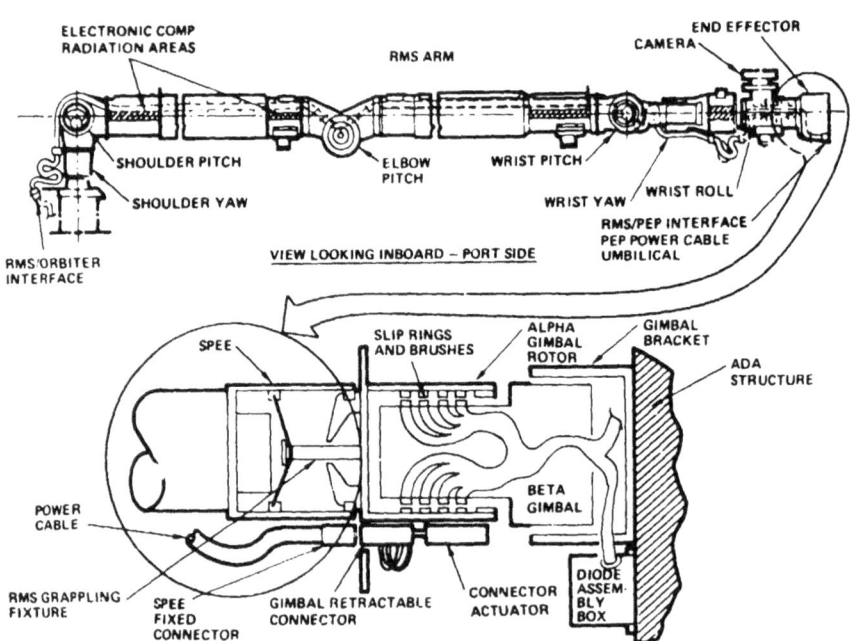

The robotic arm gimbal/slip-ring/grapple assembly.

---

[1] A slip-ring is an electromechanical device that facilitates the transmission of power and electrical signals from a stationary structure to a rotating structure.

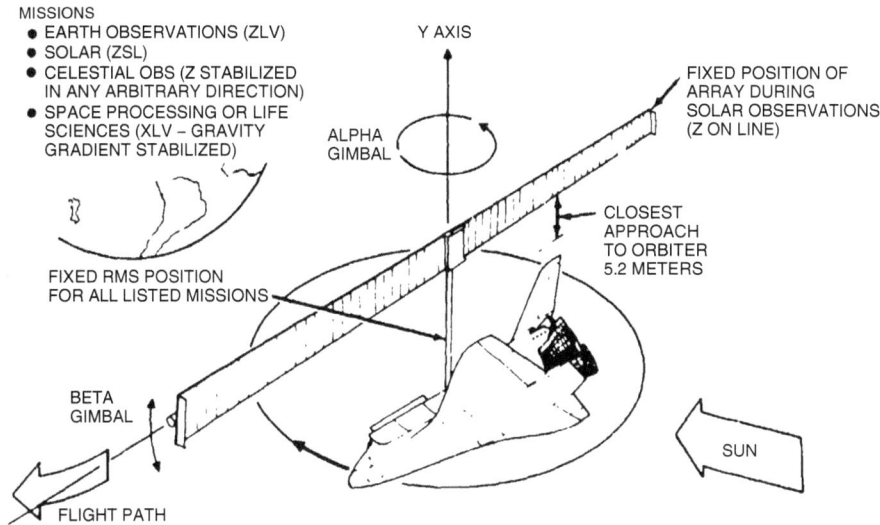

MISSIONS
● EARTH OBSERVATIONS (ZLV)
● SOLAR (ZSL)
● CELESTIAL OBS (Z STABILIZED IN ANY ARBITRARY DIRECTION)
● SPACE PROCESSING OR LIFE SCIENCES (XLV – GRAVITY GRADIENT STABILIZED)

Y AXIS

ALPHA GIMBAL

FIXED POSITION OF ARRAY DURING SOLAR OBSERVATIONS (Z ON LINE)

CLOSEST APPROACH TO ORBITER 5.2 METERS

FIXED RMS POSITION FOR ALL LISTED MISSIONS

BETA GIMBAL

SUN

FLIGHT PATH

The PEP solar array orientation.

In sunlight, the fuel cells would be either switched off or operated at a controlled minimum (idle) load in parallel with the solar arrays. The paralleling method was the preferred one due to its advantages in terms of system simplicity, reliability, voltage regulation, and minimum array size. The attitude of the Orbiter would be managed by the vernier thrusters of the RCS because of their superior propellant economy and small dynamic loading on the extended and flimsy solar arrays. The primary thrusters would be available as a backup, but they would consume more propellant and impose greater dynamic loads on the arrays.

Although it may appear to be a rather complex system, the basic performance of a PEP-Orbiter looked pretty good. For instance, a payload power requirement of 7 kW (equivalent to the baseline Orbiter) could be supplied for 12 days in a 28.5° orbit (as compared to 6 days without the PEP), or for 20 days in a 55° orbit, or 48 days in a polar orbit. The greater the altitude, the more continuous sunlight would be available, thereby increasing the period of power generation and reducing consumption of the cryogens for the fuel cells. The performances also looked good if a reduction in the number of cryogen tank sets for the Orbiter was considered. For instance, with only two tank sets, the power provided by PEP would allow a flight time of 6.5 to 12 days, as compared to just 5 days granted by the baseline complement of four tank sets. And the steeper the inclination, the more pronounced the gain in mission duration due to the increased amount and exposure time to sunlight. Offloading two tank sets meant a net payload gain of at least 3,500 pounds.

Several other configurations were considered, and their performances compared with the Orbiter baseline and with each other. It was evident that PEP could expand the flight envelope of the Orbiter either by extending the mission duration to several weeks or by allowing heavier flights to take place.

Further good news was that the PEP system would have a negligible impact on mission turnaround in comparison to the 39 hours required to install each cryogen tank set. In fact, as the PEP hardware could be carried anywhere in the payload bay using standard payload attachment points, fitting would be easy, quick, and readily accommodate the needs of the other payloads of a mission. Obviously, it would not be cheap, but the projected $47 million to develop PEP would be rapidly recouped by savings in Orbiter maintenance. In fact, the power available from the PEP meant that each fuel cell would be exposed to a much lower load in-flight, reducing wear and tear, extending its useful life, and shrinking the refurbishment costs. By having fewer fuel cells or reducing the amount of electricity that they had to supply would mean dissipating less waste heat through the Orbiter's radiators, thereby improving their life and reducing maintenance.

By 1984 the PEP had evolved into the so-called Power Module, which was an "orbital solar panel farm" at which an Orbiter could dock to draw power. A decade later, the expectation was that these orbital solar farms would be able to supply an Orbiter with at least 100 kW for extended periods. Such free-flying platforms could also provide services for standalone payloads, such as for materials processing, left there by one Orbiter and collected by a later mission. A large solar farm would also be able to provide attitude and control functions for both standalone payloads and a docked Shuttle. In this manner, the projected $139 million development cost of the Power Modules would be fully exploited.

THE 25 KW POWER MODULE

The issue of expanding the flight envelope of the Orbiter was also investigated by rival Marshall Space Flight Center in Huntsville, Alabama. Its proposal was similar to JSC's Power Module by being a free-flying platform but the design made use of the Apollo legacy. Meant to be delivered into orbit aboard the Shuttle, the primary structure was based on the Apollo Telescope Mount (ATM) that was developed for Skylab,[2] connected to a tubular forward truss that carried a single-axis gimbaled solar array. It would also have two docking ports to afford simultaneous coupling with an Orbiter in sortie mode, such as a Spacelab flight, and support a free-flying platform. Docking and payload berthing would be executed by the robotic arm of the Orbiter. Electric power would be generated from the single-axis tracking array, aimed by Sun sensors. Batteries would provide power for the night side of the orbit and would be recharged in daylight. Refurbished and upgraded Control Moment Gyros (CMG) commanded by an ATM computer would manage the attitude of the platform, aided by the RCS of the Orbiter when necessary. The onboard computer would also take care of communications and data handling functions

---

[2] Skylab, the first space station orbited by NASA, consisted of a Saturn V third stage externally fitted with solar panels and radiators and internally with living and working quarters for a crew of three. The ATM, was a boxy structure derived from the descent stage of an Apollo lunar lander. It was in line with the longitudinal axis of Skylab for launch, and then turned 90° on-orbit. It provided the mounting support and avionics for a suite of four solar telescopes and four solar panels.

via a data link with TDRS satellites during free-flying operations and via umbilical when docked with an Orbiter. Heat regulation was based on a thermal radiator similar to those of the Orbiter. The curved profile would enable the radiator to be folded against the module for ascent in the Shuttle payload bay. Once deployed on-orbit, the radiator would dissipate heat transported by Freon 21 coolant loops. Since this radiator was only for the free-flyer's own equipment, any berthed payloads would require their own heat rejection systems.

As the baseline power output of this free-flyer was 25 kW it was known as the 25 kW Power Module, but it was presumed future versions would provide 100 kW or more.

EXTENDED DURATION ORBITER

Sadly, neither the PEP nor the 25 kW Power Module ever left the drawing boards. Apart from the constant issue of funding, the main obstacle was the looming space station. The delay in starting to design and fabricate a permanent outpost meant the Shuttle had to serve as a miniature interim space station. Spacelab had been created to undertake research of the kind that was intended ultimately to be done aboard a space station. If a space station had been included in the program from the start, the Shuttle would have simply ferried astronauts and apparatus to and from the station, where most of the real work would be done. However, all space station work had been frozen by the mid-1970s so that NASA could focus on developing the Shuttle. By the early 1980s the space station was back on the agenda, and in 1984 President Reagan ordered it be built within a decade. If all went to plan, by the early 1990s there would no longer be any need for extended duration Shuttle missions.

But as the saying goes, "The rest is history." The decade envisaged by Reagan passed without any space station hardware being launched, and on-orbit assembly did not start until the end of the 1990s. In the meantime, NASA was obliged to find a way to sustain an Orbiter in space for longer than the baseline.

On April 2, 1990, NASA announced that it had signed an agreement with the Space Transportation System Division of Rockwell to develop a pallet to carry fuel cell cryogens in the payload bay as a major part of the Extended Duration Orbiter (EDO) program. It was to sustain Shuttle missions lasting up to 16 days. The terms of the agreement required Rockwell to fund the design and construction work, with delivery not later than December 1991. The agency would reimburse the company in three yearly installments.

The EDO pallet was precisely what JSC and MSFC had both ruled out in the late 1970s when developing methods for extending the in-space time of the Orbiter. It added four sets of cryogen tanks for a total of 368 pounds of liquid hydrogen and 3,125 pounds of liquid oxygen. They were to be carried on a 3,500-pound, 15-foot-diameter structural pallet at the rear of the payload bay. Internal connections and associated control panels and avionics would feed cryogens to the fuel cells as the flight progressed. Technological advancements and an improved understanding of how the spacecraft's resources were managed had improved the case for additional cryogens rather than a solar generator.

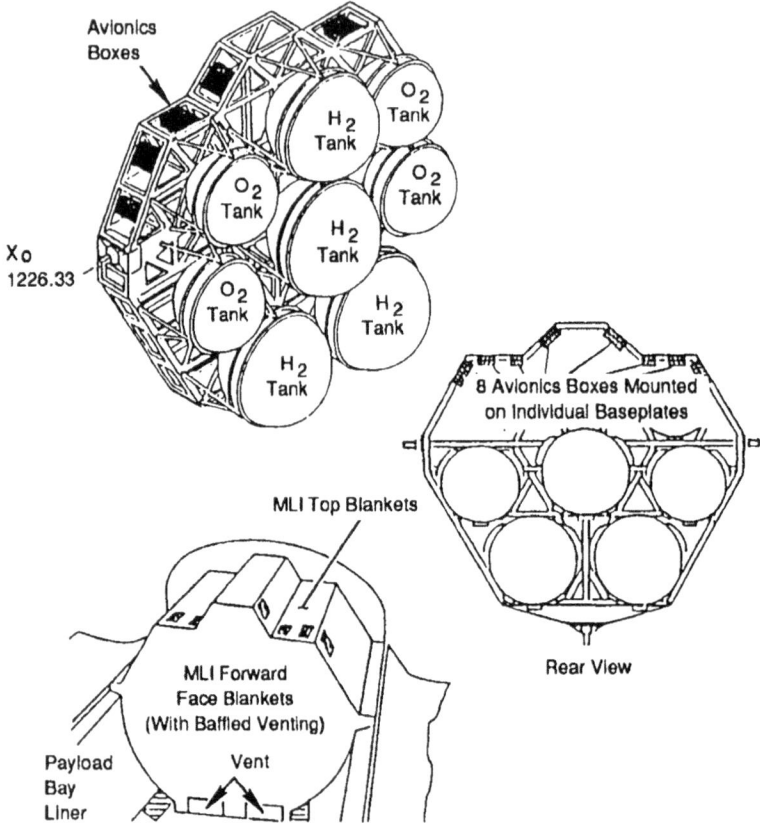

The Extended Duration Orbiter pallet structure for storing cryogens.

The agreement with Rockwell included the development of other modifications that would be essential for an extended duration. One was the Regenerable Carbon Dioxide Removal System (RCRS) which was intended to reduce the use of lithium hydroxide canisters for scrubbing carbon dioxide (CO_2) from the air onboard. As at least two such canisters were required per day, a 16-day mission would require a prohibitive number of canisters which added weight and also required a substantial amount of storage.

The RCRS system would remove CO_2 by passing cabin air through one of two identical solid amine resin beds. This resin is a polyethylenimine sorbent coating on a porous polymeric substrate. On exposure to CO_2-laden air, the resin combines with water vapor in the air to form a hydrated amine that reacts with the CO_2 to produce a weak bicarbonate bond. Water is required by the process because dry amine is unable to react directly. While one bed was active, the other one was to be regenerated by thermal treatment and vacuum venting. This latter requirement precluded using the RCRS for ascent or re-entry, so lithium hydroxide canisters still had to be carried for these mission phases.

An EDO pallet installed in the payload bay of an Orbiter.

This innovative and vital system required thorough testing in an Environmental Test Article (ETA), a high-fidelity simulator of the Orbiter crew compartment and its environmental control system. Putting the RCRS through its paces identified various faults, but these were fixed. It also removed some unknowns, as Henry Rotter, one of the leading designers, recalls, "We had been doing a lot of testing…to verify that this solid amine would not harm humans. Then we realized we did not know whether humans could harm the solid amine. We decided we'd better test humans against it, so we put seven guys in the ETA with cots and stuff so they could sleep in there and locked them up at 14.7 psi[3] for 7 days…The carbon dioxide performance did not change during that time. Then we did

---

[3] This was the nominal pressure maintained in the crew compartment during flight.

seven women for 7 days…We saw no change, no effect to the solid amine, and so we flew that in the EDO."

Additional nitrogen tanks were installed next to the existing ones in the payload bay in order to maintain the air in the crew cabin on longer missions.

An Improved Waste Collection System (IWCS) was also designed to withstand the heavy workload imposed by a crew of seven astronauts during a 16-day mission. This toilet had unlimited capacity, was more comfortable and sanitary than the earlier model, and eliminated many of the mechanical problems.

Finally, thanks to a new airlock storage bay, and space regained by reducing to a minimum the number of lithium hydroxide canisters, it was possible to add some 127 cubic feet of storage space to the crew cabin.

Altogether, the pallet and the other Orbiter modifications constituted the so-called EDO-kit. *Columbia* was the first to be retrofitted with it during a major modification at Rockwell's Orbiter Assembly and Modification Facility in Palmdale, California, that ran from August 1991 to February 1992.[4] It lifted off on June 15, 1992, for the first EDO flight which was scheduled to last 13 days and be devoted to microgravity research.

EDO FIRST FLIGHT: STS-50

Astronaut Bonnie J. Dunbar was one of the chief architects of STS-50. Recalling her first flight as a mission specialist on STS-61A with Spacelab-D1, she says, "MAN Technologies, a materials company in Germany, had sponsored some experiments. I think it was directional solidification of binary alloys that you cannot produce in 1-g. Some of it was very fundamental research that would allow them insights into, for instance, their turbine blade processes. Of course, you know that you aren't actually going to produce them in space but gravity really obstructs our ability to understand some processes because it produces convection and turbulent flow." The experiments were to study the processes in weightlessness.

Given Dunbar's background in materials processing, it was only natural that she should ask NASA whether the space station would have facilities such as furnaces for materials research and development. The reply was disconcerting, as she recalls, "The answer I got back was that, well…we really weren't building any facilities for the station. All we had were middeck-level things, so I said, 'Well, that won't allow us to really utilize the station or to do this leading research.'"

With Skylab, the agency had taken its first steps in the field of metals directional solidification, proving that microgravity does affect how materials perform and react, transfer heat, and so on. To Dunbar, it really "seemed that it was a real loss of our investment and scientific discoveries of the future if we didn't build facilities for the station." Fortunately, she found a powerful ally in fellow astronaut Sally Ride. "We saw a link between our

---

[4] *Endeavour* was built with the crew cabin modified for extended flights and with the plumbing to operate the pallet carried in the payload bay, so it was already EDO compliant. *Atlantis* and *Columbia* underwent the retrofit. *Atlantis* never used it. *Discovery* was not retrofitted. Only *Columbia* and *Endeavour* flew with the pallet.

ability to operate facilities in microgravity and exploration of the Moon and Mars, which are in fractional gravity, where let's say you're going to do in situ resource utilization. As a materials person, it was pretty evident to me that if you don't understand what is happening at zero, how are you going to extrapolate to one-sixth and one-third? I can't just put a mining operation up on the Moon and expect to get the same chemical reactions, heat transfers, et cetera; it's just not going to work that way."

At the request of the NASA Administrator, Sally Ride was at that time leading an Exploration Strategic Plan aimed at defining future goals for space exploration. She asked Dunbar to create and lead the Microgravity Materials Research Task Group to investigate how to get ready to exploit the space station for materials research. After discussing the issue with all NASA centers and drafting possible ways to fill the gap, Dunbar and her team made several important recommendations. The chief one was that dedicated Spacelab missions be planned in which prospective materials research facilities and procedures for the space station could be trialed and improved prior to being permanently installed on the orbiting outpost. As the reader may recall, one of the reasons behind the Spacelab concept was the possibility of a rapid turnaround of experiments or hardware which could be tweaked or improved upon in the light of flight experience and reflown for further investigations. In its report, delivered at the end of 1987, the Microgravity Materials Research Task Group advocated a series of Spacelab flights exclusively for materials research.

Five years later, STS-50 lifted off with a Spacelab pressurized module carrying a number of furnaces to grow crystals in microgravity and experiments to observe the behavior of fluids and combustion processes. Bonnie Dunbar was a member of the crew, "We ended up with a flight of brand-new hardware that had two purposes: one, to do science that would advance the state of the art, and the other as a testbed for the space station." Having been appointed payload commander, Dunbar's role carried an increased degree of responsibility, "It was my job as payload commander to take the requirements for the crew training, divide them out among the crew, and work with the crew activity planners on the timeline. It was a wonderful experience. We had a 13-day flight of two shifts, working around the clock. When we did a shift handover, it was a source of pride for me, which may not mean much to many people, but based on the fact that we were able to train as well as we did at Marshall, we never had a shift handover that was any longer than 15 minutes late. There had always been a problem in Spacelab flights with not getting everything done in a shift, being late, or having people stay up late. We never had that on our flight. I was really pleased with the way it was executed and with the ground support, both in Johnson and Marshall."

After 47 flights, it appeared that NASA had finally mastered the art of flying the Shuttle. "I feel almost guilty saying this," Dunbar observes, "but it was an extremely smooth flight. There were no failures of anything. The only event we had, was one of the furnaces had a large power switch and a switch guard was not narrow enough to preclude someone's toe from getting in and turning the furnace off. That was a great disappointment to the investigator, but we were able to load some backup samples in. That was, as I recall, the only big event. Everything else was extremely successful."

In fact, among other things, STS-50 provided several powerful lessons for how to do research on the space station. For example, since it is not possible to foresee how an experiment will behave in microgravity, blocks of time should be allotted to allow the astronauts in space and the scientists on the ground to react in real time to what is going on, and thereby take decisions while the experiment is in progress. This was in contrast to the usual way of planning a mission timeline down to almost every single minute in order to make the most of the time on-orbit.

Another significant lesson was how much time ought to be devoted to pre-flight training. One school of thought held that for space station flights, a basic preparation in how to operate the research facility would suffice because there will be more time available on a long duration mission. Being a long duration flight, at least by Shuttle standards, STS-50 proved this to be incorrect. As Dunbar notes, "As it turned out, we have much more research than time, and so where do you make that trade? I maintain that time is money, and there is a cost-per-hour on the station. If I'm going to invest in the training time, I would rather invest on the ground pre-flight than use valuable resource up there. I think we proved that in our Spacelab flights."

With the advent of fast communications and digital media that astronauts on the International Space Station now enjoy, it might seem that this reasoning is debatable. Whenever a new procedure or experiment is to be tried, a video tutorial can be easily uplinked. But as Dunbar notes, "You still need to have some sort of core training. I would never take an astronomer, for example, that's never dissected a rat, and then go ask them to extract the part of the inner ear of a rat based on a training film. That's a hands-on lab skill that you'd want to train people to. But I think we exercised some of those limitations on STS-50."

A major contributor to the success of STS-50 was the flawless performance of the EDO pallet. About 24 hours into the flight, *Columbia*'s three fuel cells began to draw upon the additional cryogen tanks. The new CO_2 removal system performed well for the first 25 hours, then experienced six shutdowns which led to its deactivation and a return to using lithium hydroxide canisters, a large stock of which were carried as a backup measure. After 4 days of troubleshooting, the system was reactivated and it worked satisfactorily for the remainder of the flight.

The EDO package was flown 14 times, all but one on *Columbia*.[5] On STS-80 it allowed *Columbia* to perform the longest mission in the Shuttle program. Launching on November 19, the five person crew returned on December 7 after a flight lasting 17 days 15 hours 53 minutes.

The EDO package lived up to the expectations of its designers, but undeniably it was a massive piece of equipment which sacrificed payload mass for extended orbital duration. As is common in any branch of engineering, planners have to compromise. Nevertheless, we can reasonably speculate that if either the PEP or the 25 kW Power Module had been available, the space agency would have been able to make much longer flights than STS-80, perhaps lasting up to a month.

By the late 1990s the need for long duration missions faded as the International Space Station began taking shape. With the exception of STS-125, the final servicing of the Hubble Space Telescopes, all flights subsequent to the loss of *Columbia* were to the

---

[5] This total includes the ill-fated STS-107 mission, where the EDO was lost together with *Columbia* and its crew. The only flight by the EDO aboard another vehicle was the ASTRO-2/STS-67 mission by *Endeavour* in 1995.

growing outpost. With each flight delivering either a component of the station or a load of spares and supplies, there was simply no room left for an EDO pallet in the payload bay. In fact, it was not necessary. ISS-bound missions did not require an extended period. Once the delivery was accomplished, there was nothing more to do. However, as the Shuttle program drew to a close, NASA recognized there would still be some room for improvement.

An Orbiter could transfer power from its 28-volt direct current electrical system to the 120-volt direct current grid of the ISS via an Assembly Power Converter Unit (APCU). During the early years of assembly, when the ISS had a limited capacity for generating electricity, this unit played a vital role in supporting the increased energy consumption that a Shuttle visit imposed. However, it was a one-way flow. By May 2007 both *Endeavour* and *Discovery* received the Station-to-Shuttle Power Transfer System (SSPTS) upgrade that involved replacing the APCU with the Power Transfer Unit (PTU) which could feed power in either direction. In the meantime, the SSPTS upgrade was also made on the ISS side.

Between January and February 2007 three spacewalks by ISS residents lay down the external cables that were required for the connection between the Orbiter and ISS power systems. Following reduction from 120 volts to 28 volts of direct current, up to 8 kilowatts could be routed from the ISS to the Orbiter thereby reducing the rate at which the fuel cells consumed their cryogens. Based on the flight configuration, the SSPTS allowed an Orbiter to stretch its docked mission phase from 6–8 days to 9–12 days in order to gain more time to conduct logistics transfers, additional experiments, and detailed Orbiter inspections (if required).

STS-118 in 2007 was the first mission to test the SSPTS upgrade. It performed so well that 3 days after being brought online the mission management team approved a 3-day extension that allowed a fourth spacewalk to be added in order to "get ahead" with some assembly tasks that would assist future flights. In a sense, this was a return to the PEP and Power Module strategy of having the Orbiter powered by sunlight.

12

Adding New Capabilities

NIGHT OPERATIONS

Sooner or later, it was bound to happen. Designed to serve the commercial, military, and scientific needs of the United States, the Shuttle had to be sufficiently flexible to launch at any time of the day, including during the night.

At 6:32 GMT on August 30, 1983, *Challenger* lit up the dark Floridian sky riding upon what looked like the Sun. Within the protective shell, the five crew members of the STS-8 mission were thoroughly enjoying the show.

Aviation regulations require the lights in the cockpit and in the passenger cabin of an aircraft taking off or landing in darkness be dimmed in order to accustom the eyes to ambient conditions, so that in an emergency evacuation people are not stunned by a sharp contrast in illumination. Respectful of this safety precaution, the STS-8 crew trained with dimmed lights during all their simulation runs. In fact, in the case of an emergency evacuation prior to lift-off, they would have been able to run through the structure of the launch pad without wasting precious time adapting their eyes to night vision. That was the theory. However, at the moment of lift-off the good intention to maintain night vision vanished. As the astronauts would report post-flight, "It was day inside the cockpit" and "it looked like you were in the middle of a glowing ball," basically like "looking into a cloud with light." That overwhelming light was the two tongues of fire rushing from the solid rocket boosters.

The need for a night launch derived from the requirements of the payload for this mission. In the aft portion of the payload bay was a cradle enclosing INSAT-1B on a PAM perigee-kick motor. Once in geosynchronous orbit, the Indian communications satellite was to provide telephone and data links, direct television broadcasting, and comprehensive weather services. In order to position the satellite above the equator at the longitude of the Indian subcontinent, orbital mechanics solutions dictated that the launch must occur at night. It would be only the second night launch of the American human space program.[1]

---

[1] The first night launch of an American manned spacecraft was when Apollo 17 illuminated the sky over Florida in the early morning hours of December 7, 1972.

© Springer International Publishing AG 2017
D. Sivolella, *The Space Shuttle Program*, Springer Praxis Books,
DOI 10.1007/978-3-319-54946-0_12

Despite NASA declaring the Shuttle operational after only four orbital flight tests, STS-8 had the flavor of a test flight. As pilot Dan Brandenstein notes, "The fact that we launched at night meant that we would end up landing at night."[2] It was another first for the program, and something which astronaut Loren Shriver says was greatly needed, "To be a fully operational program, you are going to eventually land at night somewhere."

All Shuttle pilots are fully experienced test pilots, well trained in landing in any conditions. Despite the airplane-like appearance of the Orbiter, landing at night set a challenge that promised a unique piloting experience. To better understand the issues, it is necessary to recall that when an aircraft lands at night the runway must be lit up along both sides for its entire length. And the touchdown area must be clearly visible to the pilots, so they can be sure they are about to put the wheels down on the runway rather than on a nearby corn field. Modern aircraft have landing lights, usually on the wing leading edges and on the nose landing gear. In making their approach, the pilots can execute corrections to ensure they are correctly lined up with the runway and will touch down at the desired point.

The heat of atmospheric re-entry meant the entire surface of the Orbiter had to be covered by the blankets and tiles of the thermal protection system. This left no room to install landing lights on the outer mold line. It would have been possible to install lights on the landing gear, but they would have been useless because the procedures for landing the Shuttle dictated that the gear not be lowered until the vehicle was a few hundred feet from the runway threshold, and a few seconds of illumination prior to touching down at 300 mph would not have allowed final adjustments to be made.

As Brandenstein, continues: "Dick Truly[3] and I kind of looked at each other and said, 'Oooh. This is going to be interesting.'" Shriver, who was in charge of the effort to develop a solution to the shortage of landing illumination, says, "The runway itself has lights along the edge. It has threshold lights or approach lights. There are a lot of cues, but there is nothing that illuminates the touchdown zone like the landing lights of an airplane do. So we had to figure out a way to supply some of that lighting onto the touchdown zone, and far enough ahead that the commander could get the visual cues he'd normally have." The most satisfactory solution was found in xenon lights, which are very bright. As Shriver says, "We found that certain arrangements of these lights in groups of four or groups of two and angled across the touchdown zone, not only headed you in the right direction, they also supplied the light."

Once the right combination of lights and reflectors was identified, a final tweak remained. So far, all flights had landed on the dry lakebed at Edwards in California, a rather dusty surface, as one might easily imagine. While this had not been a problem in daylight, landing at night on a dusty surface would create a serious and dangerous situation. As Shriver observes, "It became apparent that once you come in, if the light sources are behind you…the wingtip vortexes and [other effects of] touching down and rolling out stir up a huge amount of dust on the lakebed. If the lighting sources are behind you and the dust billows up behind you, it starts to cut out the light in the rest of the touchdown zone. The dust is soon going to block out all the light." It was therefore decided that for night

---

[2] A night landing is defined as a landing which occurs no later than 15 minutes prior to sunrise.

[3] A previous X-15 pilot and pilot on STS-2, Dick Truly was now commander of STS-8.

landings the Shuttle would touch down on a concrete runway. As Brandenstein says, "We landed on the runway using the lighting system that we'd devised, and that worked great."

Including STS-8, the program saw a total of 26 night landings, six of which were at Edwards Air Force Base.

A night landing by a Shuttle, clearly showing the illumination provided by the threshold and runway lights.

"THE CONSERVATIVE THING TO DO"

On April 4, 1997, *Columbia* and the STS-83 crew of seven began the planned 16-day Microgravity Science Laboratory mission to undertake 33 experiments in a Spacelab module to investigate the behavior of metals, alloys, proteins, fluids, and combustion processes in space.

Of particular interest were three Protein Crystal Growth Experiments aimed at unlocking the high-order structure[4] of proteins related to human physiology, such as insulin and

---

[4] In general terms, a protein is a sequence of a many amino acids. There are 20 different amino acids. The linear sequence in which the amino acids are organized is defined as the protein's primary structure. However, Brownian motion and electrostatic interactions between different parts of the protein chain can form secondary, tertiary and even quaternary structures. Such high-order structures define how the amino acid sequence is folded, and account for the specific behavior of the protein. An incorrect folding has the potential to create a "faulty" protein whose behavior can seriously damage cellular functions.

HIV-Reverse-Transcriptase.[5] Overall, they were expected to produce some 1,500 protein crystal samples for analysis on Earth in terms of cancer, diabetes, alcoholism, Alzheimer's, and AIDS. The advantage of growing proteins in space is that the absence of gravity provides structures of higher quality than those that can be obtained on Earth. In fact, on the ground such structures tend to buckle and deform under their own weight, altering their three-dimensional organization.

Although an outbreak of fire is one of the most serious hazards for a spacecraft, the process of combustion is the all-time favorite topic of study for space scientists. By eliminating gravity, researchers can gain a better understanding of fuels, ignition, and flames. This research offers the prospect of making internal combustion engines more efficient and less polluting. At the time of STS-83, the American Petroleum Institute estimated the US had an annual expenditure on crude oil of $200 billion. By increasing fuel efficiently by just 1%, it would be possible to save 100 million barrels per year, equivalent at that time to $5.5 million per day.

Metals and alloys also express their most natural behavior in weightlessness. In particular, undercooling can be studied. This occurs when a drop of liquid is rapidly cooled below its freezing point, yet remains fluid. An understanding of undercooling could lead to better aircraft, car, and truck engines, stronger construction materials, better foundry methods, and better welding, casting and soldering techniques.

All of the experiments were activated by the third flight day, but it was also clear that the mission would have to be cut short. Shortly after orbital insertion, fuel cell #2 started to display signs of possible degradation.

A fuel cell generates electricity by combining oxygen and hydrogen to initiate a chemical reaction with an electrolyte of potassium hydroxide. On one side of the cell, hydrogen reacts with hydroxide ions (OH^-) to produce water and electrons ($2H_2 + 4OH > 4H_2O + 4e$). The electrons (e) leave the cell as current to power the electrical loads of the Orbiter. As the electrons complete the circuit by returning to the cell they react with oxygen and water on the other side of the cell ($O_2 + 2H_2O + 4e > 4OH$) to produce new OH^- ions to replenish those consumed in the hydrogen reaction. The net result is two hydrogen molecules and one oxygen molecule are consumed to generate a flow of four electrons, two water molecules and heat. The number of electrons that leave the cell determines the amount of reactants consumed by the fuel cell. This also provides a way of checking whether there are any leaks in the distribution system. If a fuel cell is shut down, or if no loads are applied to it, a continuing consumption of the reactants from the supplying cryogen tanks[6] is clear evidence of a leak.

Each Orbiter had three fuel cells, all of them located in the forward portion of the mid-fuselage, beneath the floor of the payload bay. As with many subsystems, the fuel cells

---

[5] Reverse transcriptase (RT) is a protein responsible for assembling a DNA sequence that is complementary to a sequence of RNA. This protein is used by so-called retroviruses, such as HIV, which introduce into the cell a malign sequence of RNA that is read by the RT and translated into a corrupted DNA sequence which then reaps havoc in the cell and leads to the well-known symptoms of AIDS.

[6] Since oxygen and hydrogen can exist as liquids only at low temperatures fairly close to absolute zero, the adjective "cryo" (or cryogenic) is applied to any component designed to handle them.

were designed to be reusable with minimal maintenance between flights. In this case, they had a useful life of around 2,000 hours of on-line operation. Each fuel cell was 14 inches tall, 15 inches wide, and 40 inches long, which were fairly modest dimensions considering that it could supply 10 kW of maximum continuous power in nominal situations, 12 kW continuous in off-nominal situations (for example, if one or more fuel cells had failed), and a peak of 16 kW for 10 minutes.

A fuel cell was split into a power section where the chemical reactions took place and an accessory section that monitored the reactant flows, removed waste heat and water, and controlled the temperature. The power section contained 96 cells grouped in three sub-stacks of 32 cells. Manifolds ran the length of each substack, distributing the reactants and coolant fluid. The reaction in each cell was facilitated by an oxygen electrode (cathode) and a hydrogen electrode (anode) that were separated by a porous matrix saturated with potassium hydroxide electrolyte. Each cell produced electrical current at a voltage of 1.15 volts with no load applied, or 0.9–1.0 volts under normal loads. With the cells connected in series, a substack produced a total voltage ranging from 28 volts (the nominal voltage of the electrical loads) to 32 volts; i.e. the number of cells times the 0.9–1.0 volts that each cell was able to provide.

During normal fuel cell operation, the hydrogen and oxygen were diffusely mixed to generate electricity. The matrix, which was a fibrous asbestos blotter device, acted to contain the electrolyte in order to limit the mixing between the reactants. But after many hours of exposure to the caustic electrolyte and the heat generated within the cell, a manufacturing flaw in the matrix or impurity in the matrix fibers could cause a pin hole to develop and permit direct combination of the reactant molecules at that point. The heat from this localized reaction would cause the hole to burn and enlarge. Left to propagate uncontrollably, the reaction would cause an explosion. Early in the Shuttle program there was no way to monitor the onset of this "crossover" condition because the degraded output of that cell would be picked up by the remaining healthy cells.

This occurred 2.5 hours into the STS-2 mission. The nominal 5-day flight had to be reduced to 2 days, this being the minimum duration mission allowed by the flight rules. In addition to being disappointing, this also raised an issue that could seriously jeopardize the health of the astronauts. As pilot Joe Engle points out, "We also had a problem with our water, in that the membrane that failed on the one fuel cell allowed excess hydrogen into our drinking water supply,[7] making it very bubbly. Whenever we'd go for a drink, a large percentage of the volume was hydrogen bubbles in the water, and they didn't float to the top like bubbles would in a glass [on Earth] and get rid of themselves, because in zero gravity they don't; they just stay in solution. We had no way to separate those out, so the water had an awful lot of hydrogen in it and once you got that into your system it was like when you drink Coke real fast and it's still bubbly [with carbon dioxide]; you want to belch and get rid of that gas. That was the natural physiological reaction, but anytime you did that you'd regurgitate water. It wasn't a nice thing, so we didn't drink any water. We were dehydrated when it was time to come back."

---

[7] A useful by-product of fuel cells is water. It was collected and stored into four water tanks named with letters of the alphabet. Usually, Tank A stored purified water which the crew would use in meal preparation and for personal hygiene.

Starting with STS-9, a third substack was added to each fuel cell to produce more power because the missions were getting more demanding. With the expansion of the fuel cells, the likelihood of individual cell failures by "wear out" or age increased. So a Cell Performance Monitor (CPM) was designed to detect imminent failures such as crossover in individual cells. It compared each half substack voltage and calculated a delta-volts. As all the cells in a substack had to produce the same current, both halves of a substack had to be at about the same voltage. The critical threshold was defined as a delta of 300 millivolts.

This is what NASA thought was happening to fuel cell #2 on *Columbia*. In fact, the CPM noticed fuel cell #2 go out of limits and restabilize. The available data was insufficient to permit Mission Control to exclude the possibility of a crossover, so at 9:00 am on April 6, the decision was taken to curtail the mission and return to Earth within the next 2 days. Although one fuel cell was sufficient to power the Orbiter, for redundancy the flight rules required three good fuel cells for re-entry, so it would be the first attempt to return with only two operating.

At 9:12 am, Canadian astronaut Chris Hadfield, serving as Capcom in Mission Control, read up the bad news from the Mission Management Team, "The MMT had all players in on the meeting right through from the factory. The consensus is they just do not understand the behavior of fuel cell 2. Even though your efforts have done a good job toward stabilizing the problem, it is significantly out of family, so we will shorten the mission." Speaking for all of his crew, mission commander Jim Halsell professionally responded, "That's certainly a disappointment but we know you guys put your best effort forward and you're doing the right thing. We appreciate all the work that's gone into that."

Disappointment is a feeling that rarely clings to an astronaut, and in the limited time available the crew resolved to carry out as much useful science as possible. On the other hand, they had trained hard for more than 18 months and the investigators had spent even more time designing and preparing their experiments. The task was to do the best in the circumstances.

In order to route as much power as possible to Spacelab and its experiments, all non-essential utilities were powered down, including the onboard lighting. In fact, watching the post-flight video presentation it is possible to see members of the crew working by the illumination of small torches, a scene reminiscent of Apollo 13. The Spacelab controllers in Huntsville worked with the astronauts to reduce the run times of experiments and to realign the schedule to maximize the overall scientific return.

In the Combustion Module, which was designed to support the so-called Laminar Soot Processes investigation of flame shape, the type and amount of soot generated in various conditions, and the temperature of the soot components, provided the first ever observation of a steady non-buoyant flame. Combustion scientist and principal investigator Gerard Faeth of the University of Michigan, enthusiastically reported, "Scientists have gotten their first glimpse of the concentration and structure of soot from a fire burning in micro-gravity. It's a real first and the pictures we saw today will probably find their way into textbooks of the future." Much of the energy from fire is expelled in the soot. "Researchers are gaining a better understanding of the role that soot plays in combustion, and how it is produced by different fuels. Soot has a lot of negative attributes and that is why we are concerned about it. It is a pollutant. It is harmful to public health. It is the major source of difficulties in fires in homes. It has carbon monoxide associated with it, which is toxic and in that role soot is responsible for the deaths of about 4,000 people a year in the United States, and is responsible for the fire injuries of about 25,000 people."

Another first in combustion research was achieved on April 7, when payload commander Janice Voss made several runs of the Droplet Combustion Experiment. As principal investigator Dr. Forman Williams of the University of California at San Diego reported, "Six burns were successful, and for the first time we're burning free droplets." The experiment was designed to study the burning rates of flames, flame structures, and conditions in which flames are extinguished. "We can't get this kind of data in ground-based experiments. We've burned at two different atmospheres of oxygen concentration and calculated the burning times of free fuel droplets at each," Williams explained. Combustion of fuel droplets is an important aspect of heating furnaces for materials processing, heating homes and businesses, and making energy using gas turbines and gasoline powered engines. Findings from this study provided researchers with a better understanding of the combustion process leading to cleaner and safer ways to burn fossil fuels, as well as more efficient methods of generating heat and power on Earth.

Continuing the combustion research, the Structure of Flame Balls at Low Lewis-number (SOFBALL) study in the Combustion Module was, in the words of principal investigator Dr. Paul Ronney of the University of Southern California, "beyond my wildest dreams." The investigation was designed to ascertain the conditions in which a stable flame ball can exist and whether heat loss is responsible in some way for the stabilization of the flame ball in burning. Throughout the first experiment, a mixture of hydrogen, oxygen, and carbon dioxide burned in the facility for 500 seconds. This was a significant result, as Ronney noted, "These are the weakest flames ever burned, the lowest temperature, weakest, most diluted mixtures. They won't burn in Earth's gravity. We have long known that burning weaker mixtures increases efficiency, but not much is known about the burning limits of these mixtures." Findings from this work gave scientists a better understanding of the combustion process, and helped to improve theoretical models. "Combustion models give different results for these types of flames," Ronney said. "This is an acid test to show which, if any, current combustion modules should be used."

By the evening of April 7, Spacelab and the experiments had been deactivated, *Columbia* had been configured for re-entry, and the crew had time to answer some questions from the press. The next day *Columbia* smoothly touched down in Florida, successfully completing the first Shuttle re-entry with only two working fuel cells.

The post-mortem of the defective fuel cell revealed that several cells had suffered a slight degradation whose symptoms made it appear that a crossover had occurred. In fact, the fuel cell was far from that critical point. Unfortunately, by its very nature, the CPM was not able to tell whether the fuel cell had fallen victim of a crossover or a simple degradation. With hindsight, the STS-83 mission could have continued, but in the absence of better knowledge of the real status of the fuel cell NASA took the correct decision. Following this incident, the CPM was modified and improved to be able to report on a cell-by-cell-basis. Had another problem like this happened, NASA would have known exactly what was going on in the fuel cell, and been able to take a more informed decision. STS-83 had been prematurely recalled, but its mission was far from being over…

"THE POTENTIAL FOR ADDITIONAL FLEXIBILITY IN THE FUTURE"

In fact, NASA managers were already toying with the possibility of re-flying the full mission in the near future to finish the MSL investigations. Astronauts, the scientific community, and everyone in between were willing to try. After touchdown, STS-83 mission commander James Halsell confidently said, "We're ready to go fly. If it were up to me, I'd like to give the guys a week or two off to let them decompress from this flight and then we'll come back and start ramping up again for the next flight." And mission scientist Michael Robinson added, "From a payload standpoint, we could be ready. All the science teams say they could be ready." Then, "It's going to be tight, but again, the major-ity of the actual (experiment) samples were not processed so they don't require to be turned around. There wouldn't be facility upgrades or changes. There's no need to."

Even from an economic standpoint, it looked like the numbers would add up. As of 1995 a typical Shuttle flight cost $500 million, with most of the expense borne by testing, processing, training, mission planning, simulations and other activities on the ground. However, a re-flight would require just a handful of activities to prepare the spacecraft and the crew to execute precisely the same mission. According to NASA's Space Shuttle Manager Tommy Holloway, it would cost $50 to $60 million to return STS-83 to space. This was a fairly modest amount of cash, well within the agency's budgetary reserves. Joel Kearns, Director of the Microgravity Research Office at the Marshall Space Flight Center that had sponsored the MSL payload, was even more optimistic, "It is certainly less than half of the number Tommy was talking about, and potentially a lot less than that." He added the important point, "The faster we can do it, the cheaper it is."

On April 25, NASA announced that STS-83 would be reflown as STS-94 (the first available mission number) in early July. This gave the agency less than 3 months to get *Columbia*, its crew, and the Spacelab ready. Some outside-of-the-box thinking would be required to carry out the unusual processing flow. Normally, the Spacelab would be extracted from the Orbiter's payload bay and sent to the Operations and Checkout (O&C) building at KSC for post-flight inspections and preparations for its next mission. Ten weeks prior to launch it would then be returned to one of the three Orbiter Processing Facilities (OPF) to be installed into the Orbiter assigned for that mission. Additional sys-tems integration verification would confirm the connections between the Orbiter and the Spacelab. A month before launch, the Orbiter would be positioned on the launch pad, where a final systems verification would occur.

The normal process took about 13 months, but for STS-94 NASA had to squeeze it down to just 2 months, a 75% reduction in processing time never attempted before. Even before the announcement that the mission was to be reflown, it was decided to keep the Spacelab aboard *Columbia* and perform the checkout processing directly in the OPF build-ing. This meant that most of the Spacelab ground support equipment could not be used, as it was not designed to handle the laboratory aboard an Orbiter. Most checkouts had there-fore to be carried out at systems integration level instead of at the component level. Requirements had to be modified or deleted to accommodate the lack of access to ground support equipment. An effort was made to identify only those inspections or tests that absolutely had to be performed.

Every time a mission is concluded, the hardware is subjected to a series of post-flight inspections. Then once the hardware for the new mission has been assembled, another set of pre-flight checks must be made. As *Columbia* and Spacelab would not be separated, an analysis of the required post-flight inspections determined that while some inspections were indispensable, others could be done as part of the pre-flight inspections for STS-94 instead of in the context of the post-flight checks for STS-83, or even deferred until STS-94 returned.[8] The focus was on systems integration, but some verifications were done at component level to ensure that none of the Spacelab components had reached the end of their qualified life and would need replacement. A remote acquisition unit and an experiment input/output unit had shown anomalous behavior on STS-83, but there was not time to assess and resolve the problem. The troublesome video recorder was left in place because it was likely any anomaly could be worked around in-flight.

Columbia was also the object of inspections, because crew safety has the highest priority irrespective of how much time remains before the next flight. An assessment was made of every check that was not to be performed, to confirm that it had already been made prior to STS-83 and that no associated interfaces had been broken in the meantime. In addition, an overall systems check of the safety and reliability of all the critical components was performed ahead of STS-94 in order to confirm that these were working properly.

The hard work paid off brilliantly. Launched on July 1, 1995, STS-94 performed beyond all expectations. After landing, Tommy Holloway had only words of praise, "The mission has been executed in an outstanding way. All of the science objectives, including the highly desirables, have been accomplished. The vehicle has performed in an absolutely exemplary manner and I could not be happier." Mission commander James Halsell was also delighted, "*Columbia* has performed absolutely flawlessly for us. Days have gone by without [us] having to do an error log reset. That is our way of saying there have just been no problems whatsoever…So overall, this flight has done what the previous flight set out to do."

The successful flight of STS-94 is a clear example of how confident, and capable NASA and the Shuttle program had become in coping with unexpected problems and turning them into successful achievements. At the same time, the example set by the re-flight was a powerful lesson that, if necessary, could be applied to upcoming space station assembly flights.

As Holloway pointed out while the re-flight option was being debated, "I think it would be a very good test case of a capability we should have in place for the station, an ability to bring a payload back, an element of the station back for whatever reason, either for something like this or perhaps the interface didn't work as we expected it to at the station, and be able to turn it around in as reasonable a time as practical."

Such a situation never recurred, but we can be sure that if it had the space agency would have been able to recover promptly.

---

[8] This was a sensible choice, since some of the STS-83 post-flight inspections were similar in intent to the pre-flight checks for STS-94. This expedient saved time by avoiding duplication of the same checks.

PAYLOAD PLANNING AND INTEGRATION

The last splashdown of an Apollo capsule, concluding the Apollo-Soyuz Test Project (ASTP) on July 24, 1975, enabled NASA to devote all its resources and talents to the development of the Space Shuttle, whose projected first flight was fast approaching. The small group of employees who had made ASTP such a success in international politics[9] initially found themselves idle. However, there was a critical aspect of the Shuttle program that had been overlooked and now required attention.

Since all expendable launch vehicles in the US stable were to be phased out once the Shuttle became operational, the latter clearly had to be capable of accommodating a large variety of payloads, each with its own requirements, on a routine basis in the same manner as a freight airline.[10] This would clearly not be an easy task, because it presented the space agency with a steep learning curve.

The former ASTP team of several dozen people was renamed the Shuttle Payload Integration and Development Program Office (SPIDPO) and set to work developing and implementing the strategies designed to allow the Shuttle to become an effective orbital delivery truck. NASA was no stranger to the discipline of payload planning and integration, but the Shuttle introduced a whole new series of issues that had not been faced before. For instance, all previous manned space programs had been self-contained within NASA, with each of the agency's field centers being assigned to a piece of the program, with only a few external connections with industry. Also, only a few small, simple payloads had been flown on manned missions. Now NASA had a spacecraft with a large capacity for cargo which was to be filled up on each mission by a different mixture of payloads, most of which would be complicated, expensive, unrelated to one other, and, worst of all, developed outside the agency and without its oversight.

As the Shuttle was to carry commercial satellites mated to their upper stages, the SPIDPO group decided that, based on their sizes, a maximum of four satellites could be carried per flight. This raised the critical issue of how to physically interface each payload with the Orbiter. The most suitable place was a connector plate at the base of the forward bulkhead of the payload bay, which housed in excess of 100 connectors for power, commands, and data transfer to and from the Orbiter. Each payload would be manually hard wired to the connector plate by stringing cables along the bay. But this was a complicated approach that could readily lead to misplacing a cable, whose recertification would stretch out an already time consuming process. In fact, unless a given Orbiter were always assigned to deliver four commercial satellites, the payload bay would require to be configured for the manifested payload, each time involving a rewiring scheme. The issue was peculiarly exacerbated by Spacelab, whose different configurations and experiments

---

[9] The Apollo-Soyuz Test Project of 1975 marked the first time a US and a Soviet spacecraft rendezvoused in space and conducted docked operations. It was the first example of international cooperation between the two main protagonists of the Cold War. Its legacy eventually led to the creation of the International Space Station.

[10] Several economic analyses carried out in the early 1970s had predicted that the Shuttle would be much cheaper than any existing expendable launcher if it could be flown at least once, if not twice, a week.

would demand full access to all of the resources the Orbiter had to offer. Hence, any Spacelab would require an intricate web of cables to the connector plate.

A more elegant solution was eventually devised. As Glynn S. Lunney, leading the SPIDPO effort, recalls, "We ended up dividing all the connectors into four equal sets of connectors at four stations in the bay, which had one-fourth of all of the utilities, the electrical services, that we were providing for the Shuttle, so that if you were this size customer, you got one set. If you were twice as big as that, you got two sets. And if you were a full cargo bay dedicated flight, you could have all the services."

A similar approach was adopted when writing the so-called Space Transportation System User Charge Policy to determine the fee NASA should charge a commercial customer. The issue was whether the price should be based on the volume or mass of the transported payload. Considering that a satellite might take up the entire bay but have a relatively small mass, NASA elected to partition the payload bay into quarters and bill the customer based on how many quarters the payload would occupy. When necessary, weight would also be considered. Simple though this sounds, other factors played an influential role in setting the real pricing. In particular, competition from the French launcher Ariane forced NASA to cut their prices in order to win as many customers as possible, because that was how the Shuttle was supposed to justify itself in economic terms. In fact, unless the Shuttle flew once or twice a week, the program would not be able to yield an economic return. Hence, NASA had to do everything it could to prevent losing contracts, even if this meant operating at a loss (so to speak).

Manifesting was another challenge for Lunney's team. It required learning how to mix and match different payloads in an Orbiter. As Lunney admits, "Mixed cargoes presented a brand new challenge that nobody had dealt with very much before." First of all, the planners had to avoid interference between the payloads, not only in terms of physical intrusion of one into the operating envelope of a neighboring payload, but also by thermal, electromagnetic and radio frequency interference. Other factors were also important, such as a payload's deployment/operating altitude, orbital inclination, period of operation in the mission timeline and so on. As Lunney remembers, "We had to go through a process of assuring ourselves, and the payload people, that we'd planned something that could be run and could be achieved, and would not fall apart just by doing that."

The unprecedented reusability of the Shuttle forced an unforeseen shift in attitude on both NASA and its paying customers in terms of launch services. Until that time, when a customer such as a telecommunications company or the military purchased a launch service, they de facto gained complete control on the launch vehicle itself. By owning the launcher, the customer dictated their requirements. This was a reasonable agreement for an expendable use-once rocket, but the reusability of the Shuttle and the presence of crew meant that now NASA was the owner, with the customers only leasing a launch service in compliance with requirements imposed by NASA.

As Lunney recalls, "People came in with this mindset that they owned the launch vehicle, and the people who operated it should do whatever they say. So it was kind of like a control shift for them. It wasn't so obvious to us at the time…It took a little grumbling and working with people before it began to dawn on us just how they had previously worked, what their expectations were, why this was different, and why in some respects we were having some friction on subjects that would derive from that mentality." In response,

SPIDPO prepared a large number of documents to clearly explain the Shuttle environment and interface, to help the customers to design their payloads in a manner that would be compatible with the Orbiter without endangering crew safety. As Lunney says, "We gave them a book of requirements…They would have the mechanical requirement and an electrical requirement and a radio frequency requirement…We organized that document into logical groupings of subjects and we basically then documented that and evolved it over a number of years."

For previous manned space programs the space agency had been responsible for the development and construction of almost every piece of hardware, but the Shuttle was to carry equipment whose development had been conducted without supervision by NASA. The payloads would be developed by companies all across the world. As Lunney wryly put it, "You don't know what they are doing." This is why payload safety panel reviews were held at significant milestones during payload development and construction to verify adherence to accommodation and integration requirements. For the same reason, NASA demanded that payloads be designed with three layers of redundancy, and be able to sustain two failures without threatening the safety of the Orbiter and its crew.

Although the Shuttle made it more difficult for customers to design and fly their payloads, NASA did its best to ease the pain. Interpretation documents were written to clarify most of the accommodation and integration requirements so that customers would fully understand how to design their payloads to pass the safety review panels without delays. Cargo integration reviews were established in which NASA showed how they would take care of payloads and how they would meet the requirements for payload deployment and/or operation. The position of Payload Integration Manager (PIM) was introduced to liaise between NASA and its customers. For each mission, a PIM would be nominated to orchestrate the mix of payloads assigned to the flight by coordinating all of the activities surrounding payload approval, safety reviews, cargo integration, and so on.

SPIDPO was set up at a time when it was believed the Shuttle would operate with the regularity of a freight airline. It continued to serve the program even when it was apparent that only a handful of flights could be handled per year, and even after the carriage of commercial satellites was canceled in the wake of the loss of *Challenger*. Throughout three decades of Shuttle operations, SPIDPO proved to be of paramount value in making each flight a success. At the same time, it allowed NASA and the worldwide aerospace community to learn how to operate a reusable spacecraft.

13

The Legacy of the Shuttle Program

INTRODUCTION

On September 25, 2005, *USA Today* published a one-page article by Traci Watson entitled "NASA Administrator says Space Shuttle was a Mistake." It recounted that Michael D. Griffin, the Administrator of NASA since April, had said in conversation that the Shuttle had been a mistake and, "It is now commonly accepted that [it] was not the right path."[1]

This strong claim was made at a delicate moment of the Shuttle program. In fact, on February 1, 2003, *Columbia* was tragically lost when, less than 15 minutes from its scheduled return to Florida, it disintegrated in the skies over Texas and Louisiana. The accident investigation board soon identified multiple reasons for this loss. Chief among them was a piece of foam that detached from the external tank during ascent and hit the leading edge of the Orbiter's left wing. The resulting 20-inch hole in the wing had later allowed the hot plasma of re-entry to penetrate the Orbiter and cause a catastrophic structural failure. As the vehicle broke up, all seven crew members lost their lives. As a consequence, NASA strove to implement a multifaceted program of upgrades and improvements to enhance safety and reduce the risk of a recurrence.

Two years later, on July 26, 2005, *Discovery* was launched for the return-to-flight mission, which was to deliver much needed items to the International Space Station. Analysis of the ascent imagery revealed that once again foam had detached from the external tank. Although none of the debris hit *Discovery*, it was a near miss which, in the wake of *Columbia*, the agency could not afford. The Shuttle fleet was grounded for another year to further modify the external tank. It was during this second hiatus that Griffin made his criticism of the Shuttle program. Given the historical context, with two Orbiters lost, his assertion might seem reasonable, but can we really call the Shuttle a mistake?

---

[1] It is worth noting that Griffin expressed the same sentiment towards the International Space Station.

© Springer International Publishing AG 2017
D. Sivolella, *The Space Shuttle Program*, Springer Praxis Books,
DOI 10.1007/978-3-319-54946-0_13

THE FAILURES OF THE SPACE SHUTTLE

The losses of *Challenger* and *Columbia* are without a doubt the most serious failures of the Shuttle program. Both were claimed by catastrophic structural failures which were attributed to flaws in the design that dated back to the inception of the program. These accidents claimed the lives of fourteen brave astronauts, who, although aware of the perils of flying in space, had placed their trust in the competence of NASA.

On January 26, 1986, an unusually cold morning for winter in the Sunshine State, *Challenger* lifted off for its tenth mission. The flight abruptly ended some 72 seconds later. The accident was investigated by the Rogers Commission, which found that the joints between the four segments that made a solid rocket booster had a fatal flaw. At low temperatures, such as those experienced the night prior to the launch, the rubber O-rings tended to lose their elasticity, preventing them from hermetically sealing the joints.[2] As a result, the hot, highly pressurized gas inside the right-hand booster was able to leak out of a joint in the vicinity of the lower attachment of the booster to the external tank. This "blowtorch" both burned a hole at the bottom of the external tank and severed the load-bearing strut. The tank breach dumped liquid hydrogen and the broken strut permitted the booster to pivot about its upper attachment and breach the oxygen tank at the top of the ET, leading to an uncontrolled explosion and structural failure.[3]

Most importantly, the Commission unearthed a pattern of wrongdoing at the heart of NASA dating back more than a decade. It started when NASA sold the Shuttle as the only launcher that the nation would ever need. The much-touted reusability of the new spacecraft, it was argued, would make access to space cheap, even profitable. There would be no need for wasteful expendable launchers. But there was a catch. As was established by a number of studies, profitability and low operating costs would require the Shuttle to operate like a commercial airline. Specifically, it would have to fly missions at least once per week. The absurdity of the proposition was evident, but the need for funding and political backing overcame common sense. The future space airline started to receive contracts for the Shuttle to deploy commercial and military satellites, and this led (as shown in Chapter 2) to the design and development of two types of upper stages.

After only four missions, the flight test program was declared complete and the Shuttle became operational. Once again, the insanity of such a proposition is evident when it is considered that any modern airliner or military aircraft requires thousands of hours of

[2] It was Nobel Prize winner Richard P. Feynman, famous for making complex concepts in physics clear to the general public, who exposed this problem. At a public hearing of the Rogers Commission, he demonstrated how a piece of the same rubber as was used on the O-rings would lose its integrity under freezing temperatures. For the experiment, the brilliant physicist simply used a glass of water with ice! You can search on YouTube for a video of his presentation. It is amusing to watch this genius at work.

[3] Technically, *Challenger* did not explode. In fact, while the external tank did explode as a consequence of hydrogen and oxygen mixing, the Orbiter suffered a structural break up due to the acceleration produced by the exploding external tank and the ensuing loss of control. This is akin to a person suffering broken bones or split organs when struck by the blast of an explosion. In this case, the damage incurred by the body is a consequence of the blast, not the explosion of the body itself.

ground and flight testing prior to being approved for service. The Shuttle, being a complex flying machine, ought to have undergone the same verification but lack of funds and the urgency to fulfill the commercial contracts conspired to end the testing much earlier.

The pressure to launch on time obliged NASA to overlook safety by adopting the "normalization of deviance" attitude. The term was coined by Columbia University's Professor Diane Vaughan when she analyzed the *Challenger* accident. As she writes, normalization of deviance "means that people within the organization become so much accustomed to a deviant behavior that they don't consider it a deviant, despite the fact that they far exceed their own rules for the elementary safety."[4]

In effect, this translates into accepting that because the occurrence of an anomaly (or deviant behavior) did not have any impact on safety, it is not necessary to rectify that anomaly, which becomes a part of the normal (hence the "normalization" effect) behavior of the system.

In the case of *Challenger*, it was well known that the O-rings did not work as they were intended. In fact, several recovered boosters had shown that some O-rings had suffered damage to such an extent that if the joint design had worked as advertised it should have never happened. A *Challenger*-like accident could well have happened sooner. But this anomaly was not deemed worthy of detailed analysis because it had never caused any significant issue. As a matter of fact, the whole joint design had not been adequately tested, only computer simulations and ground testing had been used to analyze the static loads. Owing to lack of money and time, the more demanding dynamic loads of the ascent were barely investigated. A better analysis would have revealed that the flexibility of the joints would permit hot gas to leak out even if the O-rings had not been subjected to cold temperatures.

Challenger's normalization of deviance was further compounded by the pressure that NASA managers placed on the engineers responsible for the solid boosters, who were concerned that the unusually chilly temperatures could have compromised the integrity of the O-rings.[5] Their technical opinions were cast aside by a management more interested in fulfilling their commercial obligations than in assuring safety.[6] It was a direct consequence of NASA having sold the Shuttle as an airliner.

---

[4] *The Challenger Launch Decision* by Diane Vaughan is a must-read for anyone who wants to understand the normalization of deviance, and how this can affect safety at any level within an organization. The issue is so fundamental that the *Challenger* accident is routinely cited as an example when studying and explaining human factors issues.

[5] You can gain a sense of how cold it was during the night prior to the launch by the impressive pictures available on the internet which show large and long icicles hanging from the metal framework of the launch pad structure. As opposed to the pad being located in the Sunshine State, one could be forgiven for thinking it was in Scandanavia!

[6] *Truth, Lies and O-Rings* by Allan J. McDonald is another must-read book on the *Challenger* accident, as the author was one of the few engineers that kept opposing to the launch decision even after Morton Thiokol, the manufacturer of the solid boosters, had yielded to the pressure from NASA.

The accident also highlighted how the Orbiter lacked an escape system for saving a crew in the event of losing a vehicle.[7] But to be fair, NASA never had enough resources to add a proper emergency system, and the mass saved by not having one helped to achieve the payload capability required to retain political support.

Eventually, the Shuttle program recovered stronger and safer than ever.[8] When *Discovery* launched on September 29, 1988, it marked the return of NASA to human space flight with an ethos which put crew safety first and foremost. The space airline services were handed over to expendable launchers.

But memory has the tendency to fade away and over time old and flawed customs slowly resurfaced. When *Columbia* missed its appointment with the Kennedy Space Center on February 1, 2003, at the end of a highly successful 16-day mission, it was again self-examination time for NASA. The *Columbia* Accident Investigation Board determined that during the ascent a piece of foam fell off the external tank and struck the left wing of the Orbiter, opening a sizeable hole in one of the leading edge panels. The crew were unaware of it, but they were doomed. Inevitably, during re-entry the hot plasma penetrated the wing, melting metal and causing the wing to collapse. The resulting loss of control resulted in the breakup of the vehicle.

It transpired that once again the normalization of deviance was at the heart of the problem. As in the case of the flawed booster joint, the fragility of the foam covering the external tank was well-known. But changes to the design of certain features of the external tank that were promoting foam detachment had not been implemented since (surprise!) thus far no harm had been done. The vulnerability of the Orbiter's thermal protection was also well understood. The silica tiles plastering the underside and the carbon-carbon panels which formed the nose and the leading edges of the wings were excellent at protecting the aluminum structure from melting during re-entry but were poor at dealing with mechanical stress. It was, however, the best solution that NASA could come up with, given the size of the Orbiter and the limits on technology that precluded the adoption of a hot structure.[9] Initial studies for on-orbit repairs of the thermal protection system were dismissed when

---

[7] For the first two missions, the two-man crews could use ejection seats to escape a crippled Orbiter if there was a problem at launch. When larger crews began to be carried the ejection mechanisms of the front seats were disabled. The ejection seats were later removed. After *Challenger*, an escape system with a pole was added in the crew cabin. In case of an emergency, the ingress/egress hatch would be blown off and the pole extended into the airflow. Then one by one, the astronauts would exit the cabin using the pole as a guide to avoid striking the wing. Once free from the Orbiter, they would open their parachute and seek a safe place to land. Although crews trained for this scenario in swimming pools and in the open sea, nobody ever believed that the system would work. In fact, it could be used only when in the lower atmosphere and with the Orbiter in a stable attitude. It was really only an expedient to give members of the public a false sense that their astronauts were now better off, but quite frankly the mass taken up by this system would have been better used for additional payload.

[8] The joints of the solid rocket boosters were greatly improved, and no further issues of this kind were ever detected. For an in-depth description of the joint design prior to and after *Challenger*, refer to Chapter 4 of my previous book *To Orbit and Back Again: How the Space Shuttle Flew in Space*.

[9] Refer to Chapter 1 for a discussion on the "hot structure" concept initially proposed for the Shuttle.

the Shuttle was declared operational. They were resumed as a matter of urgency after the loss of *Columbia*.[10] Once again, the program recovered and flew to its conclusion in 2011 with no other major issues. One aspect of the Shuttle often advanced by its detractors was it was too complex, and thus too expensive to operate. Specifically, it failed to deliver the promised cheap access to space. However, we must understand that complexity is not a bad thing, as long as it is thoroughly understood, properly managed and implemented in a stepwise manner.

The mistake that NASA made was to jump from the "simple" Apollo capsule to a Boeing B737-like spacecraft, without any intermediate steps. It was as if the Wright brothers, after making the historic flight of the Wright Flyer I on December 17, 1903, had set out to develop a Boeing B747! The Queen of the Skies[11] came about some six decades later, after thousands of technical innovations had been implemented and millions of hours of flying experience and operations had been accumulated. NASA would have been better to start with a smaller spacecraft, maybe only for delivering astronauts and small payloads to a space station that was assembled using rockets of the Saturn V class. A spaceship similar to the European Space Agency's Hermes or the Dream Chaser that Sierra Nevada Corporation is currently developing privately would have been the right size for the job. Orbital tugs and space stations could then be employed for research, manufacturing, and satellite servicing. With experience in operating a reusable spacecraft, upgraded versions could be rolled out with enhanced reliability and safety. Operational costs would have been better contained, and would have fallen over time. If the need arose, a larger, more complex reusable spacecraft could have been developed. Costs, reliability, and safety, would have all been better than the Shuttle. Maintenance proved to be so expensive and lengthy prior to a flight mainly because no one knew how to operate a reusable spacecraft, let alone one the size of a narrow-body airliner. NASA had to learn from scratch, often pursuing a trial and error process.

With sufficient cash, NASA would have been able to build space stations, expand human activities to Mars, and jump start a space-based manufacturing industry. With forethought, Saturn V launchers could have been kept in production, or more modern heavy lifting rockets developed from scratch. This would have significantly reduced the size of the Orbiter to that appropriate for transporting people and small payloads.

Since the benefits of space explorations have never been fully understood by the general public and politicians, NASA was only barely able to develop the Shuttle. To obtain the necessary funding, they were forced to sell it as a jack-of-all-trades. All of a sudden, complexity increased appreciably. They were also compelled to accept the requirements set by the USAF for a voluminous payload bay to carry their heavy and cumbersome spy satellites and delta wings to maneuver extensively during re-entry. These requirements added more complexity, and once again NASA was not allowed to learn from operating a smaller vehicle. On the other hand, the alternative was not to have a Shuttle at all. The agency faced a rock-and-a-hard-place conundrum which had no simple solution.

---

[10] For more details on changes to the external tank and on the thermal protection system repair techniques developed after *Columbia*, refer to Chapter 5 and Chapter 8 of my previous book *To Orbit and Back Again: How the Space Shuttle Flew in Space*.

[11] The Boeing B747 is widely referred to in this manner because it is one of the most beautiful and important aircraft ever built.

Furthermore, the need for a large payload delivery capability coupled with a system that was already more complex than anything done before did not permit the addition of a suitable emergency escape system for the crew.

But history cannot be made with "what ifs," and wishful thinking will not change the past. While we cannot deny the failures of the Shuttle program, at the same time we should also celebrate its legacy of real accomplishments that probably will not be matched by any other program for a long time to come.

THE ACHIEVEMENTS OF THE SPACE SHUTTLE

The idea of a reusable spacecraft to lower the cost of access to space is a good one. It is a ludicrous proposition to build expensive and complex spacecraft and launchers as throw-away items. Every means of transportation ever created by humankind, from a primitive canoe to a jet fighter, has the intrinsic property of reusability over a large number of cycles. But the space launcher industry did not follow this tendency and, apart from a few exceptions in recent years,[12] is still pursuing the throw-away ethos. Hence it was a bold step for NASA to take when, even before the end of the Apollo program, it proposed a fully reusable space transportation system. Due to limitations placed on the design, it achieved a partially reusable configuration. *Columbia* flew 27 times,[13] *Challenger* 9 times,[14] *Discovery* 39 times, *Atlantis* 33 times and *Endeavour* 25 times. The fact that the program tallied an outstanding 134 flights is a tremendous achievement. Considering that each Orbiter had been designed for an effective life of 100 missions, the program could have gone much further, in particular because after the loss of *Columbia* missions suffered fewer and fewer issues.

The fact that the Orbiter is by far the largest spacecraft yet to see service[15] made possible several other achievements.

First of all, NASA had to learn to operate and manage a space program that was far more complex than anything it had attempted before. With Apollo the agency had to train small crews[16] for a small number of missions that all had basically the same objective of landing on the Moon.[17] With the Shuttle, NASA opened the gates to a program comprising a vast number of missions, each different from the other, carried out by an average crew of

---

[12] Such as SpaceX and Blue Origins.

[13] This number includes the last fatal mission, STS-107.

[14] The last flight is not considered, as it did not reach orbit.

[15] The International Space Station and its Soviet/Russian predecessor Mir, are obviously much bigger but in this context the term spacecraft refers to a vehicle that commutes from and to orbit. It is also worth mentioning that the Soviet Shuttle Buran was slightly bigger than its American counterpart. However, it flew only once and was unmanned. It is therefore safe to say that the NASA Shuttle is the biggest reusable spacecraft ever operated.

[16] Only three astronauts per flight.

[17] The early Apollo missions were directed either to prove procedures and hardware in low Earth orbit or to circumnavigate the Moon, rather than to land on the Moon. However, they were all aimed at building confidence and experience for the landing missions, which were the real goal of the Apollo program.

seven, some of whom were not professional astronauts. It was necessary to learn how to simultaneously plan several complex missions, design, build, and certify different types of hardware, maintain and prepare for flight a fleet of five Orbiters, and train crews and ground support teams. Even if the yearly flight manifest was reduced after *Challenger*, the logistic and programmatic effort was kept at the same level to assure mission success in each instance.

The tiny Apollo capsule could house a crew of three, but the Shuttle afforded an unprecedented opportunity for up to seven astronauts to live and work on missions lasting up to 2 weeks.[18] Among other things, this meant the US space program was finally opened to women.[19] For instance, Sally K. Ride became the first American women to fly in space, Kathryn D. Sullivan was the first American women to make a spacewalk, and Eileen M. Collins became the first female Shuttle pilot and later the first female Shuttle commander.

The Orbiter afforded NASA the opportunity to allocate seats to non-professional astronauts. Along with the payload specialist class of spacefarer,[20] private citizens were offered seats. As the Shuttle had been funded with taxpayer money, it was only reasonable that several of them would be selected to participate in a mission. NASA had in mind to fly teachers, artists, poets, musicians, and other people who would be able to convey the experience of space flight in a manner that would be closer to the general public.[21] The first to participate in this so-called "citizen in space" program was Sharon Christa McAuliffe, an elementary school teacher from Concord in New Hampshire. Assigned to *Challenger*/ STS-51L, she was meant to deliver a number of lessons from space. Unfortunately, she perished along with her six colleagues when *Challenger* was lost. That curtailed the "citizen in space" program. In fact, NASA understood that the Shuttle was nothing like an airliner, and that only astronauts with comprehensive training should bear the risks.[22]

The ability to fly a large crew also allowed room for non-American professional astronauts. On STS-9 in 1983, Ulf Merbold of West Germany[23] became the first of a long list of

---

[18] As related in Chapter 7, STS-61B was an exception because it had a crew of eight astronauts. Luckily, the presence of the Spacelab laboratory avoided uncomfortable overcrowding. It is also interesting to remember that STS-71, the first Shuttle to dock with the Russian Mir station, lifted off with seven astronauts but returned with eight. The longest mission ever flown by a Shuttle was STS-80. In that case the almost 18-day-long flight was accomplished with a reduced crew of five.

[19] The reader may recall that Russian Valentina Tereshkova became the first woman in space when she was launched on June 16, 1963. However it was a propagandistic flight more than anything else. In fact, ever since the Soviet/Russian space program has allowed such a small number of female astronauts that it can be clearly defined as a "for men only" program.

[20] This was discussed at length in Chapter 7.

[21] The NASA professional astronauts did not share the same enthusiasm for this initiative, as it gave away flight opportunities to people who did not have to go through the astronaut selection process and the same degree of training.

[22] After the accident, payload specialists were still allowed on certain flights, but compared to the payload specialists pre-*Challenger* they received better training and they were indeed indispensable for the execution of the mission.

[23] Refer to Chapter 7.

foreign professionals to fly on the Shuttle. If you read the names of all the crews flown through the entire program, you will not have any problem recognizing people from Italy, Canada, Spain, Holland, Germany, Japan, United Kingdom and so on.[24]

The international flavor of STS-9 was extended to the large pressurized cylinder in *Columbia*'s payload bay, because this was the first flight for the made-in-Europe Spacelab. For two decades this allowed the Shuttle to become a mini-research station to help to prepare for the International Space Station. Spacelab also opened NASA to significant collaborations with international partners. In this regard, one of the most quintessential cases was the partnership with the Italian Space Agency that flew the Tethered Satellite System on STS-46 and STS-75.[25]

An even more international flavor was added with the Shuttle-Mir program. In an effort to reduce the cost of construction of its long overdue space station, NASA was obliged to partner with Russia and exploit that nation's know-how in long-term space flight, mastered by missions dating back to the late 1970s. The joint program started with *Discovery* rendezvousing with the Mir station in February 1995 to validate the approach to a range of just 10 meters. This cleared the way for Robert "Hoot" Gibson to dock *Atlantis* to the former Soviet space complex on June 29, 1995. It was sort of a déjà vu moment. During the Apollo-Soyuz Test Project mission, Thomas P. Stafford had shaken hands with the Soviet Alexi Leonov. Now, some 20 years later, the ever-smiling Gibson shook hands with his Russian counterpart Anatoly Solovyev. While the ASTP was fundamentally a political gesture of apparent friendship between the two rival superpowers and did not have any follow-on, this time the cooperation was to last. Shuttles docked with Mir a total of nine times, and American astronauts lived in the outpost for months at a time. In this manner, NASA gained its first hands-on experience in long-duration space flight since Skylab, in the process discovering that it had forgotten many of the lessons learned from that time.[26]

Finally, in December 1998, *Endeavour* flew the STS-88 mission that connected together the first two components of the International Space Station and initiated the most complex construction project ever attempted in space. In fact, it is questionable whether the modular configuration of the ISS was the best choice. It might have been easier and faster to assemble a station from a number of units similar to Skylab, but it must be recognized that production of the Saturn V ended well before the final Moon landing and therefore the

---

[24] It is fair to recall that the Soviet space program had already opened the doors to foreigners well before the Shuttle, but few of these people were professional cosmonauts. Rather they were private citizens of communist countries who were given the chance to fly in space as a sign of friendship and support towards that country. In fact, the Soviet/Russian program did not initiate an age of true international cooperation. Nowadays, the Russian Soyuz capsules do routinely ferry non-Russian cosmonauts, but it is because there is currently no other means of reaching the International Space Station.

[25] Refer to Chapter 10 for an in-depth examination of these fascinating missions.

[26] Skylab, America's first space station, consisted of a Saturn V third stage fitted with living and working quarter plus additional external modules. It was visited only three times, with the longest mission lasting 84 days. Although NASA considered using an early Shuttle mission to attach a small propulsion module to boost the orbit of the station for possible reuse, the station fell into the atmosphere before this could be attempted. No further experience in multi-month missions was pursued up to the Shuttle-Mir joint program.

Shuttle was the largest launcher NASA could use. Besides, from its inception the Shuttle was meant to be capable of assembling large orbiting infrastructures. We can therefore confidently affirm that the Shuttle lived up to those expectations.

CONCLUSION

After this brief overview of the failures and achievements of the Shuttle program, we must decide whether the Shuttle was a mistake. It is my firm opinion that it was *not* a mistake.

Although the idea of the Space Transportation System was faulty, the Shuttle was a remarkable flying machine and will remain one of the most sophisticated creations of humankind for a long time. With hindsight, it is always easy to point at every flaw and defect, but it is important to discern precisely why the Shuttle happened to be so faulty. This mostly boils down to the lack of adequate funding and political backing that forced NASA and the US aerospace industry to accept onerous requirements and settle for the least expensive solutions, even though it was readily apparent they were not the best in engineering terms. In this regard, the selection of solid rocket boosters is prominent. The aerospace industry is also to blame. If it had embraced the ethos of satellite servicing and refurbishment, the Shuttle would have had a long log of flights to accomplish. The more you fly it, the better and cheaper it becomes to operate your spacecraft. The USAF must also to bear its share of the blame. The requirements that it imposed were barely exploited in reality. If it had supported the program with real money and confidence, the Shuttle would have had a significant number of missions to perform.

In this context, we can easily understand that the Shuttle as we know it today was indeed the best solution available to NASA. As a little kid that grows into adulthood, the Shuttle matured mission by mission, and the achievements of 30 years in service include: reusability; the ability to perform a broad variety of mission objectives; the opening of international collaborations; flights for women, minorities, and foreigners; and assembling the International Space Station. These achievements are just a small part of the legacy left to us by the Shuttle and its crews. No other space program has been able to match such success.

Today, the mission of the Shuttle is not finished. The three surviving Orbiters are exhibits at prestigious locations across the United States. *Discovery* is at the National Air and Space Museum in Washington DC, *Atlantis* is at the Kennedy Space Center in Florida, and *Endeavour* is at the California Science Center in Los Angeles. The fact they are permanently grounded does not preclude them from inspiring the future generations of explorers who, one day, we hope, will return to the Moon and, farther still, set foot on Mars.

About the Author

Davide Sivolella was born in Pinerolo, Italy, in 1981. As a child, he developed a fascination with all kinds of flying machines, especially those which travel above the atmosphere. This passion for astronautics led to bachelor's and master's degrees in Aerospace Engineering from the Polytechnic of Turin (Italy). Since 2009 he has been employed as a specialist in aircraft structural repairs for civil airliners in the United Kingdom. He thinks of aircraft as spacecraft that fly low and slow. His first book *To Orbit and Back Again: How the Space Shuttle Flew in Space* was born from a life-long passion for the Space Shuttle program. In addition to a fondness for human space exploration, he enjoys cooking, traveling, and landscape photography. He currently lives near London with his Spanish wife Monica.

© Springer International Publishing AG 2017
D. Sivolella, *The Space Shuttle Program*, Springer Praxis Books,
DOI 10.1007/978-3-319-54946-0

References

Astronautix.com. (n.d.). *Industrial Space Facility*. [online] Available at: http://www. astronautix.com/i/industrialspacefacility.html [Accessed 15 Mar. 2017].

Applications of Tethers in Space: Workshop Proceedings, Volume 1. (1983). NASA-CP-2364. [online] Williamsburg, Virginia: NASA. Available at: http://www. ntrs.nasa.gov [Accessed 15 Mar. 2017].

Cassutt, M. (2009). *Secret Space Shuttles*. [online] Air & Space Magazine. Available at: http://www.airspacemag.com/space/secret-space-shuttles-35318554/ [Accessed 15 Mar. 2017].

Dressler, G., Matuszak, L. and Stephenson, D. (2003). Study of High-Energy Upper Stage for Future Shuttle Missions. In: *39th AIAA Joint Propulsion Conference and Exhibit*. [online] Available at: http://www.ntrs.nasa.gov [Accessed 15 Mar. 2017].

Dtic.mil. (2006). *Stealth satellites: Cold War myth or operational reality?*. [online] Available at: http://www.dtic.mil/dtic/aulimp/citations/gsa/2006_153341/132330.html [Accessed 15 Mar. 2017].

Erickson, M. (2005). *Into the Unknown Together. The DOD, NASA, and Early Spaceflight*. 1st ed. Ft. Belvoir: Defense Technical Information Center.

Evans, B. (2007). *Space Shuttle Challenger: Ten Journeys Into The Unknown*. 1st ed. Chichester: Springer-Praxis.

Evans, B. (2010). *Space Shuttle Columbia: Her Missions and Crews*. 1st ed. Chichester: Springer-Praxis.

Forum.nasaspaceflight.com. (n.d.). *Industrial Space Facility*. [online] Available at: http:// forum.nasaspaceflight.com/index.php?topic=33781.0 [Accessed 15 Mar. 2017].

Frochlich, W. (1983). *Spacelab: An international short-stay orbiting laboratory*. NASA-EP-165. [online] NASA. Available at: http://www.ntrs.nasa.gov [Accessed 15 Mar. 2017].

Harland, D.M. (2004). *The Story of the Space Shuttle*. 1st ed. Berlin: Springer-Verlag.

Harland, D.M. and Catchpole, J. (2002). *Creating the international space station*. 1st ed. London: Springer.

© Springer International Publishing AG 2017
D. Sivolella, *The Space Shuttle Program*, Springer Praxis Books,
DOI 10.1007/978-3-319-54946-0

Harland, D.M. and Lorenz, R. (2006). *Space systems failures*. 1st ed. Berlin [u.a.]: Springer.

Inertial Upper Stage User's Guide. (1980). [online] Boeing. Available at: http://www.nasaspaceflight.com [Accessed 15 Mar. 2017].

Jackson, T., Pido, J. and Zimmerman, P. (1991). *Navigation of the TSS-1 mission*. [online] Huston, Texas: Rockwell Space Operation Company. Available at: http://www.ntrs.nasa.gov [Accessed 15 Mar. 2017].

Jsc.nasa.gov. (2017). *Oral History Project Participants - complete list*. [online] Available at: https://www.jsc.nasa.gov/history/oral_histories/participants.htm [Accessed 15 Mar. 2017].

Lardas, M. (1989). Shuttle tethered operations: The effect on orbital trajectory and inertial navigation. [online] Houston, TX, United States: McDonnell-Douglas Astronautics Co. Available at: http://www.ntrs.nasa.gov [Accessed 15 Mar. 2017].

Lord, D. (1987). *Spacelab: An International Success Story*. NASA-SP-487. [online] NASA. Available at: http://www.ntrs.nasa.gov [Accessed 15 Mar. 2017].

Manned Maneuvering Unit User's Guide. (1978). NASA-CR-151864. [online] Martin Marietta. Available at: http://www.ntrs.nasa.gov [Accessed 15 Mar. 2017].

Marshall, L. and Geiger, R. (1995). *Deployer Performance Results for the TSS-1 Mission*. NASA-CR-202595. [online] Martin Marietta Astronautics. Available at: http://www.ntrs.nasa.gov [Accessed 15 Mar. 2017].

Molczan, T. (2011). *Evaluation of the Opportunity to Launch Prowler on STS 38*. [online] Available at: http://www.satobs.org/ [Accessed 15 Mar. 2017].

National Reconnaissance Program's Planned Use of the Space Shuttle. (1981). [online] National Reconnaissance Office. Available at: http://www.nro.gov/foia/ [Accessed 15 Mar. 2017].

National Space Transportation System Reference. Volume 1: Systems and Facilities. (1988). [online] NASA. Available at: http://www.ntrs.nasa.gov [Accessed 15 Mar. 2017].

Nsarchive.gwu.edu. (n.d.). *The Spy Satellite So Stealthy that the Senate Couldn't Kill It*. [online] Available at: http://nsarchive.gwu.edu/NSAEBB/NSAEBB143/ [Accessed 15 Mar. 2017].

O'Connor, B. and Stevens, J. (2016). *Tethered Space Satellite-1 (TSS-1): Wound About a Bolt*. MSFC-CS1007-1. [online] NASA. Available at: http://www.ntrs.nasa.gov [Accessed 15 Mar. 2017].

Orbital service module systems analysis study documentation. Volume 1: Executive summary. (1978). NASA-CP-151877. [online] NASA. Available at: http://www.ntrs.nasa.gov [Accessed 15 Mar. 2017].

Ordahl, C. (1982). The MDAC Payload Assist Module. In: *AIAA 9th Communication Satellite Systems Conference*. [online] McDonnell Douglas. Available at: http://www.nasaspaceflight.com [Accessed 15 Mar. 2017].

PAM-D User's Requirements Document. (1983). MDC G6626E. [online] McDonnell Douglas Astronautics Company. Available at: http://www.nasaspaceflight.com [Accessed 15 Mar. 2017].

Portanova, P. (1983). *DOD Space Shuttle Operations at Vandenberg Air Force Base Launch and Landing Site*. [online] The Aerospace Corporation. Available at: http://www.nasaspaceflight.com [Accessed 15 Mar. 2017].

Powers, B., Shea, C. and Mcmahan, T. (1992). *The first mission of the Tethered Satellite System*. NASA-TM-107955. [online] NASA. Available at: http://www.ntrs.nasa.gov [Accessed 15 Mar. 2017].

Power Extension Package (PEP) system definition extension, orbital service module systems analysis study. Volume 1: Executive summary. (1979). NASA-CR-160321. [online] McDonnell Douglas. Available at: http://www.ntrs.nasa.gov [Accessed 15 Mar. 2017].

Ryan, R., Mowery, D. and Tomlin, D. (1993). *The dynamic phenomena of a tethered satellite: NASA's first Tethered Satellite Mission, TSS-1*. NASA Technical Paper 3347. [online] NASA. Available at: http://www.ntrs.gov [Accessed 14 Mar. 2017].

Renewing Solar Science: The Solar Maximum Repair Missions. (1985). NASA-EP-206. [online] NASA. Available at: http://www.ntrs.nasa.gov [Accessed 15 Mar. 2017].

Repairing Solar Max: The Solar Maximum Repair Mission. (1984). [online] NASA. Available at: http://www.ntrs.nasa.gov [Accessed 15 Mar. 2017].

Science.ksc.nasa.gov. (2017). *STS-83*. [online] Available at: https://science.ksc.nasa.gov/shuttle/missions/sts-83/mission-sts-83.html [Accessed 15 Mar. 2017].

Science.ksc.nasa.gov. (2017). *STS-94*. [online] Available at: https://science.ksc.nasa.gov/shuttle/missions/sts-94/ [Accessed 15 Mar. 2017].

Shuttle Crew Operation Manual. (2008). [online] NASA. Available at: https://www.nasa.gov/centers/johnson/news/flightdatafiles/ [Accessed 15 Mar. 2017].

Sivolella, D. (2014). *To Orbit and Back Again*. 1st ed. New York, NY: Springer New York.

Spaceflightnow.com. (2016). *'Slick 6:' 30 years after the hopes of a West Coast space shuttle – Spaceflight Now*. [online] Available at: https://spaceflightnow.com/2016/02/08/astronaut-interview-30-years-after-the-hopes-of-a-west-coast-space-shuttle/ [Accessed 15 Mar. 2017].

Spacelab News Reference. (1983). NASA-TM-102974. [online] ESA, NASA. Available at: http://www.ntrs.nasa.gov [Accessed 15 Mar. 2017].

Spacelab-1. (1983). NASA-TM-85197. [online] NASA. Available at: http://www.ntrs.nasa.gov [Accessed 15 Mar. 2017].

Spacelab-2. (1985). NASA-EP-217. [online] NASA. Available at: http://www.ntrs.nasa.gov [Accessed 15 Mar. 2017].

Spacelab-3. (n.d.). NASA-EP-203. [online] NASA. Available at: http://www.ntrs.nasa.gov [Accessed 15 Mar. 2017].

Space Transportation System User Handbook. (1982). [online] NASA. Available at: http://www.ntrs.nasa.gov [Accessed 15 Mar. 2017].

Strozier, J., Sterling, M., Schultz, J. and Ignatiev, A. (2001). Wake vacuum measurement and analysis for the wake shield facility free flying platform. *Vacuum*, 64(2), pp. 119–144.

STS investigators' guide. (1989). [online] NASA. Available at: http://www.ntrs.nasa.gov [Accessed 15 Mar. 2017].

STS-5 Press Information. (1983). [online] Rockwell International Space Transportation & Systems Group. Available at: http://www.ntrs.nasa.gov [Accessed 15 Mar. 2017].

STS-5 Space Shuttle Mission Press Kit. (1982). [online] NASA. Available at: http://www.ntrs.nasa.gov [Accessed 15 Mar. 2017].

STS-6 Press Information. (1983). [online] Rockwell International Space Transportation & Systems Group. Available at: http://www.ntrs.nasa.gov [Accessed 15 Mar. 2017].

STS-6 Space Shuttle Mission Press Kit. (1983). [online] NASA. Available at: http://www.ntrs.nasa.gov [Accessed 15 Mar. 2017].

STS-8 Press Information. (1983). [online] Rockwell International Space Transportation & Systems Group. Available at: http://www.ntrs.nasa.gov [Accessed 15 Mar. 2017].

STS-9 Press Information. (1983). [online] Rockwell International Space Transportation & Systems Group. Available at: http://www.ntrs.nasa.gov [Accessed 15 Mar. 2017].

STS-34 Press Information. (1989). [online] Rockwell International Space Transportation & Systems Group. Available at: http://www.ntrs.nasa.gov [Accessed 15 Mar. 2017].

STS-35 Space Shuttle Mission Press Kit. (1990). [online] NASA. Available at: http://www.ntrs.nasa.gov [Accessed 15 Mar. 2017].

STS-37 Space Shuttle Mission Press Kit. (1991). [online] NASA. Available at: http://www.ntrs.nasa.gov [Accessed 15 Mar. 2017].

STS-39 Press Information. (1991). [online] Rockwell International Space Transportation & Systems Group. Available at: http://www.ntrs.nasa.gov [Accessed 15 Mar. 2017].

STS-41B Press Information. (1984). [online] Rockwell International Space Transportation & Systems Group. Available at: http://www.ntrs.nasa.gov [Accessed 15 Mar. 2017].

STS-41C Space Shuttle Mission Press Kit. (1984). [online] NASA. Available at: http://www.ntrs.nasa.gov [Accessed 15 Mar. 2017].

STS-45 Press Information. (1992). [online] Rockwell International Space Transportation & Systems Group. Available at: http://www.ntrs.nasa.gov [Accessed 15 Mar. 2017].

STS-46 Press Information. (1992). [online] Rockwell International Space Transportation & Systems Group. Available at: http://www.ntrs.nasa.gov [Accessed 15 Mar. 2017].

STS-46 Space Shuttle Mission Press Kit. (1992). [online] NASA. Available at: http://www.ntrs.nasa.gov [Accessed 15 Mar. 2017].

STS-49 Press Information. (1992). [online] Rockwell International Space Transportation & Systems Group. Available at: http://www.ntrs.nasa.gov [Accessed 15 Mar. 2017].

STS-50 Press Information. (1992). [online] Rockwell International Space Transportation & Systems Group. Available at: http://www.ntrs.nasa.gov [Accessed 15 Mar. 2017].

STS-51 Space Shuttle Mission Press Kit (1993). [online] Rockwell International Space Transportation & Systems Group. Available at: http://www.ntrs.nasa.gov [Accessed 15 Mar. 2017].

STS-51A Space Shuttle Mission Press Kit. (1985). [online] NASA. Available at: http://www.ntrs.nasa.gov [Accessed 15 Mar. 2017].

STS-51B Space Shuttle Mission Press Kit. (1985). [online] NASA. Available at: http://www.ntrs.nasa.gov [Accessed 15 Mar. 2017].

STS-51D Press Information. (1985). [online] Rockwell International Space Transportation & Systems Group. Available at: http://www.ntrs.nasa.gov [Accessed 15 Mar. 2017].

STS-51F Space Shuttle Mission Press Kit. (1985). [online] NASA. Available at: http://www.ntrs.nasa.gov [Accessed 15 Mar. 2017].

STS-51I Space Shuttle Mission Press Kit. (1985). [online] NASA. Available at: http://www.ntrs.nasa.gov [Accessed 15 Mar. 2017].

STS-54 Press Information. (1993). [online] Rockwell International Space Transportation & Systems Group. Available at: http://www.ntrs.nasa.gov [Accessed 15 Mar. 2017].

STS-57 Press Information. (1993). [online] Rockwell International Space Transportation & Systems Group. Available at: http://www.ntrs.nasa.gov [Accessed 15 Mar. 2017].

STS-60 Space Shuttle Mission Press Kit. (1994). [online] NASA. Available at: http://www.ntrs.nasa.gov [Accessed 15 Mar. 2017].

STS-60 Space Shuttle Mission Report. (1994). NASA-CR-197233. [online] NASA. Available at: http://www.ntrs.nasa.gov [Accessed 15 Mar. 2017].

STS-61A Space Shuttle Mission Press Kit. (1985). [online] NASA. Available at: http://www.ntrs.nasa.gov [Accessed 15 Mar. 2017].

STS-61B Space Shuttle Mission Press Kit. (1985). [online] NASA. Available at: http://www.ntrs.nasa.gov [Accessed 15 Mar. 2017].

STS-67 Space Shuttle Mission Press Kit. (1995). [online] NASA. Available at: http://www.ntrs.nasa.gov [Accessed 15 Mar. 2017].

STS-69 Space Shuttle Mission Press Kit. (1995). [online] NASA. Available at: http://www.ntrs.nasa.gov [Accessed 15 Mar. 2017].

STS-69 Space Shuttle Mission Report. (1995). NSTS-37402. [online] NASA. Available at: http://www.ntrs.nasa.gov [Accessed 15 Mar. 2017].

STS-72 Space Shuttle Mission Press Kit. (1996). [online] NASA. Available at: http://www.ntrs.nasa.gov [Accessed 15 Mar. 2017].

STS-75 Space Shuttle Mission Press Kit. (1996). [online] NASA. Available at: http://www.ntrs.nasa.gov [Accessed 15 Mar. 2017].

STS-80 Space Shuttle Mission Press Kit. (1996). [online] NASA. Available at: http://www.ntrs.nasa.gov [Accessed 15 Mar. 2017].

STS-80 Space Shuttle Mission Report. (1997). NASA-TM-112252. [online] NASA. Available at: http://www.ntrs.nasa.gov [Accessed 15 Mar. 2017].

STS-83 Space Shuttle Mission Press Kit. (1997). [online] NASA. Available at: http://www.ntrs.nasa.gov [Accessed 15 Mar. 2017].

STS-87 Space Shuttle Mission Press Kit. (1997). [online] NASA. Available at: http://www.ntrs.nasa.gov [Accessed 15 Mar. 2017].

STS-94 Space Shuttle Mission Press Kit. (1997). [online] NASA. Available at: http://www.ntrs.nasa.gov [Accessed 15 Mar. 2017].

S. T. Wu, (1978). *UAH/NASA Workshop on The Uses of a Tethered Satellite System*. NASA-CR-161836. [online] Huntsville: The University of Alabama, NASA. Available at: http://www.ntrs.nasa.gov [Accessed 15 Mar. 2017].

Thespacereview.com. (2013). *The Space Review: On the trail of "The Curse of Slick-6"*. [online] Available at: http://www.thespacereview.com/article/2349/1 [Accessed 15 Mar. 2017].

Thespacereview.com. (n.d.). *The Space Review: A lighter shade of black: the (non) mystery of STS-51J*. [online] Available at: http://www.thespacereview.com/article/1536/1 [Accessed 15 Mar. 2017].

Tethered Satellite System (TSS-1) (STS Flight 46). Volume 2; System Description. (1991). [online] Martin Marietta Astronautics. Available at: http://www.ntrs.nasa.gov [Accessed 15 Mar. 2017].

The 25 kW power module evolution study. Part 3: Conceptual designs for power module evolution. Volume 1: Power module evolution. (1979). NASA-CR-161145. [online] NASA. Available at: http://www.ntrs.nasa.gov [Accessed 15 Mar. 2017].

Tsiao, S. (2008). *"Read you loud and clear!"*. 1st ed. Washington, DC: National Aeronautics and Space Administration, NASA History Division, Office of External Relations.

Upper Stage Alternatives for the Shuttle Era. (1981). NASA-TM-84137. [online] NASA. Available at: http://www.ntrs.nasa.gov [Accessed 15 Mar. 2017].

Williams, D. and Johnson, B. (2003). *EMU Shoulder Injury Tiger Team Report*. NASA/TM-2003-212058. NASA.

Wired, I. (2006). *I Spy*. [online] WIRED. Available at: https://www.wired.com/2006/02/spy-3/ [Accessed 15 Mar. 2017].

Index

© Springer International Publishing AG 2017
D. Sivolella, *The Space Shuttle Program*, Springer Praxis Books,
DOI 10.1007/978-3-319-54946-0